Qualitative Methods in Quantum Theory

Qualitative Methods in Quantum Theory

A. B. Migdal
I. V. Kurchatov Atomic Energy Institute
Moscow, Russia

Translated from the Russian edition by
Anthony J. Leggett
University of Sussex

Advanced Book Program

CRC Press
Taylor & Francis Group
Boca Raton London New York

CRC Press is an imprint of the
Taylor & Francis Group, an **informa** business

First published 1977 by Westview Press

Published 2018 by CRC Press
Taylor & Francis Group
6000 Broken Sound Parkway NW, Suite 300
Boca Raton, FL 33487-2742

CRC Press is an imprint of the Taylor & Francis Group, an informa business

Visit the Taylor & Francis Web site at
http://www.taylorandfrancis.com

and the CRC Press Web site at
http://www.crcpress.com

A CIP catalog record for this book is available from the Library of Congress.

ISBN 13: 978-0-7382-0302-7 (pbk)

ADVANCED BOOK CLASSICS

David Pines, Series Editor

Anderson, P.W., *Basic Notions of Condensed Matter Physics*
Bethe H. and Jackiw, R., *Intermediate Quantum Mechanics, Third Edition*
Cowan, G. and Pines, D., *Complexity: Metaphors, Models, and Reality*
de Gennes, P.G., *Superconductivity of Metals and Alloys*
d'Espagnat, B., *Conceptual Foundations of Quantum Mechanics, Second Edition*
Feynman, R., *Photon-Hadron Interactions*
Feynman, R., *Quantum Electrodynamics*
Feynman, R., *Statistical Mechanics*
Feynman, R., *The Theory of Fundamental Processes*
Gell-Mann, M. and Ne'eman, Y., *The Eightfold Way*
Khalatnikov, I. M. *An Introduction to the Theory of Superfluidity*
Ma, S-K., *Modern Theory of Critical Phenomena*
Migdal, A. B., *Qualitative Methods in Quantum Theory*
Negele, J. W. and Orland, H., *Quantum Many-Particle Systems*
Nozières, P., *Theory of Interacting Fermi Systems*
Nozières, P. and Pines, D., *The Theory of Quantum Liquids*
Parisi, G., *Statistical Field Theory*
Pines, D., *Elementary Excitations in Solids*
Pines, D., *The Many-Body Problem*
Quigg, C., *Gauge Theories of the Strong, Weak, and Electromagnetic Interactions*
Schrieffer, J.R., *Theory of Superconductivity, Revised*
Schwinger, J., *Particles, Sources, and Fields, Volume I*
Schwinger, J., *Particles, Sources, and Fields, Volume II*
Schwinger, J., *Particles, Sources, and Fields, Volume III*
Schwinger, J., *Quantum Kinematics and Dynamics*
Wyld, H.W., *Mathematical Methods for Physics*

CONTENTS

Editor's Foreword

Perseus Publishing's *Frontiers in Physics* series has, since 1961, made it possible for leading physicists to communicate in coherent fashion their views of recent developments in the most exciting and active fields of physics—without having to devote the time and energy required to prepare a formal review or monograph. Indeed, throughout its nearly forty year existence, the series has emphasized informality in both style and content, as well as pedagogical clarity. Over time, it was expected that these informal accounts would be replaced by more formal counterparts—textbooks or monographs—as the cutting-edge topics they treated gradually became integrated into the body of physics knowledge and reader interest dwindled. However, this has not proven to be the case for a number of the volumes in the series: Many works have remained in print on an on-demand basis, while others have such intrinsic value that the physics community has urged us to extend their life span.

The *Advanced Book Classics* series has been designed to meet this demand. It will keep in print those volumes in *Frontiers in Physics* that continue to provide a unique account of a topic of lasting interest. And through a sizable printing, these classics will be made available at a comparatively modest cost to the reader.

The late Arkady B. Migdal was one of the great theoretical physicists of our time, who made significant contributions to nuclear physics, condensed matter physics, astrophysics, and particle physics. He combined keen physical insight with a deep understanding of physics and physical phenomena and believed it truly important that students learn more about quantum theory than about the mathematical manipulation of formulae and equations. It was for that reason that he developed the series of lectures on qualitative methods in quantum theory that are contained in this book. Migdal's originality and careful attention to pedagogy make his book, *Qualitative Methods in Quantum Theory*, required reading for every scientist interested in learning and applying quantum theory. I am most pleased that its publication in *Advanced Book Classics* will now make it readily available to the future generation of scientists, who may be expected to profit greatly from reading it.

David Pines
Cambridge, England
May, 2000

TRANSLATOR'S NOTE

In the interests of rapid and economical production most of the equations in this book have been photocopied directly from the Russian version. Generally, notations which may be unfamiliar to English-speaking readers have been changed to conform to the English usage, but there are a few cases where this has proved awkward or impossible; the most frequent ones are the following:

A simple sequence of two vectors (indicated by boldface) denotes the scalar or dot product.

The notation [a, b] with a, b boldface denotes the vector product of a and b.

∇_r^2, ∇_R^2 denotes the Laplacian with respect to r and R respectively.

sh indicates hyperbolic sine. (sinh)

Sp indicates the trace of a matrix.

PREFACE

The solution of most problems in theoretical physics begins with the application of the qualitative methods which constitute the most attractive and beautiful characteristic of this discipline. By "qualitative methods" we mean dimensional estimates and estimates made by using simple models, the investigation of limiting cases where one can exploit the smallness of some parameter, the use of the analytic properties of physical quantities, and finally the derivation of consequences from the symmetry properties, that is, the invariance relative to various transformations (e.g. Lorentz or isotopic invariance). However, as experience in the classroom shows, it is just these aspects of theoretical physics which are most difficult for the beginner.

Unfortunately, the methods of theoretical physics are usually presented in a formal, mathematical way, rather than in the constructive form in which they are used in scientific work. The object of this book is to make up this deficiency, that is, to teach to beginning students of the subject the correct approach to the solution of scientific problems. This goal largely determines the character of the presentation; the general results are always obtained first in special cases or with extremely simplified models. It seems to me that a formal exposition, which leaves no traces of a gradual approach to the problem, no traces of the "sweat" involved, can often leave the beginner in scientific research with a sense of something lacking. I have therefore endeavoured as far as possible to indicate the general method of approach to the problem, especially at the first stage of the work. Of course, this means I have had to sacrifice rigour in the exposition, and in return disclose some "trade secrets", that is, the little tricks which shorten the derivation of the results.

A common mistake of beginners is the desire to understand everything completely right away. In real life understanding comes gradually, as one becomes accustomed to the new ideas. One of the difficulties of scientific research is that it is impossible to make progress without clear understanding, yet this understanding can come only from the work itself; every completed piece of research represents a victory over this contradiction. Similar difficulties will inevitably occur in reading this book; I hope that by the time it is read to the end they will be overcome.

Each of the six chapters of the book begins with a detailed introduction, in which the physical meaning of the results obtained in the chapter is explained in a simple way. The first three chapters are devoted to dimensional and model-based estimates in atomic physics, the applications of various types of perturbation theory and the quasiclassical approximation respectively. These chapters are a revised version of the book by A. B. Migdal and

V. P. Krainov "Approximation Methods in Quantum Mechanics" (Nauka, 1966 : translation published by W. A. Benjamin, Inc., New York, 1969). The fourth chapter is devoted to various problems solution of which requires the use of the analytic properties of physical quantities. The fifth chapter develops the graphical method and its application to the many-body problem. Finally, the sixth chapter is devoted to questions connected with the interaction of elementary particles at short distances; in this problem of quantum field theory it is precisely the application of qualitative methods which plays the main role.

The author is deeply grateful to A. A. Migdal, A. M. Polyakov and B. A. Khodel' for numerous discussions and suggestions, and to V. P. Krainov for his help in the selection of material for the first three chapters. He also thanks his friends and students G. Zasetskii, D. Voskresenskii, N. Kirichenko, O. Markin, I. Mishustin, G. Sorokin and A. Chernoutsan for help in the preparation of the manuscript.

A. B. Migdal

CHAPTER 1

DIMENSIONAL AND "MODEL" APPROXIMATIONS

No problem in physics can ever be solved exactly. We always have to neglect the effect of various factors which are unimportant for the particular phenomenon we have in mind. It then becomes important to be able to estimate the magnitude of the quantities we have neglected. Moreover, before calculating a result numerically it is often necessary to investigate the phenomenon qualitatively, that is, to estimate the order of magnitude of the quantities we are interested in and to find out as much as possible about the general behaviour of the solution.

To this end we first consider the problem in the most simplified form possible. For instance, in the case of a particle moving in a Coulomb field, we replace the problem by that of its motion in a square well potential with an appropriately chosen depth and width depending on the particle energy, and so on. Moreover, we should consider all the limiting cases in which the solution is simplified. For instance, if it is required to solve the problem of scattering of particles of arbitrary energy, we should first consider the limits of small and large energy and trace out how the corresponding expressions match up in the intermediate-energy region. The aim of this chapter is to instruct the reader in the art of obtaining approximate solutions from dimensional estimates with the aid of a simplified model of the phenomenon to be investigated.

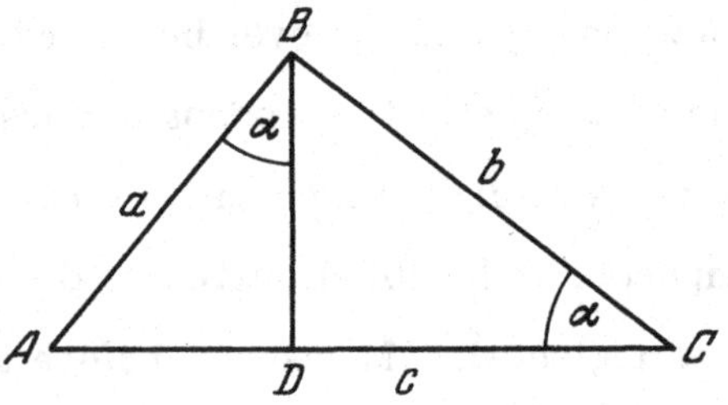

Fig. 1

In some cases dimensional techniques actually enable one to get quantitative rather than merely qualitative results. For instance, we can prove Pythagoras' theorem purely from dimensional considerations (see Fig. 1). It follows by dimensional reasoning that the area of the triangle ABC can depend only on the square of the hypotenuse, c^2, multiplied by some function $f(\alpha)$ of the angle α. The same applies to the areas of the two similar triangles ABC and BCD, but for these the hypotenuse is respectively the sides AB, BC of the large triangle. Hence

$$c^2 f(\alpha) = a^2 f(\alpha) + b^2 f(\alpha).$$

which proves the theorem.

As a second example we consider the problem of finding the resistive force when a body moves in a viscous medium with arbitrary velocity. We start with the limiting case of small velocities; then the resistive force will be determined by the viscosity of the medium. The parameter which defines the notion of "small" as opposed to "large" velocities may be found by forming a quantity with the dimensions of velocity from the viscosity, the density of the medium and the dimensions of the body. We assume that the body has all its dimensions approximately equal; then for the purposes of dimensional estimates it can be characterized, just like a sphere, by a single length R. From the viscosity η, density ρ, length R and velocity v we can form only one dimensionless combination, the so-called Reynolds number

$$\mathrm{Re} = \frac{vR}{\nu},$$

where $\nu = \eta/\rho$. Since the momentum current is given by $\eta \nabla v$, the order of magnitude of the force acting on unit surface area is $P \sim \eta v/R$. (The estimate of the velocity gradient as v/R is made as follows: at the surface of the body the velocity of the liquid is v, while far from the body (at distances of order R) the liquid is at rest. Hence $\nabla v \sim \Delta v/R \sim v/R$.) If we estimate the surface area of the body as $S \sim 4\pi R^2$, then the total resistive force is given by

$$F \sim 4\pi\eta vR.$$

We may note that an exact solution of the problem for a spherical body gives in the case of small velocity

$$F = 6\pi\eta vR,$$

In the case of arbitrary velocity this expression must be multiplied by some function of the dimensionless parameter Re:

$$F = 6\pi\eta vR\Phi\left(\frac{vR}{\nu}\right).$$

Now let us consider the limit of very large velocities. In this case the resistive force is independent of the viscosity and is determined by the momentum which is transferred per unit time to the column of liquid lying in front of the body; the base area of this column is just the cross-sectional area of the body, and so we find

$$F \sim \pi R^2 \rho v^2.$$

Thus, for large velocities the function $\Phi(x)$ satisfies the estimate $\Phi(x) \sim \frac{1}{6} x$.

The rough character of the solution for all velocities is determined by the interpolation formula

$$F \sim 6\pi\eta v R \left(1 + \frac{1}{6} \frac{vR}{\nu}\right).$$

According to this estimate, the transition from one regime to the other should take place for $Re \sim 6$. In reality, the transition to the turbulent regime, where the resistive force is independent of viscosity, takes place for $Re \sim 100$. Here we have run up against a rather unusual case - usually, the transition from one limiting case to the other is characterized by a value of the relevant dimensionless parameter of order unity.

Another example we shall give relates to the possibility of constructing a theory which will connect gravitation and electrodynamics. Such a theory, if it existed, would have to fix the value of a dimensionless parameter relating the gravitational constant g to the quantities which characterize electromagnetic processes, viz. the electron charge e and mass m, the speed of light c and Planck's constant $\hbar$. From these quantities it is possible to construct two dimensionless ratios:

$$\alpha = \frac{e^2}{\hbar c} = \frac{1}{137}, \qquad \xi = \frac{gm^2}{\hbar c} = 2 \cdot 10^{-45}.$$

As already mentioned, the dimensionless parameters which occur as a result of solution of the equations usually turn out to be of order unity. Thus the quantity ξ must enter in such a way that one obtains a number of order unity, e.g.

$$\alpha \ln (1/\xi) \sim 1.$$

It is indeed in just such a form that the parameters α and ξ come into those estimates which give some hope of establishing a connection between gravitation and electrodynamics.

We shall give one more example of the way in which order-of-magnitude estimates help one to orient oneself in complicated problems. Let us answer the question: beyond what strengths of the electric field $\mathcal{E}$ and magnetic field $\mathcal{H}$ do Maxwell's equations in free space become nonlinear ? The reason for the nonlinearity is the perturbation of the vacuum by the external field. Let us then construct a quantity with the dimensions of field from the quantities which characterize the vacuum fluctuations of the electron-positron field. Since $e\mathcal{E}$ has the dimensions of energy divided by length, we find

$$e\mathcal{E}_C \sim mc^2 \Big/ \frac{\hbar}{mc}, \qquad \mathcal{E}_C \sim \frac{m^2c^3}{e\hbar}.$$

It is clear from this expression that the critical field $\mathcal{E}_c$ is determined by the particles with the smallest mass, that is, by the electron-positron field. We see that the quantity $\mathcal{E}_c$ is the field strength at which the potential difference across a Compton wavelength is of the order of the energy necessary for pair creation. Substitution of the numerical values of e, m, $\hbar$ and c gives $\mathcal{E}_c \sim 10^{16}$ V/cm.

1. ORDER-OF-MAGNITUDE ESTIMATION OF MATHEMATICAL EXPRESSIONS

Before explaining techniques for estimating physical quantities, we first review a rather simpler sort of estimation problem, namely the estimation of mathematical expressions. The basic idea involved consists in determining the region of the variables which gives the principal contribution to the result, separating the part of the mathematical expression which is fast varying in this region from the slowly varying part, and also using the asymptotic forms of the expression.

Estimation of a derivative. In the simplest case, when the important region of variation of the function $F(x)$ is characterized by a single length ℓ, then the order of magnitude of the derivative $F'(x)$ is simply $F(\ell)/\ell$. For instance, if $F(x) = \exp(-x^2/\ell^2)$, then the derivative $F'(x) = -(2x/\ell^2)\exp(-x^2/\ell^2)$, so that $F'(\ell) \sim F(\ell)/\ell$. However, for x much larger than ℓ this estimate is clearly invalid. For a power function $F(x) = x^n$ the "region of appreciable change" is defined by the variable x itself. In fact we have

$$F'(x) = nx^{n-1} \sim nF(x)/x.$$

In some cases the relevant length ℓ is different for different regions of variation of the variable x. Then in each region of x the derivative $F'(x)$ is of order $F(x)/\ell(x)$, where $\ell(x)$ is the length over which $F(x)$ changes appreciably in this region. For instance, suppose $F(x)$ has the form shown in fig. 2. Then we have $F'(x) \sim F(x_1)/\ell_1$ for $x \sim x_1$, but $F'(x) \sim F(x_2)/\ell_2$

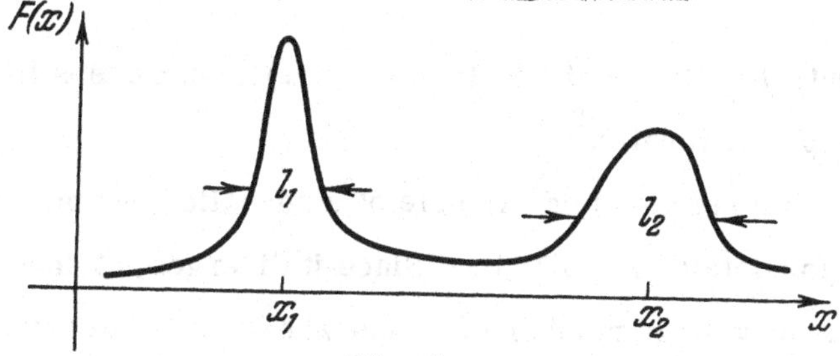

Fig. 2

for $x \sim x_2$. In more complicated cases, if F(x) can be sketched, even roughly, the best way of estimating its derivative is from the graph.

<u>Estimation of integrals.</u> We will demonstrate some methods of estimating integrals by various examples:

1. Often one can obtain approximate values of integrals by expanding the integrand in a power series. For instance, we write

$$\int_0^x \exp(-t^2)\,dt = \int_0^x (1 - t^2 + t^4/2 - \dots)\,dt =$$
$$- x - \frac{x^3}{3} + \frac{x^5}{10} - \dots .$$

This integral converges for all x. To estimate the integral we may restrict ourselves to the first few terms of the series; the resulting estimate will of course be appropriate only for $x \lesssim 1$.

How can we estimate this integral for large x? Integrating repeatedly by parts, we obtain

$$\int_0^x = \int_0^\infty - \int_x^\infty = \frac{\sqrt{\pi}}{2} - \left[\frac{\exp(-x^2)}{2x} - \int_x^\infty \frac{\exp(-t^2)\,dt}{2t^2}\right] =$$
$$= \frac{\sqrt{\pi}}{2} - \exp(-x^2)\left[\frac{1}{2x} - \frac{1}{4x^3} + \dots\right];$$

where the n-th term of the series has the form $(-1)^n (2n-1)!!/2^{n+1} x^{2n+1}$. It is easy to see that the series is divergent: for $n \to \infty$ the factorial $(2n-1)!!$ increases faster than the power term x^{2n+1}.

This series is an example of a so-called asymptotic series (for details see p.153). Since it diverges, it does not pay to take a very large number of terms when one is estimating the integral; this actually makes it <u>less</u> accurate. How do we find the optimum number of terms to keep ? We notice that for large x the terms of the series first decrease in absolute magnitude and then subsequently begin to increase. The optimum number of terms is evidently defined by the requirement that the remainder of the series should be a minimum. It is easy to see that the remainder is of the order of the (n + 1)-th term of the series. Hence the correct prescription is to sum as far as the smallest term of the series. The condition for the minimum to be reached at the n-th term can be approximated by setting the n-th and (n+1)-th terms equal:

$$\frac{(2n-1)!!}{2^{n+1}x^{2n+1}} \sim \frac{(2n+1)!!}{2^{n+2}x^{2n+3}}.$$

Hence we find $n \sim x^2$.

<u>Problem.</u> Show that, in estimating the integral $\int_x^\infty t^{-1}\exp(-t)dt$ in the case $x \gg 1$, the optimum number of terms to keep in the asymptotic series is equal to x.

2. Many integrals can be estimated by separating out the most important part of the integrand. Consider the following examples:

Case 1) $I(x) = \int_0^x \frac{\exp(t^2)}{\sqrt{x^2 - t^2}} dt.$

If $x \ll 1$, then the exponential $\exp(t^2)$ in the integrand is approximately equal to 1. Consequently

$$I(x) \approx \int_0^x \frac{dt}{\sqrt{x^2 - t^2}} = \int_0^1 \frac{dz}{\sqrt{1 - z^2}}.$$

Since this integral contains no parameters, we have $I(x) \sim 1$. (An exact calculation of the above integral gives $\int_0^1 \frac{dz}{\sqrt{1-z^2}} = \frac{\pi}{2}$)

If $x \gg 1$, then, in view of the exponential increase of the factor $\exp(t^2)$, the principal contribution to the integral comes from the region near the point $t = x$. If we write $\xi = x - t$, then we have

$$I(x) = \int_0^x \exp(x^2 - 2\xi x + \xi^2) \frac{d\xi}{\sqrt{2\xi x - \xi^2}}.$$

The region of ξ for which the integrand is large is concentrated near the lower limit, and its width is of order $1/2x$. In this region we have $\xi^2 \sim 1/4x^2 \ll 1$, hence $\exp(\xi^2) \approx 1$ and so

$$I(x) \approx \exp(x^2) \int_0^x e^{-2\xi x} \frac{d\xi}{\sqrt{2\xi x}} \approx \frac{\exp(x^2)}{2x} \int_0^\infty e^{-z} \frac{dz}{\sqrt{z}} = \frac{\sqrt{\pi}}{2x} \exp(x^2).$$

For $x \sim 1$ the two expressions for $I(x)$ are of the same order of magnitude (namely unity), as of course they must be. Thus the estimates given form a good description of $I(x)$ over the whole range of the variable x.

Case 2) $I(\alpha, \beta) = \int_0^\infty e^{-\alpha x^2} \sin^2 \beta x \, dx, \quad \alpha > 0.$

We rewrite this integral in the form

$$I(\alpha, \beta) = \frac{1}{\sqrt{\alpha}} \int_0^\infty e^{-z^2} \sin^2\left(\frac{\beta}{\sqrt{\alpha}} z\right) dz.$$

For $z > 1$ the integrand decreases fast, so that the important region of integration is $0 < z < 1$. If $\beta \gg \sqrt{\alpha}$, the function $\sin(\beta z/\sqrt{\alpha})$ oscillates many times in the region of z which is important. Therefore, we may replace $\sin^2(\beta z/\sqrt{\alpha})$ by 1/2, and the integral $I(\alpha, \beta)$ is approximately given by

$$I(\alpha, \beta) \approx \frac{1}{2\sqrt{\alpha}} \int_0^\infty e^{-z^2} dz = \frac{\sqrt{\pi}}{4\sqrt{\alpha}}.$$

If $\beta \ll \sqrt{\alpha}$, then in the region of z which is important we have $\sin \beta z/\sqrt{\alpha} \approx \beta z/\sqrt{\alpha}$. Consequently,

$$I(\alpha, \beta) \approx \frac{\beta^2}{\alpha^{3/2}} \int_0^\infty e^{-z^2} z^2 dz = \frac{\sqrt{\pi}\, \beta^2}{4\alpha^{3/2}}.$$

For $\beta = \sqrt{\alpha}$ the two expressions coincide and are equal to $(1/4).\ (\pi/\alpha)^{\frac{1}{2}}$. We may note that the exact value of the integral we have estimated is

$$I(\alpha, \beta) = \frac{1}{4} \sqrt{\frac{\pi}{\alpha}} (1 - \exp(-\beta^2/\alpha)).$$

It is easy to verify that in the limits $\beta \gg \sqrt{\alpha}$ and $\beta \ll \sqrt{\alpha}$ this expression reduces to the appropriate forms given above. For $\beta = \sqrt{\alpha}$ we have $I(\alpha, \beta) = (1/4)\sqrt{\pi/\alpha}\ (1 - 1/e)$, which, though not numerically equal to our estimated value, is of the same order of magnitude.

Case 3) $I(\alpha, a) = \int_0^\infty \frac{e^{-\alpha x}}{x + a} dx, \quad a, \alpha > 0.$

Let us change the variable of integration, putting $x = az$. Then $I(\alpha, a)$ takes the form

$$I = \int_0^\infty \frac{e^{-\beta z}}{z+1}\, dz,$$

where $\beta = \alpha a$. The region in which the integrand is appreciably different from zero is defined by $z \lesssim 1/\beta$.

Suppose we have $\beta \gg 1$. Then $z \ll 1$ in the important region and so

$$I \underset{\beta \gg 1}{\approx} \int_0^\infty e^{-\beta z} dz = \frac{1}{\beta}\,. \tag{1.1}$$

For $\beta \ll 1$, then in the important region $z \gg 1$. Consequently,

$$I \underset{\beta \ll 1}{\approx} \int_1^{1/\beta} \frac{dz}{z} = \ln \frac{1}{\beta}\,. \tag{1.2}$$

Actually this integral can be expressed in terms of the exponential integral $\mathrm{Ei}(x)$[1]:

$$\int_0^\infty \frac{e^{-\alpha x}}{x+a}\, dx = -\,e^{\beta}\mathrm{Ei}\,(-\beta), \quad \beta = \alpha a.$$

If $\beta \gg 1$, then $\mathrm{Ei}(-\beta) \approx e^{-\beta}/(-\beta)$, and we obtain formula (1.1). For $\beta \ll 1$, we have $\mathrm{Ei}(-\beta) \approx -\ln \beta$, which agrees with (1.2).

Case 4) $I(a) = \int_{-1}^{1} \frac{f(x)\, dx}{\sqrt{x^2 + a^2}}$;

where it is assumed that the function $f(x)$ is varying appreciably on a scale of x of order unity.

---

1 P. Morse and F. Feshbach, Methods of Theoretical Physics, (McGraw-Hill Book Company, New York, 1953).

In the case $a \ll 1$ the principal contribution to the integral comes from the region near the origin. Since the function $f(x)$ is smoothly varying in this region, we can replace it by $f(0)$ and take it out from under the integral. Then we get

$$I(a) \approx f(0) \int_{-1}^{1} \frac{dx}{\sqrt{x^2+a^2}} = 2f(0) \ln \frac{1+\sqrt{1+a^2}}{a} \approx$$
$$\approx 2f(0) \ln \frac{1}{a} .$$

In the case $a \gg 1$ we get

$$I(a) \approx \frac{1}{a} \int_{-1}^{1} f(x)\, dx.$$

Problems. Estimate the following integrals in the limits $a \gg b$ and $a \ll b$:

(1) $\int_0^\infty e^{-bx^2} \sin ax^2 dx, \quad b > 0.$

(2) $\int_0^\infty \exp(-x/a) \frac{dx}{\sqrt{x(x+b)}}, \quad a, b > 0.$

(3) $\int_0^\infty \frac{\sin(x/a)}{x(x^2+b^2)} dx, \quad a, b > 0.$

Solutions.

(1) $a \gg b$: $\sqrt{\pi/8a}$; $a \ll b$: $a\sqrt{\pi/16b^3}$.

The exact value of the integral is

$$\frac{\sqrt{\pi}}{2\sqrt[4]{a^2+b^2}} \sin\left(\frac{1}{2} \operatorname{arctg} \frac{a}{b}\right) .$$

(2) $a \gg b$: $\ln(a/b)$; $a \ll b$: $\sqrt{\pi a/b}$.

The exact value is

$$\exp(b/2a) \cdot K_0(b/2a)$$

where K_o is the Macdonald function.

(3) $a \gg b$: $\pi/2ab$; $a \ll b$: $\pi/2b^2$.

The exact value is

$$\frac{\pi}{2b^2}[1 - \exp(-b/a)].$$

The method of steepest descents* Consider the integral $I = \int_0^\infty g(t) \exp f(t) dt$, where $f(t)$ is a function which has a sharp maximum for some value $t_o > 0$ of t. Suppose that near t_o the function $g(t)$ is slowly varying. Then we can replace the function ge^f near the maximum by a simpler function; to do this we expand f in a Taylor series around its maximum t_o:

$$f(t) = f(t_0) + \frac{1}{2}(t - t_0)^2 f''(t_0) + \ldots$$

Assume that $|f''(t_o)| \gg t_o^{-2}$; this is just the mathematical expression of the assumption that f has a sharp maximum. In fact, the values of $(t-t_o)^2$ which are important in the integral I are of order $1/f''(t_o)$, as we shall see below (eqn. 1.3); thus, $(t-t_o)^2/t_o^2 \ll 1$.

---

* This method is discussed in more detail in Chapter 3 (p.170).

This condition makes it legitimate to omit the higher terms in the Taylor series written above for f(t) (cf. below).

We have

$$I \approx g(t_0) \int_{-\infty}^{\infty} \exp\left[f(t_0) - \frac{1}{2}(t-t_0)^2 |f''(t_0)|\right] dt = \tag{1.3}$$
$$= \sqrt{\frac{2\pi}{|f''(t_0)|}}\, g(t_0)\, e^{f(t_0)}.$$

Here we replaced the limits of integration by $[-\infty, \infty]$, since the integrand decreases exponentially in the region

$$\delta t > \frac{1}{\sqrt{|f''(t_0)|}} \ll t_0.$$

Let us now estimate the correction given by the subsequent terms in the Taylor series. If we keep only the cubic term and expand the exponential of it in a series, then the first term of the expansion gives no contribution, since the integrand is odd. Therefore we consider the fourth-order term in the Taylor expansion of f, namely $f^{(IV)}(t_0)(t-t_0)^4/4!$ If we now further expand the exponential of this quantity in a series, we find that the correction is of order $f^{(IV)}/(f'')^2$ relative to the expression (1.3). If the function f(t) is characterised by a single parameter, then estimating the order of magnitude of the derivatives of f, we find

$$f^{(IV)}/(f'')^2 \sim 1/f(t_0).$$

Thus, the condition for applicability of the method of steepest descents is $f(t_0) \gg 1$; this is equivalent to the assumption $|f''(t_0)| \gg t_0^{-2}$ made above.

If in (1.3) we were to make the substitution $t-t_0 = i\xi$, the integrand would become an increasing exponential. In other words the point t_0 is a saddle point in the complex t plane and the direction of integration is the direction of steepest descent from the saddle point (see Fig. 3). Hence the name, saddle-point method or method of steepest descents. We have actually considered the special case in which the direction of steepest descent coincides with the real axis; one can also consider the general case in which the direction of steepest descent makes an arbitrary angle with the real axis.

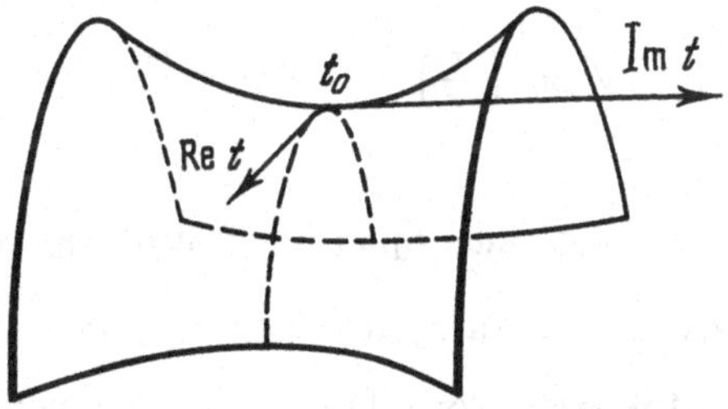

Fig. 3

Let us use the method of steepest descents to obtain an asymptotic expression, for large x, for the gamma function

$$\Gamma(x+1)=\int_0^\infty \exp(-t+x\ln t)\,dt.$$

We write $-t + x \ln t = f(t)$. Then the condition that $f'(t)$ be zero gives us the saddle-point t_0:

$$f'(t_0)=-1+\frac{x}{t_0}=0,$$

whence we have $t_0 = x$. Since $f(t_0) = x \ln x - x$, the condition of applicability of the method of steepest descents, viz. $f(x) \gg 1$, means that $x \gg 1$. We further have $f''(t_0) = -x/t_0^2 = -1/x$. Using (1.3), we therefore obtain

$$\Gamma(x+1) \underset{x \gg 1}{\approx} \sqrt{2\pi x}\,(x/e)^x. \tag{1.4}$$

This asymptotic formula is called the Stirling formula. To estimate its accuracy we use the relation $\Gamma(x+1) = x\Gamma(x)$, and write the (unknown) exact expression for $\Gamma(x+1)$ in the form

$$\Gamma(x+1) = \sqrt{2\pi x}\left(\frac{x}{e}\right)^x [1 + \varphi(x+1)],$$

where φ is a so far unknown function. With the help of the above recurrence relation we obtain, for large x, $\varphi(x+1) - \varphi(x) \approx -1/12x^2$. For $x \gg 1$ the difference $\varphi(x+1) - \varphi(x)$ is approximately equal to $\varphi'(x)$, and we accordingly find $\varphi(x) = 1/12x$. Thus, for $x \gg 1$, we have

$$\Gamma(x+1) \approx \sqrt{2\pi x}\left(\frac{x}{e}\right)^x \left[1 + \frac{1}{12x} + O\left(\frac{1}{x^2}\right)\right].$$

It is interesting that this formula is actually very accurate even for small values of x. In fact, we can check its accuracy by using the fact that for integral values of x the gamma-function $\Gamma(x+1)$ is just $x!$ For instance, even for $x = 1$ we get

$$\sqrt{2\pi}\,\frac{1}{e}\left(1 + \frac{1}{12}\right) = 0{,}9990 \approx 1! = 1,$$

and for $x = 2$,

$$\sqrt{4\pi}\,\frac{1}{e^2}\left(1+\frac{1}{24}\right) = 1{,}9990 \approx 2! = 2.$$

Problems

(1) Evaluate the integral $\int_0^\infty \cos\left(\frac{1}{3}t^3 + xt\right) dt$ for $x \gg 1$ by the method of steepest descents.

(2) Evaluate the integral $\int_0^\infty x \exp\left(-ax - \frac{b}{\sqrt{x}}\right.$. Show that the method of steepest descents is valud provided $ab^2 \gg 1$.

Solutions

(1) $\frac{\sqrt{\pi}}{2\sqrt[4]{x}} \exp\left(-\frac{2}{3}x^{3/2}\right)$.

(2) $\frac{b}{a}\sqrt{\frac{\pi}{3a}} \exp\left(-\frac{3}{2}\sqrt[3]{2ab^2}\right)$.

Properties of Integrals of Oscillating Functions. Estimates of the Higher Terms in Fourier Series Expansions.

We shall consider some examples which illustrate the properties of integrals of oscillating functions

Example 1. $I = \int_{-\infty}^{\infty} \frac{e^{i\omega t}\,dt}{\sqrt{1+t^2}}, \quad \omega \gg 1.$

The singularities of the integrand occur on the imaginary axis: $t = \pm i$. We can calculate the integral by deforming the contour

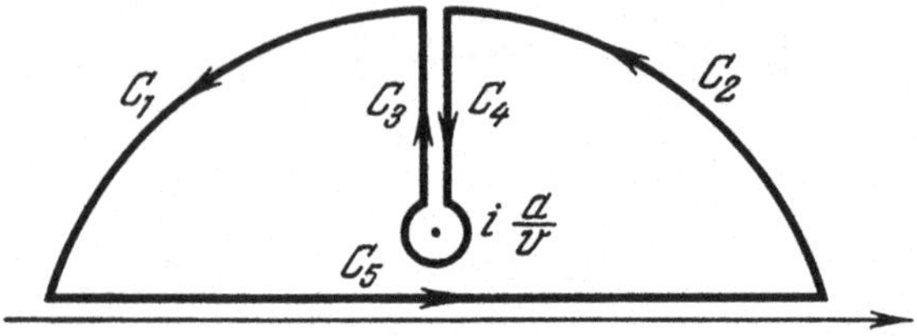

Fig. 4

into the upper half-plane (see Fig. 4). The contributions from C_1 and C_2 vanish when the contour is moved away to infinity, so that the integral is just the contributions from the parts $C_3 + C_4 + C_5$ which go around the branch point $t = i$. The denominator of the integrand $\sqrt{1 + t^2}$ changes sign when we go around the branch point; therefore the contributions from C_3 and C_4 are equal. The contribution from C_5 tends to zero as the radius of the circle is decreased; this is easy to see if we make the substitution $t = i + re^{i\varphi}$, and let r tend to zero. Then we get

$$\int_{C_5} \sim \int_0^{2\pi} \frac{re^{i\varphi} d\varphi}{\sqrt{re^{i\varphi}}} \sim \sqrt{r} \to 0.$$

Introducing the variable of integration y by $t = i(1 + y)$ and calculating the integral over C_3, we find

$$I \simeq 2e^{-\omega} \int_0^{\infty} e^{-\omega y} \frac{dy}{\sqrt{2y}} = \sqrt{\frac{2\pi}{\omega}}\, e^{-\omega}.$$

Thus the integral is exponentially small for large ω.

Example 2. $$I = \int_{-\infty}^{\infty} \frac{e^{i\omega t} dt}{\sqrt{(a^2 + t^2)(b^2 + t^2)}}, \quad \omega \to \infty.$$

In this case there are two branch points in the upper half-plane, $t = ia$ and $t = ib$. ; we assume that $a > b$. First of all we can reduce the number of independent parameters in I: measuring t, a, ω and I in units of the appropriate powers of b, we get

$$I = \int_{-\infty}^{\infty} e^{i\omega t} \frac{dt}{\sqrt{(a^2+t^2)(1+t^2)}}.$$

We shift the contour of integration into the upper half-plane (see Fig. 5). In Fig. 5 there is a cut from the point i to the point ia. By considerations analogous to those used in the previous example,

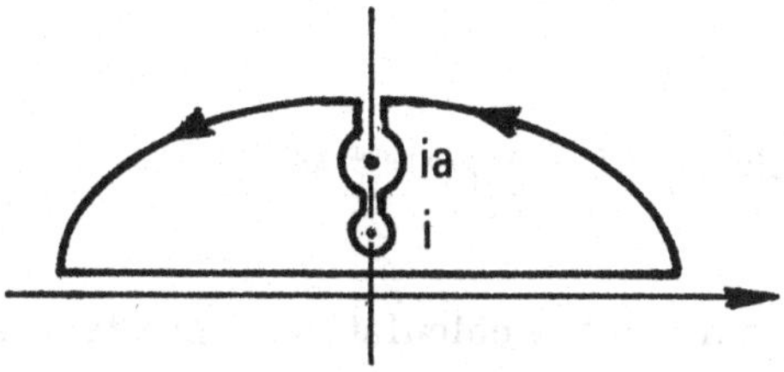

Fig. 5

it is easy to see that the required integral is just twice the integral along the cut. By changing the variable of integration ($t = i(1+y)$) we get

$$I \simeq 2e^{-\omega} \int_0^{a-1} \frac{e^{-\omega y}dy}{\sqrt{2y(a^2-1-2y)}}.$$

In calculating this integral there are two possibilities. If $a-1 \gg \omega^{-1}$, then the important region of integration is cut off by the exponential factor $e^{-\omega y}$ and is of order ω^{-1}. Consequently we have

$$I \simeq 2e^{-\omega} \int_0^{\infty} \frac{e^{-\omega y}dy}{\sqrt{2y(a^2-1)}} = \sqrt{\frac{2\pi}{\omega(a^2-1)}}\, e^{-\omega}.$$

If on the other hand $a-1 \ll \omega^{-1}$, then within the interval of integration we have $e^{-\omega y} \approx 1$ and so

$$I \simeq 2e^{-\omega} \int_0^{a-1} \frac{dy}{\sqrt{2y(a^2-1-2y)}} = 2\, e^{-\omega} \arcsin \sqrt{\frac{2}{a+1}}.$$

In the first case the two singularities of the integrand are distant from one another, in the second they are close together. We see that in both cases the exponentially small factor is determined by the singularity closest to the real axis, whereas the pre-exponential factor depends appreciably on the position of both singularities.

Notice that in the limit $a \to 1$ we get from the last expression $I \to \pi e^{-\omega}$. We can easily get this result immediately by noticing that for $a \to 1$ the two square-root singularities coalesce to give a simple pole.

Example 3. $I = \int_{-1}^{1} f(x)\, e^{i\omega x} dx.$

This integral occurs in the calculation of scattering amplitudes (cf. Section 2.1). Here the limits of integration are finite. Integration by parts gives

$$I = f(x)\frac{e^{i\omega x}}{i\omega}\Big|_{-1}^{1} - \frac{1}{i\omega}\int_{-1}^{1} f'(x)\, e^{i\omega x} dx =$$

$$= \frac{f(1)\, e^{i\omega} - f(-1)\, e^{-i\omega}}{i\omega} + O\left(\frac{1}{\omega^2}\right).$$

Thus, when the limits of integration are finite the high Fourier components decrease, generally speaking, according to a power law, not exponentially as in the case of infinite limits. (An exception is the case in which, on repeatedly integrating by parts, we find that the contribution of $f(x)$ and all its derivatives vanishes at the limits of integration).

Example 4. $I = \int g(\mathbf{r}) \exp\left[i(kr - \mathbf{kr})\right] d\mathbf{r}$ при $kR \gg 1$,

where R is the characteristic distance over which the function $g(\mathbf{r})$ changes appreciably.

We separate out the integration over the angular variables:

$$I = \int_0^\infty r^2\, dr \int_0^\pi g(r, \theta, \varphi)\, e^{ikr(1-\cos\theta)} \sin\theta\, d\theta\, d\varphi.$$

Integrating over θ by parts, we get

$$I = \int_0^\infty r^2\, dr\, \frac{1}{ikr}\left[g(r, \theta, \varphi)\, e^{ikr(1-\cos\theta)} \Big|_0^\pi + O\left(\frac{1}{kR}\right)\right] d\varphi.$$

The contribution from the upper limit may be neglected because of the rapid oscillation of the integrand; we therefore finally get

$$I \approx \frac{2\pi i}{k} \int_0^\infty g(r, \theta = 0)\, r\, dr \left[1 + O\left(\frac{1}{kR}\right)\right].$$

These results may be generalized: the high Fourier components of a function which has no singularities on the real axis are exponentially small. In fact, if x_1 is the distance from the real axis to the nearest singularity of $f(x)$, then we may estimate

$$f_\omega = \int_{-\infty}^{\infty} f(x)\, e^{i\omega x} dx \underset{\omega x_1 \gg 1}{\sim} e^{-\omega x_1}.$$

Let us estimate the magnitude of the preexponential factor for the Fourier component f_ω. If $f(x)$ has a pole in the complex ω plane, then we have

$$f_\omega \sim f(x_1) x_1 e^{-\omega x_1}.$$

since the order of magnitude of the residue c_1 of the function $f(x)$ at the pole is easily seen from dimensional considerations to be given by $c_1 \sim x_1 f(x_1)$. If $f(x)$ has a branch point of the root type, then as we saw in the first two examples (pp. 17, 18), we have

$$f_\omega \sim \frac{f(x_1)\, x_1}{\sqrt{\omega x_1}}\, e^{-\omega x_1}.$$

In the case of finite limits we get by integration by parts (see Example 3 above)

$$f_\omega = \int_{a_1}^{a_2} f(x)\, e^{i\omega x} dx \underset{\omega\to\infty}{\simeq} \frac{f(a_1)}{i\omega}\left[1 + O\left(\frac{1}{\omega}\right)\right].$$

The same type of estimate also applies to integrals between infinite limits in cases where f(x) has a discontinuity on the real axis. To see this, we can divide the integral into two regions, that from $-\infty$ to the discontinuity $(x = a)$ and that from the discontinuity to ∞. If we then calculate each of these integrals separately, then, just as in the example just considered, we find they decrease according to a power law. In this case

$$f_\omega \approx i\,\frac{\Delta f}{\omega},$$

where Δf is the magnitude of the discontinuity.

It is easy to see that if the nth derivative of f(x) has a discontinuity, while all lower derivatives are continuous, then

$$f_\omega \approx \frac{\Delta f^{(n)}}{(-i\omega)^{n+1}},$$

where $\Delta f^{(n)}$ is the discontinuity of $f^{(n)}(x)$. To see this, we write

$$f_\omega = \int_{-\infty}^{a} f(x)\, e^{i\omega x} dx + \int_{a}^{\infty} f(x)\, e^{i\omega x} dx.$$

and integrate by parts $n + 1$ times.

Example 5. To elucidate these results we consider an example which will illustrate how the exponential behaviour of the high Fourier components turns into power law behaviour when a singularity of the function approaches the real axis. Consider the integral

$$I = \int_{-\infty}^{\infty} \frac{e^{i\omega x}}{1 + e^{\alpha x}} dx, \quad \omega = \omega_0 - i\delta, \ \alpha, \delta > 0, \ \delta \to 0.$$

The choice of sign of δ makes the integral convergent; at the end of the calculation we may let δ tend to zero.

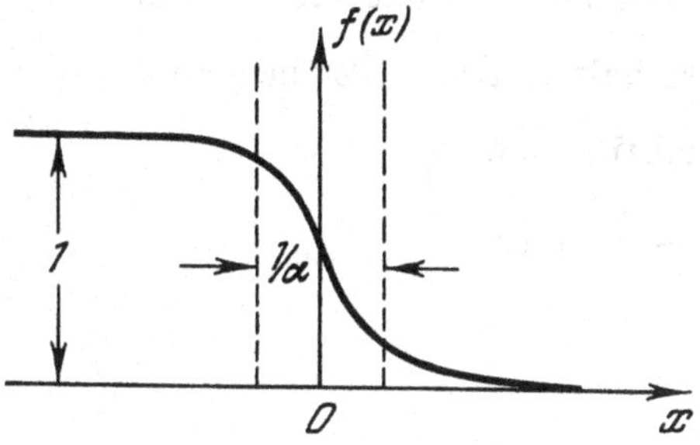

Fig. 6

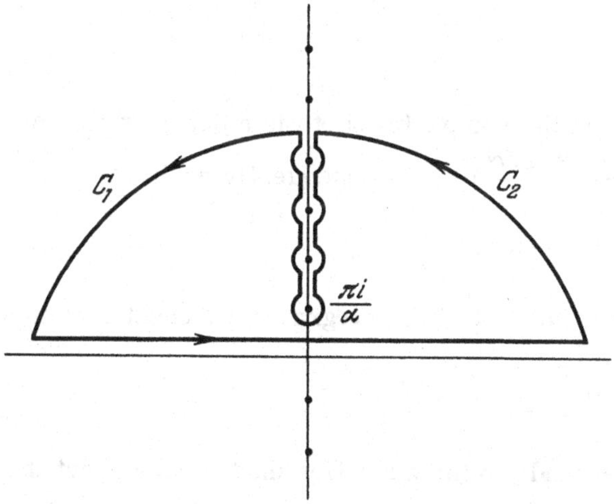

Fig. 7

The function $f(x) = (1 + e^{\alpha x})^{-1}$ has the form shown in Fig. 6. When α tends to infinity, this function tends to a step function. The function $f(x)$ has simple poles at the points $x_k = (2k+1)\pi i/\alpha$, where k is an integer. When α tends to infinity, each of these poles approaches the real axis.

We shall calculate the integral by moving the contour of integration into the upper half-plane (Fig. 7). The contributions from C_1 and C_2 vanish when the contour is shifted out to infinity, and the integral therefore reduces to the sum of residues at the poles in the upper half-plane. To find the residue we expand $e^{\alpha x}$ in a series around the pole x_k:

$$e^{\alpha x} = -1 - \alpha (x - x_k).$$

Then we have

$$I = \sum_{k=0}^{\infty} \oint e^{i\omega x} \frac{dx}{-\alpha (x - x_k)} =$$
$$= -\frac{2\pi i}{\alpha} \exp\left(-\frac{\pi\omega}{\alpha}\right) [1 + e^{-2\pi\omega/\alpha} + e^{-4\pi\omega/\alpha} + \dots].$$

The series in the square brackets is a geometrical series whose sum is $(1 - e^{-2\pi\omega/\alpha})^{-1}$. Consequently we have

$$I = -\pi i/\alpha \, \mathrm{sh} \frac{\pi\omega}{\alpha}.$$

If $\omega/\alpha \gg 1$, this integral is exponentially small:

$$I \underset{\omega/\alpha \gg 1}{\approx} -\frac{2\pi i}{\alpha} e^{-\pi\omega/\alpha}.$$

This corresponds to the assertion made above about the exponential decrease of the high Fourier components of a function with no singularities on the real axis. On the other hand when $\alpha \to \infty$, we get

$$I \underset{\omega/\alpha \ll 1}{\approx} -\frac{i}{\omega}.$$

Thus when the singularity tends to the real axis, the Fourier components decrease only according to a power law.

Problem: Estimate the high Fourier components of the function $f(x) = x\, e^{-|x|}$, which occurs in the theory of the dipole photoeffect (p. 57)

Solution: $f_\omega \underset{\omega\to\infty}{\approx} \frac{2i}{\omega^3}$

Methods of Approximate Solution of Differential Equations

In this section we shall give some examples of the qualitative solution of differential equations.

Example 1. The quasiclassical approximation (this method will be considered in detail in Section 3.1).

In many problems of theoretical physics it is necessary to solve a differential equation of the form

$$\varphi'' + k^2(x)\,\varphi = 0 \qquad (1.5)$$

The nature of the solution of (1.5) is critically dependent on the sign of k^2. For if k^2 is positive, then the solution has an oscillatory character; e.g. for k = constant, $\varphi = \exp(\pm ikx)$. If on the other hand $k^2 < 0$, then the solution is exponentially increasing or decreasing; for k = const., $\varphi = \exp(\pm|k|x)$. Even when k is a function of x the behaviour of the solution of (1.5) is qualitatively similar (Fig. 8).

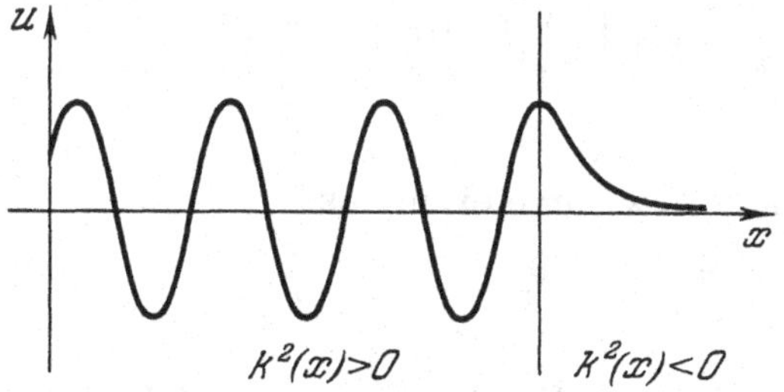

Fig. 8

If $k^2(x)$ is large and positive for the values of x which are of interest, the approximate solution of Eq. (1. 5) has the form

$$\varphi = f(x) \exp\{\pm i \int k(x)dx\}, \tag{1.6}$$

where f(x) is a slowly varying function. We can see this as follows: in this case $\varphi' \approx \pm ik(x)\varphi$ and $\varphi'' \approx -k^2(x)\varphi$. If k(x) is a sufficiently smooth function of x, then the solution (1. 6) satisfies Eq. (1. 5) approximately. We can actually find the next-order correction term (see Section 3. 1 for more details). Since $\varphi' = (f'/f)\varphi \pm ik\varphi$ and $\varphi'' \simeq -k^2\varphi \pm ik'\varphi \pm 2ik(f'/f)\varphi$, then by substituting φ and φ'' in (1. 5) we get

$$2kf' + fk' = 0$$

or

$$\frac{f'}{f} = -\frac{k'}{2k}$$

whence

$$f \sim \frac{1}{\sqrt{k}}$$

Thus the approximate solution of Eq. (1. 5) has the form

$$\begin{aligned} \varphi &\approx \frac{1}{\sqrt{k}} \exp\left\{\pm i \int k(x)\,dx\right\}, \quad k^2 > 0, \\ \varphi &\approx \frac{1}{\sqrt{|k|}} \exp\left\{\pm \int |k(x)|\,dx\right\}, \quad k^2 < 0. \end{aligned} \tag{1.7}$$

Consider now some examples:

(a) $\varphi'' + \alpha x \varphi = 0$

This is the equation satisfied by the so-called Airy function. First of all, suppose $x > 0$. Then $k^2 = \alpha x > 0$, and so according to (1. 7) the solution for large x has the form

$$\varphi \underset{x\to\infty}{\approx} \frac{1}{\sqrt[4]{\alpha x}} \exp\left[\pm \frac{2}{3} i \sqrt{\alpha x^3}\right]. \tag{1.8}$$

The solution for $x < 0$ can be obtained by analytic continuation of (1.8) (for further details on this point see Section 3.1).

$$\varphi \underset{|x|\to\infty}{\approx} \frac{e^{-\pi i/4}}{\sqrt[4]{\alpha |x|}} \exp\left[\pm \frac{2}{3} \sqrt{\alpha |x|^3}\right].$$

The extra phase factor $\exp(-\pi i/4)$ results from going around the branch point $x = 0$.

(b) $\varphi'' + (\alpha - \beta x^2)\varphi = 0.$

If $\beta x^2 \gg \alpha$, then $\int |k|\, dx \approx \sqrt{\beta} \int x\, dx = (1/2)\sqrt{\beta}\, x^2$. Thus the asymptotic solution for large x has the form

$$\varphi \underset{x\to\infty}{\approx} \frac{1}{\sqrt[4]{\beta x^2}} \exp\left[\pm \frac{1}{2} \sqrt{\beta x^2}\right].$$

Problems.

1. Find an asymptotically valid estimate for the solution of the equation $xu'' + (\gamma - x)u' - \alpha u = 0$ (the confluent hypergeometric equation) by first reducing the equation to self-adjoint form with the help of a suitable change of the variable u.

2. Using the quasiclassical approximation, obtain an asymptotic expression (valid for $x \gg 1$) for the Bessel function $J_n(x)$. Simplify it for the case $x \gg n$.

Solutions.

1. $u \sim x^{\alpha-\gamma} e^x$ for $x \to \infty$, $u \sim |x|^{-\alpha}$ for $x \to -\infty$.

2. $J_n(x) \underset{x\to\infty}{\sim} (x^2 - n^2)^{-1/4} \sin\left(\sqrt{x^2 - n^2} + n \sin^{-1}\left(\frac{n}{x}\right) + C_n\right),$

where C_n is a constant determined by the type of Bessel function. For $x \gg n$ we have $J_n(x) \sim x^{-1/2} \sin(x + C_n)$.

Example 2. Approximate solutions to differential equations can also be obtained by omitting the small terms in the equations and carrying out a systematic iteration procedure. Consider the example

$$y'' + y + \alpha y^3 = 0. \tag{1.9}$$

We assume the parameter α to be small and look for the solution in the form of successive approximations with respect to α. The solution of the unperturbed equation $y^{(0)''} + y^{(0)} = 0$ has the form $y^{(0)} = a \sin(x + \varphi)$; by an appropriate shift of the origin of the x-coordinate we can put $\varphi = 0$. If we substitute this solution in the term αy^3 of eqn. (1.9) and solve the latter directly by the iterative method, we run into difficulty: in the equation

$$y^{(1)''} + y^{(1)} = -\alpha y^{(0)^3} = -\frac{\alpha a^3}{4}(3 \sin x - \sin 3x)$$

the free (inhomogeneous) term contains an eigenfunction of the homogeneous equation, and this leads to the appearance of a spurious resonance effect. To get rid of this, we must take into account that the perturbation αy^3 alters the frequency of the vibration; in fact, we look for the zeroth-order solution in the form $y^{(0)} = a \sin \omega x$, where the frequency ω is found by requiring that there should be no such resonance. It is convenient to rewrite eqn. (1.9) in the form

$$y'' + \omega^2 y = -\alpha y^3 + (\omega^2 - 1)y. \tag{1.10}$$

and to regard the right-hand side of this equation as the inhomogeneous term. Substituting $y^{(o)}$ for y in the right-hand side of (1.10), we get

$$y^{(1)''} + \omega^2 y^{(1)} = -\frac{\alpha a^3}{4}(3\sin\omega x - \sin 3\omega x) + (\omega^2 - 1)\, a\sin\omega x. \tag{1.11}$$

We now require that the coefficient of $\sin \omega x$ on the right-hand side of (1.11) should vanish. This gives $\omega \approx 1 + 3\alpha a^2/8$.

We next look for the particular integral of the inhomogeneous equation (1.11) in the form

$$y^{(1)} = \gamma \sin 3\omega x.$$

Substituting this solution in (1.11), we find $\gamma = \alpha a^2/32$. Thus,

$$y \approx a\sin\omega x + \frac{\alpha a^3}{32}\sin 3\omega x,$$

where $\omega = 1 + 3\alpha a^2/8$. The condition for the solution we have found to be applicable has the form $\alpha a^2 \ll 1$. If we were to continue the iteration process we would obtain for ω a power series in the parameter αa^2, and for y a series containing terms of the form $a(\alpha a^2)^k \sin(2k+1)x$.

Example 3. Let us suppose that the solution is known in certain regions of the variable x. In this case we can construct approximate formulae for the solution over the whole range of x. As an example, consider the Thomas-Fermi equation

$$\varphi'' = \frac{1}{\sqrt{x}}\varphi^{3/2}, \quad \varphi(0) = 1, \quad \varphi(\infty) = 0. \tag{1.12}$$

This equation occurs when one tries to find the self-consistent field of an atom in the semiclassical approximation (cf. Section 3.2).

Let us find the approximate solution of (1.12) for $x \to \infty$.

We shall look for it in the form $\varphi = A/x^{\alpha}$. Substituting this form of solution in (1.12), we get

$$A\alpha(\alpha+1)x^{-\alpha-2} = A^{3/2}x^{-(3\alpha+1)/2},$$

whence we find $\alpha = 3$ and $A^{1/2} = \alpha(\alpha+1)$ or $A = 144$. Thus for $x \to \infty$

$$\varphi \underset{x\to\infty}{\longrightarrow} \frac{144}{x^3}.$$

We next find the next-order correction to this solution as $x \to \infty$, that is, we put

$$\varphi \underset{x\to\infty}{\approx} \frac{144}{x^3} + \psi. \tag{1.13}$$

and solve for ψ. Substituting (1.13) in (1.12), we find to the approximation linear in ψ

$$\psi'' = \frac{1}{\sqrt{x}}\frac{3}{2}\varphi^{1/2}\psi = \frac{3}{2}\cdot\frac{12}{x^2}\psi. \tag{1.14}$$

It is clear from (1.14) that ψ must be some power function of x, that is,

$$\psi = \frac{B}{x^{\beta}}$$

From (1.14) we get for β the quadratic equation

$$\beta(\beta+1) = 18$$

whence $\beta = (1/2)(-1 \pm \sqrt{73})$. Since we must have $\psi \to 0$ as $x \to \infty$, we must take the positive root $\beta \approx 3.77$. Hence

$$\varphi \underset{x\to\infty}{\approx} \frac{144}{x^3} + \frac{B}{x^{3,77}}. \tag{1.15}$$

To the same accuracy as (1.15) we can write alternatively

$$\varphi = \frac{144}{x^3\left(1+\dfrac{C}{x^{0,77}}\right)^n},$$

where n and c are for the moment undetermined constants. We now choose n so that $\varphi(0)$ shall be finite; this requires $3-0.77n=0$, or $n = 3.90$. Next we find c from the condition $\varphi(0) = 1$; this gives $144/c^{3.90} = 1$, whence $c = (12^{2/3})^{0.77}$. Thus, finally,

$$\varphi = \frac{1}{[1+(x/12^{2/3})^{0.77}]^{3.90}} . \qquad (1.16)$$

This approximate solution is in satisfactory agreement with the exact solution found by machine computation: see Fig. 9.

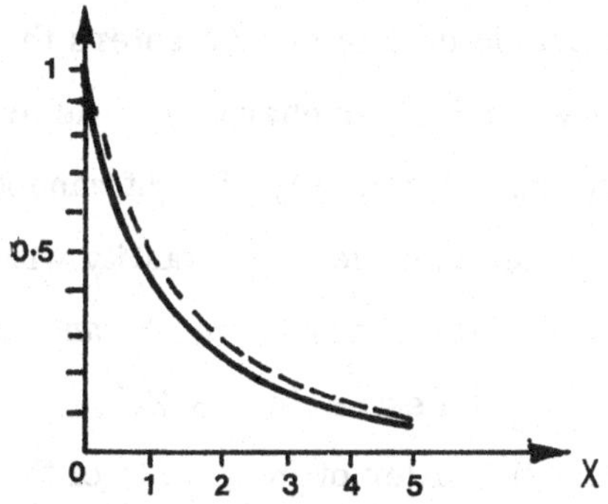

Fig. 9*

2. ATOMIC PHYSICS

In this section we shall obtain some fundamental results in atomic physics, using "model" approximations as well as dimensional ones. "Model" approximations mean those derived from limiting cases or simplified variants of the problem in question.

---

*The solid line is the approximate solution (1.16), the dashed line the exact solution.

Estimate of the Velocities and Orbit Sizes of the Inner Atomic Electrons

To estimate the velocity of an electron in one of the inner atomic orbits we must form a quantity with the dimension of velocity from the quantities characterizing the structure of the atom, namely the electronic mass m, Planck's constant $\hbar$, and the product of the electronic and nuclear charges Ze^2. (Since the effect of the other electrons can be neglected when we consider the inner atomic electrons, the electronic charge always enters the equations of motion multiplied by the nuclear charge.) The problem is a non-relativistic one, so that the velocity of light cannot enter the formulas. Then we see that the only quantity with the dimension of velocity which can be formed from m, $\hbar$, and Ze^2 is $Ze^2/\hbar$, and hence we can make the estimate $v \sim Ze^2/\hbar$.

To estimate the order of magnitude of the radius of the inner electronic orbits we use the fact that the potential energy of the electron, Ze^2/a, is of the same order as its kinetic energy $(1/2)mv^2$, that is, $Ze^2/a \sim mv^2$ or $a \sim \hbar^2/Zme^2$.

To estimate the radius of the outer orbits we must put $Z = 1$, since in this case the nuclear charge is almost completely screened by the inner electrons.

From these estimates of a and v it is obvious that a convenient system of units for atomic problems is the system in which $e = \hbar = m = 1$ (the so-called atomic units). In these units the velocity of light c and the proton mass M are given by $c = 137.2$ and $M = 1836$.

Stationary States

Consider the motion of a particle in a potential well (Fig. 10). Schrödinger's equation for this problem has the form

$$\Psi'' + k^2\Psi = 0,$$

where $k^2 = 2(E - V)$. In the classically accessible region the potential energy V is less than the total energy E. Consequently the wave function oscillates in this region [see Eq. (1.7)]. In the region where $E < V$ the wave function is exponentially decreasing. We can estimate the energy of the stationary states from the

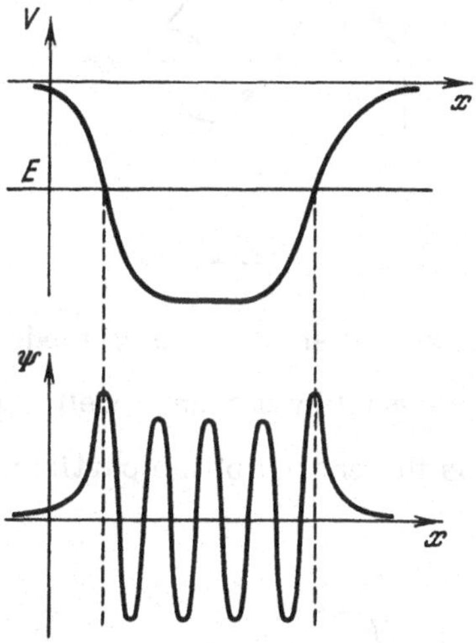

Fig. 10

condition that the characteristic length ℓ of the region in which the particle can move must contain an integral number n of de Broglie half-waves.

Note that we cannot find the stationary-state energy levels by using dimensional considerations alone, since the number n is

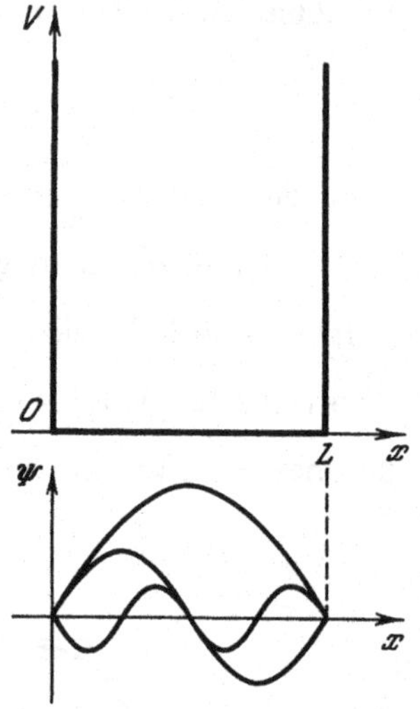

Fig. 11

dimensionless. Instead, we shall make a "model" approximation and replace the true potential by a square well (Fig. 11) with a width which depends on the energy of the particle. The de Broglie wavelength is given by

$$\lambdabar = 1/p = 1/\sqrt{2(E - V)},$$

where p is the momentum of the particle. The quantity $\lambdabar$ depends on the coordinate x; however, for purposes of estimation we may take $\lambdabar \sim 1/\sqrt{2E}$ everywhere, E being measured from the bottom of the well.

Thus the energy levels E_n of the stationary states may be approximately obtained from the relation

$$L(E_n) \approx \frac{n}{2}\lambda \sim \frac{2\pi n}{2\sqrt{2E_n}}, \tag{1.17}$$

where $L(E_n)$ is the characteristic width of the region in which the particle can move. For a square well $L(E_n)$ is equal to L, the width of the well. Hence $L = \pi n/\sqrt{2E_n}$, or $E_n = \pi^2 n^2/2L^2$. This expression satisfies Bohr's correspondence principle, which

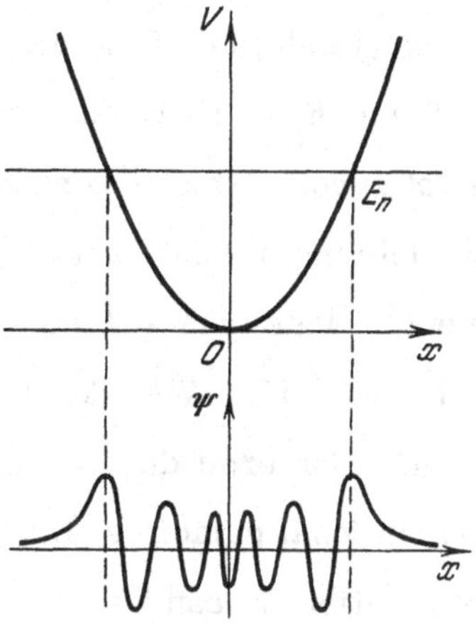

Fig. 12

tells us that for large quantum numbers the spacing between neighbouring energy levels must be equal to the classical frequency of motion (cf.p.165). In fact we have

$$\frac{dE_n}{dn} = 2\frac{\pi^2 n}{2L^2} = \frac{2\pi}{2L\Big/\frac{\pi n}{L}} = \frac{2\pi}{2L/v} = \frac{2\pi}{T_{\text{кл}}},$$

where $T_{c\ell}$ is the classical period of motion 2L/v.

Our next example is a one-dimensional harmonic oscillator with potential energy $V = (1/2)\,\omega^2 x^2$, where ω is the classical oscillator frequency. We notice that the frequency of oscillation of the wave function is greater near the origin than near the edge of the well (see Fig. 12). This is because at $x = 0$ the kinetic energy is a maximum and the wavelength $\lambda\!\!\!^{-}$ accordingly a minimum. For given energy E_n the width L of the classically accessible region is given by the condition $V(L) = E_n$. Thus, $(1/2)\omega^2 L_n^2 \sim E_n$, or $L_n \sim \sqrt{E_n}/\omega$. Then the quantization condition (1.17) becomes $\sqrt{E_n}/\omega \sim n/\sqrt{E_n}$, whence $E_n = C_1 n\omega$. It then follows from the correspondence principle that $C_1 = 1$.

As we know, exact solution of the problem leads to an oscillator energy spectrum $E_n = (n+1/2)\omega$, so that for $n = 0$ the vibrational energy is not zero. This is connected with the uncertainty principle: a particle moving in a small region of dimension d cannot have a momentum less than $\sim 1/d$. The energy E therefore is of order $E \sim d^{-2} + \omega^2 d^2$. We take the minimum of this expression; this minimization procedure corresponds to the quantum mechanical variational principle $E = \min(\psi|H|\psi)$ (in the present case the wave function ψ can be regarded as depending only on the single parameter d). The expression for E is a minimum when $d \sim 1/\sqrt{\omega}$, when $E_{min} = C_2\omega$ (actually $C_2 = 1/2$). The quantity $d^2_{min} \sim 1/\omega$ gives the order of magnitude of the squared amplitude of the zero-point vibration.

We can consider motion in a Coulomb field by a similar method (see Fig. 13). We restrict ourselves here to the case $\ell = 0$ (the case $\ell \neq 0$ will be investigated by the quasiclassical method below, p.192). The energy E is of order $(n^2/L^2) - (Z/L)$; minimizing this expression, we find $L = 2n^2/Z$, and substitution of this in the expression for the energy gives $E_n = -Z^2/4n^2$. Let us write $E_n = -C_1 Z^2/n^2$ and find C_1 from the correspondence

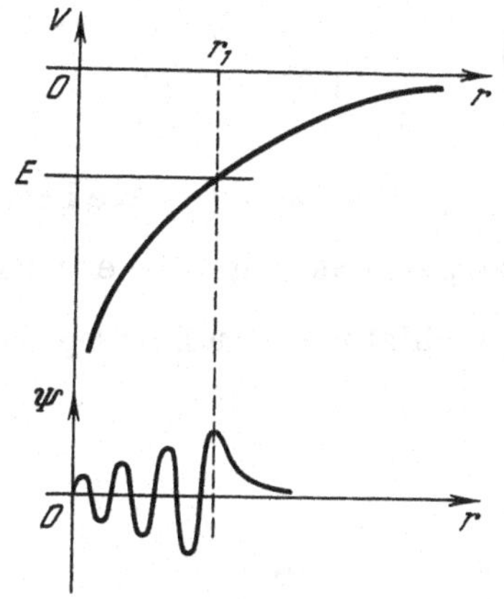

Fig. 13

principle:

$$\frac{dE_n}{dn} = C_1 \frac{2Z^2}{n^3} = \omega_{c\ell} = \frac{2\pi}{T} .$$

Here $T_{c\ell}$, the period of the classical motion, is given by

$$T_{c\ell} = 2\int_0^{r_1} \frac{dr}{v} = 2 \int_0^{-\frac{Z}{E_n}} \frac{dr}{\sqrt{2\left(E_n + \frac{Z}{r}\right)}} = \frac{2\pi Z}{(-2E_n)^{3/2}} .$$

Thus, $C_1 \cdot 2Z^2/n^3 = [\,2\pi/(2\pi Z)\,]\,(2C_1\,Z^2/n^2)^{3/2}$, whence $C_1 = 1/2$. So, finally, we get $E_n = -(1/2)Z^2/n^2$, which, as we know, coincides with the exact solution.

Let us finally investigate the more complicated case in which a particle moves in an anharmonic-oscillator field. The potential V is $\frac{1}{2}\,\omega_0^2 x^2 + \frac{1}{4}\lambda x^4$, where $\lambda > 0$. The energy E(L) may be estimated as

$$E(L) = \frac{\left(n+\frac{1}{2}\right)^2}{2L^2} + \frac{\omega_0^2 L^2}{2} + \frac{\lambda L^4}{4}.$$

The groundstate corresponds to n = 0. We have chosen the groundstate kinetic energy so as to get the exact result for the energy of a harmonic oscillator. Minimizing the above expression leads to the equation

$$-\frac{\left(n+\frac{1}{2}\right)^2}{L^3} + \omega_0^2 L + \lambda L^3 = 0,$$

or

$$\left(n+\frac{1}{2}\right)^2 = L^4(\omega_0^2 + \lambda L^2). \tag{1.17'}$$

Consider first the case of large λ. Neglecting the term in ω_o^2 in (1.17'), we obtain $\lambda L^6 = (n+\frac{1}{2})^2$. Thus, $E_n = \frac{3}{4}\lambda^{1/3}(n+\frac{1}{2})^{4/3}$. Taking the term $\omega_o^2 L^2/2$ into account by perturbation theory, we get a correction

$$\Delta E_n = \frac{\omega_0^2}{2}\frac{\left(n+\frac{1}{2}\right)^{2/3}}{\lambda^{1/3}}.$$

In the opposite case of small λ we obtain by a similar method

$$E_n = \left(n+\frac{1}{2}\right)\omega_0 + \frac{\lambda}{4}\frac{\left(n+\frac{1}{2}\right)^2}{\omega_0^2}.$$

It is straightforward to solve the cubic equation (1.17') and so get a general formula for E_n for arbitrary λ; however, since it is somewhat cumbrous we shall not give it here.

Distribution of Electric Charge in the Atom

The distribution of charge in heavy atoms may be calculated by the approximation method known as the Thomas-Fermi method (cf. also p.197). This is based on the fact that in a heavy atom most of the electrons have large quantum numbers, so that their de Broglie wavelengths are small compared with the region over which the potential changes appreciably. Then the density of electrons at any given point in the atom may be calculated in the same way as the density of free electrons in a flat-bottomed potential well.

Suppose the maximum momentum of the electrons at a given point is p_0. Then the number of states per unit volume in the interval $(\underset{\sim}{p}, \underset{\sim}{p} + d\underset{\sim}{p})$ is $d\underset{\sim}{p}/(2\pi)^3$. Since in the ground state of the atom the electrons must fill up all the states with momentum less than p_0, the number of electrons per unit volume is given by

$$n = 2\int \frac{dp}{(2\pi)^3} = 2\,\frac{4}{3}\,\pi p_0^3\,\frac{1}{(2\pi)^3}$$

(the factor of 2 is a result of the two possible spin states). We denote the electrostatic potential of the atomic field by $\varphi(r)$. By Poisson's equation

$$\nabla^2\varphi = 4\pi n,$$

i. e.

$$\nabla^2\varphi \sim p_0^3.$$

Let us relate p_0 to φ. The total energy of an electron is given by $p^2/2 - \varphi(r)$. Let ϵ_0 be the maximum value of the total energy at a given point in the atom, that is,

$$\varepsilon_0 = \frac{p_0^2}{2} - \varphi(r).$$

The quantity ϵ_0 must be constant throughout the atom (otherwise electrons would flow from points with high energy to points of lower energy). From the last two formulas we get

$$\nabla^2\varphi \sim (\varphi + \varepsilon_0)^{3/2}$$

or

$$\nabla^2\varphi \sim \varphi^{3/2},$$

if we measure φ from $-\epsilon_0$.

Let us determine the radius ℓ of the region containing the majority of the electrons of the atom. From the last formula we get the order-of-magnitude estimate

$$\frac{\varphi}{l^2} \sim \varphi^{3/2}, \text{ whence } l \sim \varphi^{-1/4}.$$

On the other hand, we can relate ℓ to the total number Z of electrons in the atom:

$$Z \sim nl^3$$

or in view of the relation $\nabla^2\varphi \sim n$,

$$Z \sim \nabla^2\varphi l^3 \sim \frac{\varphi}{l^2} l^3 \sim \varphi l.$$

Comparing this expression with the condition $\ell \sim \varphi^{-1/4}$ derived above, we find $Z \sim \ell^{-4} \cdot \ell$ or

$$l \sim \frac{1}{Z^{1/3}}.$$

The mean energy ϵ of an electron in the atom is of the order of magnitude of the electrostatic potential $\varphi(\ell)$. Since $\ell \sim \varphi^{-1/4}$, we get

$$\varepsilon \sim l^{-4} \sim Z^{4/3}.$$

The total energy E of all the electrons has the order of magnitude

$$E \sim \varepsilon Z \sim Z^{7/3}.$$

and the mean number of nodes of their wave function is of order $\ell\sqrt{\epsilon} \sim Z^{1/3}$.

Thus, the radius of the atomic K shell is of order $r_K \sim 1/Z$, the distance ℓ (the Thomas-Fermi radius) of the region containing the majority of the electrons of the atom is $\sim Z^{-1/3}$ and the radius of the outer shells is ~ 1 (the last estimate follows since the effective charge felt by the electrons of the outer shells is ~ 1).

The distribution of electric charge in an atom will be treated in more detail in Section 3.2.

The Rutherford Formula

Consider the scattering of a light particle with mass m and charge e by a heavy particle with charge Ze in the classical non-relativistic case. The scattering cross section is independent of the mass of the heavy particle and is determined only by the velocity v of the incident particle, its mass m and charge e, and the heavy-particle charge Ze. Coulomb's law implies that the cross section σ depends only on the product Ze^2. There is only one quantity with the dimensions of a length which can be formed from Ze^2, m, and v, namely Ze^2/mv^2; consequently the scattering cross section must have the form

$$\sigma(\theta) = \left(\frac{Ze^2}{mv^2}\right)^2 f(\theta).$$

We may find the angular dependence of the cross section in the limit

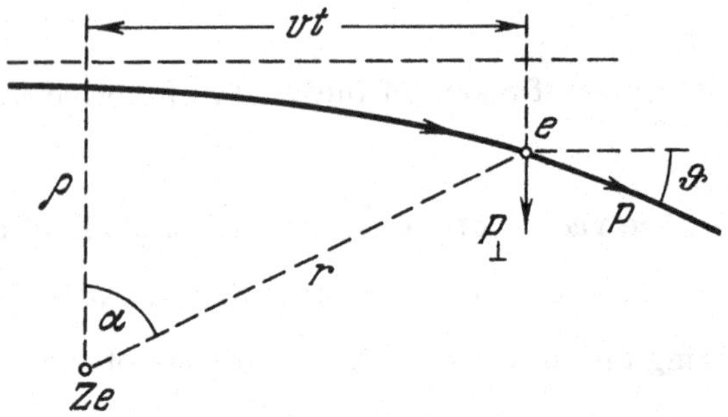

Fig. 14

of small scattering angles as follows: The angle of deflection θ is approximately given (see Fig. 14) by $\theta \approx p_\perp/p$, where p is the magnitude of the momentum of the particle and $p_\perp$ the component of the acquired momentum perpendicular to the initial direction of flight. We may calculate $p_\perp$ as follows:

$$p_\perp = \int_{-\infty}^{\infty} F_\perp \, dt = \int_{-\infty}^{\infty} \frac{Ze^2}{r^2} \cos\alpha \, dt =$$

$$= \int_{-\infty}^{\infty} \frac{Ze^2 \rho \, dt}{(\rho^2 + v^2 t^2)^{3/2}} = \frac{2Ze^2}{\rho v} \int_0^{\infty} \frac{dx}{(1 + x^2)^{3/2}} = \frac{2Ze^2}{\rho v}$$

(the integral may be performed by making the substitution $x = \tan y$), where ρ is the impact parameter (see Fig. 14). Thus,

$$\theta(\rho) = \frac{2Ze^2}{mv^2 \rho}. \tag{1.18}$$

But by definition of the differential cross section we have

$$d\sigma = \rho \, d\rho \, d\varphi = \rho \left| \frac{d\rho}{d\theta} \right| d\theta \, d\varphi = \rho \left| \frac{d\rho}{d\theta} \right| \frac{\sin\theta \, d\theta \, d\varphi}{\sin\theta},$$

or

$$\frac{d\sigma}{d\Omega} = \rho \left| \frac{d\rho}{d\theta} \right| \frac{1}{\sin\theta} . \tag{1.19}$$

From (1.18) and (1.19) we get the Rutherford formula for small scattering angles:

$$\frac{d\sigma}{d\Omega} = \left(\frac{2Ze^2}{mv^2} \right)^2 \frac{1}{\theta^4} . \tag{1.20}$$

Inapplicability of Classical Mechanics for Large Impact Parameters

In classical mechanics the total scattering cross section

$$\sigma = \int_0^\infty 2\pi\rho \, d\rho$$

tends to infinity for any potential other than those with a finite cutoff (such as a hard-sphere potential). In quantum mechanics on the other hand, the total cross section is finite for any potential which decreases with distance faster than $1/r^2$ (cf. below). We shall now show that classical mechanics must be inapplicable for sufficiently small scattering angles, that is, for sufficiently large impact parameters ρ. We shall see that as the impact parameter increases, the quantum mechanical diffraction angle θ_d decreases more slowly than the classical scattering angle $\theta_{c\ell}$. Consequently, beyond some critical value of the impact parameter ρ_1, the angle θ_d is greater than $\theta_{c\ell}$. This means that for impact parameters greater than ρ_1 quantum mechanical formulas must be used to calculate the cross section. Thus, even though in this case the de Broglie wavelength λ is much less than the characteristic dimensions of the problem, namely ρ_1, we still do not have the conditions for classical scattering; that is, the condition $\lambda \ll \rho_1$ is a necessary but not at all sufficient condition for the applicability of classical mechanics in the scattering problem.

Let us obtain a criterion for the applicability of classical mechanics. According to the estimate $\theta_{c\ell} \approx p_\perp / p$ made above, we conclude that

$$\theta_{c\ell} \sim F_\perp \Delta t / p \sim \frac{\partial V}{\partial \rho} \frac{\Delta t}{p} \sim \frac{V(\rho)}{p} \frac{\Delta t}{\rho} \sim \frac{V(\rho)}{pv}, \text{ i. e. } \theta_{c\ell} \sim \frac{V(\rho)}{E}.$$

Here we have assumed that $\partial V/\partial \rho \sim V/\rho$; E is the total energy of the particle and V the interaction potential.

To find the diffraction angle we consider the following experiment. A screen with an aperture of diameter a is placed so as to restrict the possible values of the impact parameter to within $\rho \pm a$ (see Fig. 15). If classical mechanics is to be applicable the width of the slit 2a must be much less than the impact parameter ρ, otherwise the very concept of an impact parameter loses its meaning. On the other hand, decreasing the width of the gap leads to diffraction scattering, and to be able to observe the deflection the diffraction angle must be much less than the scattering angle.

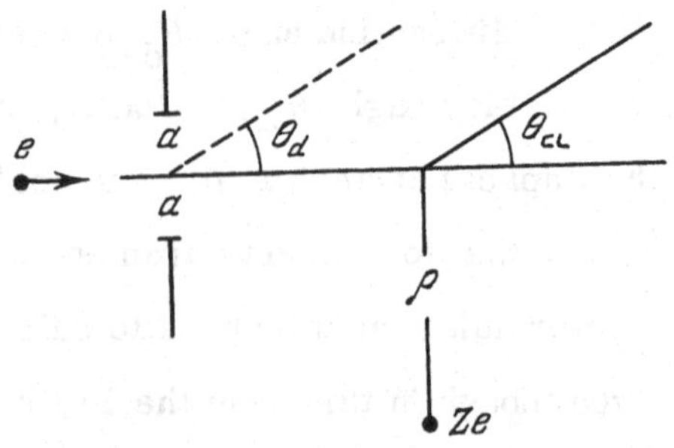

Fig. 15

It is clear from the uncertainty principle that an indeterminacy in the coordinate corresponds to an indeterminacy in the transverse component of momentum $\Delta p \sim 1/a$. Consequently, the diffraction angle is given by

$$\theta_{\text{diff}} \sim \frac{\Delta p}{p} \sim \frac{1}{pa},$$

where p is the momentum of the incident particle. For classical mechanics to be applicable we must have $\theta_{c\ell} \gg \theta_d$, or in other words $\theta_{c\ell} \gg 1/pa$; therefore

$$\theta_{c\ell} = \frac{V(\rho)}{E} \gg \frac{1}{p\rho}.$$

This is the general criterion for the applicability of classical mechanics in the scattering problem. For potentials which fall off faster than the Coulomb potential, the classical scattering angle decreases with increasing impact parameter faster than $1/\rho$. Hence we can always find an impact parameter ρ_1 such that $\theta_{c\ell} \sim \theta_d$. In the case of Rutherford scattering $\theta = 2Ze^2/p\rho v$ and the scattering will be classical for all angles provided only that $Ze^2/\hbar v \gg 1$.

Estimate of the Scattering Cross Section for Potentials Which Fall Off Faster than the Coulomb Potential

We have just seen that classical mechanics is inapplicable for impact parameters greater than the value ρ_1 defined by the condition that the classical scattering angle be equal to the diffraction angle:

$$\frac{\hbar}{\rho_1 p} = \theta_{c\ell}(\rho_1), \tag{1.21}$$

This means that the quantity ρ_1 may be found from the relation

$$\frac{\hbar}{\rho_1 p} \sim \frac{V(\rho_1)}{E}. \tag{1.22}$$

Equation (1.22) enables us to find ρ_1 for any concrete form of potential $V(\rho)$ which falls off with distance faster than the Coloumb potential (for which both sides of (1.22) are proportional to $1/\rho$, so that ρ_1 is not defined). Thus for $\rho < \rho_1$ the differential cross section may be calculated from the classical formulas.

The total cross section is given by

$$\sigma = \pi\rho_1^2 + \sigma_{\text{diff}},$$

when σ_{diff} is the diffraction cross section. If the potential falls off sufficiently fast, we may estimate σ_{diff} to be roughly equal to the cross section for scattering from a screen of radius ρ_1, that is, σ_{diff} will be of order $\pi\rho_1^2$, so that

$$\sigma \sim 2\pi\rho_1^2.$$

Let us estimate the total cross section for a power law potential $V = \alpha/r^n$. From formula (1.22) we find $1/\rho_1 p \sim \alpha/\rho_1^n E$, so that $\rho_1 \sim (\alpha/v)^{(n-1)^{-1}}$). Consequently,

$$\sigma \sim 2\pi\rho_1^2 \sim 2\pi\left(\frac{\alpha}{v}\right)^{\frac{2}{n-1}}.$$

We estimated the diffraction part of the scattering as if it were diffraction by a screen. One may show* that for this method to be valid, the potential must fall of faster than $1/r^2$; thus, the above estimate is valid for $n > 2$.

---

* L. D. Landau and E. M. Lifshitz, Quantum Mechanics, 2nd revised edition, Pergamon, Oxford, 1965, p. 473.

Resonance Effects in Scattering

Consider first the scattering from a potential having a strongly repulsive core (Fig. 16). Let the height and width of the potential barrier be V_0 and a respectively. We assume that $V_0 \gg E$, where E is the energy of the incident particle. Then the wave function must vanish at the barrier edge: $\Psi(a) = 0$. Let us consider for simplicity spherically symmetric scattering and assume that for $r > a$ the particle performs free motion, that is,

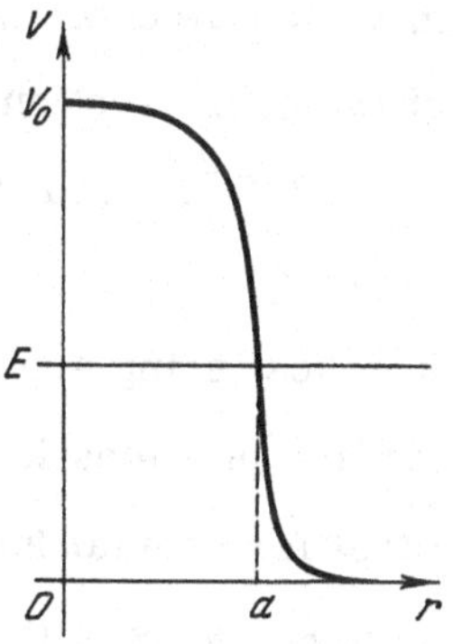

Fig. 16

its wave function is just a sum of an incident plane wave e^{ikz} and a diverging spherical wave $(f/r)e^{ikr}$, where f is the scattering amplitude. Now the plane wave has itself a spherically symmetric component, equal to $\sin kr/kr$. Thus we have

$$\int \Psi \frac{d\Omega}{4\pi}\bigg|_{r>a} = \frac{\sin kr}{kr} + \frac{f}{r}e^{ikr}.$$

Using the boundary condition $\Psi(a) = 0$, we find $\sin ka/k + fe^{ika} = 0$, or

$$f = -\frac{\sin ka}{k} e^{-ika}.$$

If the energy of the particle is such that $ka = n\pi$, where n is an integer, then $f = 0$, this is, the s-wave scattering cross section vanishes. This resonance effect is weakened in the total cross section by the fact that for $ka \gtrsim 1$ waves with $\ell \neq 0$ are also important in the scattering, so that the total scattering cross section does not tend to zero.

As is clear from the above expression for the scattering amplitude, the maxima of the cross section are obtained when $ka = (2n + 1)\pi/2$; then $\sigma = 4\pi\lambda\!\!\!^{-2}$. For small $k(ka \ll 1)$ we have $\sigma = 4\pi a^2$.

Resonance effects in scattering are applied in so-called illumination optics; the surface of a lens is coated with a layer of matter of thickness such that for normal incidence of the light the phase difference between the surfaces of the coating and of the lens is an integer times π. As a result, no light is reflected at the surface of the lens and any loss is due entirely to absorption.

An analogous effect occurs when a particle is scattered by a a potential well $(V < 0)$. Just as in the case $V > 0$, the scattering cross section assumes different values depending on the phase of the wave function at the well edge. For certain values of the phase the cross section vanishes. This effect was first observed by Ramsauer in the scattering of electrons from atoms; the classical theory of scattering cannot explain this phenomenon. Another example where classical mechanics is in outright contradiction to experiment is the capture of slow neutrons by nuclei;

the cross section for this process ($\sim 4\pi\lambda\!\!\!^{-2}$) is tens of thousands of times greater than the classical cross section ($\sigma_{c\ell} = \pi R^2$, where R is the nuclear radius).

Interaction between Atoms

Let us estimate the magnitude of the interaction of a neutral atom with an ion and with another neutral atom at large distances. The ion will produce an electric field $E = Z_1/r^2$, where Z_1 is the ionic charge. A neutral atom placed in an electric field E will acquire a dipole moment $d = \alpha E$, where α is the atomic polarizability. Thus, the interaction V of an ion with a neutral atom is given by $V = -Ed = -Z_1^2\alpha/r^4$. In atomic units $\alpha \sim 1$, since the principal contribution to the atomic polarization comes from the outermost shells, for which all quantities are of order 1 in atomic units. Thus, $V \sim -Z_1^2/r^4$.

Consider now the case when both atoms are neutral. They will then induce in one another dipole moments d_1 and d_2, and we know that the dipole-dipole interaction between atoms is given by $V \sim -d_1 d_2/r^3$. The mean value of d_1 and d_2 in the ground state is zero, since the atoms have no permanent dipole moment. However, the mean-square values d_1^2, d_2^2 are not zero, since the dipole moment operator oscillates at all possible atomic frequencies:

$$d = \sum_\omega d_\omega e^{i\omega t}.$$

A dipole moment $d_{1\omega}$ will produce a field $E_\omega \sim d_{1\omega}/r^3$. The moment $d_{2\omega}$ induced by this field is given by $d_{2\omega} = \alpha_2 E_\omega$. Substituting this expression in the interaction energy and averaging over time, we finally get

$$V \sim -\sum_{\omega} d_{2\omega}\mathscr{E}_{\omega} = -\frac{1}{r^6}\sum_{\omega} d_{1\omega}^2 \alpha_2(\omega) \sim -\frac{1}{r^6}.$$

This interaction is called the van der Waals interaction.

Ionization of Atoms

Let us estimate the energy lost in the motion of an electron with energy much greater than the energy of the electrons in the atomic shells. The incident electron, on colliding with the electrons of the atomic shells, can knock them out of the atom. For a large incident electron energy its angle of deflection θ is small, and hence the cross section for scattering by one of the shell electrons is given by the formula $d\sigma/d\Omega \sim 1/p^4\theta^4$, where p is the momentum of the incident electron.

We denote by q the change in momentum of the incident electron; q is related to the angle of deflection by $q^2/2$. The electron energy loss per unit path length is given by

$$\frac{dE}{dx} = -nZ\int \Delta E \, d\sigma,$$

where nZ is the total number of electrons per unit volume and $d\sigma$ is the differential cross section. Hence

$$-\frac{dE}{dx} \sim Zn\int q^2 \frac{1}{p^4\vartheta^4}\vartheta \, d\vartheta = Zn\frac{1}{p^2}\int\frac{dq}{q} = \frac{Zn}{p^2}\ln\frac{q_{\max}}{q_{\min}}.$$

Since the maximum energy transfer is of the order of the energy of the incident particle, we have $q_{max} \sim p$ ($\theta_{max} \sim 1$). The minimum momentum transfer corresponds to the case where the transit time τ is of the order of the reciprocal atomic frequencies, i. e. $\omega_o \tau \sim 1$; for long transit times the field of the incident electron contains no appreciable Fourier components with the atomic frequencies and the probability of ionization falls off sharply. The impact parameter ρ corresponding to this condition is defined by the relation $\rho = v\tau = v/\omega$, where v is the velocity of the incident electron. Since $(e^2/\hbar v) \ll 1$, the minimum angle θ_{min} is defined not by the classical condition but by the diffraction condition (cf. p.43), i. e. $\theta_{min} \sim 1/p\rho$ or $q_{min} = \omega_o/v$. Thus we have

$$-\frac{dE}{dx} \sim \frac{Zn}{p^2} \ln \frac{pv}{\omega_0} .$$

As we have proved above (p.40) the bulk of the electrons are to be found in a region of radius $a_o/Z^{1/3}$, and the energy per electron is of order $Z^{4/3}$. The quantity ω_o is of order $v/a \sim Z^{2/3}/Z^{-1/3} = Z$. Re-introducing dimensional quantities and introducing the ionization potential of hydrogen, I_o, we obtain

$$-\frac{dE}{dx} = C\frac{Zne^2}{mv^2} \ln \frac{E}{I_0 Z} .$$

An exact calculation gives $C = 4\pi$.

Multiple Scattering

A narrow beam of electrons passing through a medium will gradually be smeared out as a result of multiple scattering; for, although we may assume the probability of an electron being deflected to left or to right is equal, the mean-square deflection angle is not zero. We encounter a similar phenomenon in everyday life if we observe the progressive twisting of a telephone cord with time. After each telephone conversation the cord is twisted in one sense or the other with equal probability, but after a large number of conversations it ends up twisted by an angle proportional to the square root of the number of conversations.

Since the angular spreading of the electron beam is a sum of a large number of independent random processes, the angular distribution has a Gaussian form: $\Phi(\theta) = A \exp - \theta^2/\overline{\theta^2}$. The mean-square deflection angle $\overline{\theta^2}$ is proportional to the number of collisions N, which is equal to the sample thickness L divided by the mean free path ℓ. If $\overline{\theta_1^2}$ is the mean-square deviation angle in a single collision, then

$$\overline{\theta^2} = N\overline{\theta_1^2} = \overline{\theta_1^2}\, L/\ell = n\sigma L\overline{\theta_1^2} = \overline{\varphi^2} L$$

where σ is the scattering cross section, n the number of nuclei per unit volume, and $\overline{\varphi^2}$ the mean-square deviation angle per unit path length, that is, $\overline{\varphi^2} = n\sigma\overline{\theta_1^2}$. Since by definition $\overline{\theta_1^2} = (1/\sigma)\int \sigma(\theta)\theta^2\, d\Omega$, we have $\overline{\varphi^2} = n\int \sigma(\theta)\theta^2\, d\Omega$. For Rutherford scattering we get

$$\overline{\varphi^2} \sim 2\pi n \frac{Z^2}{E^2} \ln \frac{\theta_{max}}{\theta_{min}}$$

This formula is valid in both relativistic and nonrelativistic cases.

We consider two limiting cases: $Z/v \gg 1$ and $Z/v \ll 1$. In the former case the scattering may be considered classically for all angles (see p. 45). The maximum scattering angle is determined either by the condition that the impact parameter be equal to the nuclear radius R, that is, $\theta_{max} \sim V(R)/E = Z/RE$, if $Z/RE < 1$, or by $\theta_{max} \sim 1$ if $Z/RE > 1$. In the case of electron scattering the first of these conditions violates the relation $Z/v \gg 1$; for the condition $E > Z/R$ corresponds to relativistic energies, and for velocities $v \simeq c$ we have $Z/v \sim Z/c < 1$. Hence we have $\theta_{max} \sim 1$. The minimum scattering angle is determined by the atomic dimensions a: $\theta_{min} \sim Z/aE$. On estimating a from the Thomas-Fermi model, we get $\theta_{min} \sim Z/Z^{-1/3} E \sim Z^{4/3}/E$. Thus for electrons with $E \ll Z^2$ $(Z/v \gg 1)$

$$\overline{\varphi^2} \sim 2\pi n \frac{Z^2}{E^2} \ln \frac{E}{Z^{4/3}}$$

In the case $Z/v \ll 1$ the classical deflection angle is less than the quantum mechanical diffraction angle θ_d. Therefore θ_{max} and θ_{min} should be determined as follows: θ_{min} is the angle of diffraction corresponding to an aperture of the dimensions of the atom:

$$\theta_{min} \sim \frac{\lambda\!\!\!^{-}}{a} \sim \frac{Z^{1/3}}{p}$$

while θ_{max} is the angle of diffraction by the nucleus:

$$\theta_{max} \sim \frac{\lambda\!\!\!^{-}}{R} \sim \frac{1}{pR} \qquad \text{for } \frac{\lambda\!\!\!^{-}}{R} < 1$$

$$\theta_{max} \sim 1 \qquad \text{for } \frac{\lambda\!\!\!^{-}}{R} > 1$$

Thus in the case $Z/v \ll 1$ we get

$$\overline{\varphi^2} = 2\pi n \frac{Z^2}{E^2} \begin{cases} \ell n\ a/R & \lambda\!\!\!^{-}/R < 1 \\ \ell n\ a/\lambda\!\!\!^{-} & \lambda\!\!\!^{-}/R > 1 \end{cases}$$

3. INTERACTION WITH RADIATION

Zero-Point Vibrations of the Electromagnetic Field

We know that the energy of the electromagnetic field is equal to $\int (\mathcal{E}^2/8\pi + \mathcal{H}^2/8\pi)\,dr$, where $\mathcal{E}$ and $\mathcal{H}$ are respectively the electric and magnetic field intensities. If there are no free charges anywhere in space, then we can put the electrostatic potential φ equal to zero. Then $\mathcal{E} = (-1/c)\,\partial A/\partial t$, $\mathcal{H} = \text{curl } A$, where A is the the vector potential.

We decompose the vector potential into plane waves,

$$A = \sum_k A_k e^{ikr} + \text{ complex conjugate}$$

Then we get the following expression for the energy E of the electromagnetic field:

$$E = \sum_k \left\{ \frac{1}{8\pi c^2} |\dot{A}_k|^2 + \frac{k^2}{8\pi} |A_k|^2 \right\}.$$

For normalization purposes the volume in which the field is contained has been taken equal to unity.

From the above expression it is easy to see that the energy of the electromagnetic field is just the sum of energies of a set of independent harmonic oscillators. The quantity A_k plays the role of the coordinate and $\dot{A}_{\underset{\sim}{k}}$ that of the velocity, while $1/4\pi c^2$ is analogous to the mass of the harmonic oscillator. The oscillator frequency is given by $\omega_k = \sqrt{\alpha/m}$, with α the stiffness constant and m the mass; since in this case $\alpha = k^2/4\pi$ and $m = 1/4\pi c^2$, we have accordingly $\omega_k = ck$. The first term in the expression for E is the "kinetic energy" of the electromagnetic field and the second its "potential energy". Thus, the electromagnetic field in a space not containing free charges may be treated as a set of independent harmonic oscillators corresponding to all possible values of the wave vector $\underset{\sim}{k}$.

Let us apply the laws of quantum mechanics to this set of oscillators. Then the energy of the electromagnetic field will be given by

$$E = \sum_k \left(n_k + \tfrac{1}{2}\right) \omega_k,$$

where $n_{\underset{\sim}{k}}$ is the quantum number of the state of excitation of the oscillator with wave vector $\underset{\sim}{k}$ (the number of photons with wave vector $\underset{\sim}{k}$). For simplicity we have omitted the subscript λ describing the polarization of the photons; this determines the orientation of the vector $\underset{\sim}{A}$, which must lie in the plane perpendicular to $\underset{\sim}{k}$. In the ground state we have $n_{\underset{\sim}{k}} = 0$ (no photons present), and so $E = (1/2) \sum_{\underset{\sim}{k}} \omega_{\underset{\sim}{k}}$; neither the kinetic nor the potential energy of the vibration has a well-defined value. It follows that neither the electric nor the magnetic field intensity has a well-defined value either; in each case the mean value of the field is zero but the mean-square value is nonzero. This means that the electromagnetic field in vacuum is an oscillating quantity; the oscillations are called zero-point vibrations of the electromagnetic field.

We are now in a position to answer the question, why does not an atom remain in an excited state indefinitely ? After all, an excited state (we might think) is an exact stationary state of Schrödinger's equation. The explanation lies in the fact that everywhere in space exists an electromagnetic field, with which the atom interacts; as a result, the atomic states cease to be stationary states.

Let us consider, then, the interaction of a particle with the electromagnetic field. To find the change of the Hamiltonian when a field is applied, we have to replace the momentum $\underset{\sim}{p}$ of the particle by $\underset{\sim}{p} - e\underset{\sim}{A}/c$, where $\underset{\sim}{A}$ is the vector potential of the field. The kinetic energy $p^2/2m$ accordingly becomes

$$\frac{p^2}{2m} - \frac{e}{mc}pA + \frac{e^2}{2mc^2}A^2.$$

The last term in this expression is small; it needs to be taken into account in problems when first-order perturbation theory gives zero contribution, for instance, in the problem of the scattering of photons by an electron.

The operators $\underset{\sim}{A}_{\underset{\sim}{k}}$ play the role of coordinates for the corresponding field oscillators and hence have matrix elements corresponding to a change of the number of photons by one. Let us calculate the matrix element $(A_{\underset{\sim}{k}})_{01}$ for the transition from the vacuum state to the state with one photon present. To do this we find the quantity $(A_{\underset{\sim}{k}}^2)_{00} = \sum_n <0|A_{\underset{\sim}{k}}|n><n|A_{\underset{\sim}{k}}\, 0>$. Since the operator $A_{\underset{\sim}{k}}$ creates a single photon (with frequency ω_k) the only nonzero term in the sum is that with $n = 1$. Thus we have $|(A_{\underset{\sim}{k}})_{01}| = \sqrt{(|A_{\underset{\sim}{k}}|^2)_{00}}$. Now we know that for a harmonic oscillator the mean values of the kinetic and potential energies are equal and are each just half of the total energy. Hence for the ground state $\bar{T}_{\underset{\sim}{k}} = \bar{U}_{\underset{\sim}{k}} = (1/2)E_{\underset{\sim}{k}} = (1/4)\,ck$. Since $\bar{U}_{\underset{\sim}{k}} = (k^2/8\pi) \times (|A_{\underset{\sim}{k}}|^2)_{00}$, we have $(1/4)\,ck = (k^2/8\pi)\,(|A_{\underset{\sim}{k}}|^2)_{00}$, whence $(A_{\underset{\sim}{k}}^2)_{00} = 2\pi c/k$. Consequently,

$$|(A_k)_{01}| = \sqrt{\frac{2\pi c}{k}} = \sqrt{\frac{2\pi c^2}{\omega_k}}.$$

In an entirely analogous way we can find the matrix element of $A_{\underset{\sim}{k}}$ for the transition from a state with n photons to a state with n+1 photons. We easily obtain

$$(A_k)_{n,\,n+1} = \sqrt{n+1}\sqrt{\frac{2\pi c^2}{\omega_k}},$$

$$(A_k)_{n,\,n-1} = \sqrt{n}\sqrt{\frac{2\pi c^2}{\omega_k}}.$$

Thus the matrix element for the transition of a particle from a state φ_{λ_1} to a state φ_{λ_2} with the emission of a photon is given by

$$V_{\lambda_1\lambda_2} = \left(-\frac{ep}{mc}A_k\right)_{\lambda_1\lambda_2} = -\frac{e}{mc}(\varphi_{\lambda_1}|\,p\eta_{k\lambda}e^{ikr}\,|\varphi_{\lambda_2})\sqrt{\frac{2\pi c^2}{\omega_k}}. \quad (1.23)$$

To adapt this formula to the relativistic case, it is sufficient for an order-of-magnitude estimate to replace $\underset{\sim}{p}/m$ by $\underset{\sim}{v}$. Here $\underset{\sim}{p}\cdot\underset{\sim}{\eta}_{k\lambda}$ is the component of the momentum along the direction of $\underset{\sim}{A}_k$ ($\underset{\sim}{\eta}_{k\lambda}$ is a unit vector representing the polarization of the photon).

The Photoeffect

Let us consider the photoeffect, in which a photon is

absorbed by an atom and as a result an electron is ejected from it. The number of transitions per unit time is given by

$$\frac{2\pi}{\hbar}|V_{0,p}|^2\delta(E_i - E_f),$$

where E_i is the energy of the initial state, which is equal to the sum of the photon energy ω_k and the energy of the electron in the initial state, ϵ_0, and E_f is equal to ϵ_p, the energy of the ejected electron. The number of transitions by the electron per unit time into states with momenta $\underset{\sim}{p}$ in the interval $[\underset{\sim}{p}, \underset{\sim}{p}+d\underset{\sim}{p}]$ is given by

$$dW = 2\pi\int |V_{0,p}|^2\delta(\varepsilon_0 + \omega_k - \varepsilon_p)\frac{dp}{(2\pi)^3} =$$
$$= 2\pi\int |V_{0,p}|^2 \frac{p^2\frac{dp}{d\varepsilon_p}d\varepsilon_p}{(2\pi)^3}\delta(\varepsilon_0 + \omega_k - \varepsilon_p)\,d\Omega =$$
$$= \frac{1}{(2\pi)^2}\int |V_{0,p}|^2\frac{p^2}{v}\,d\Omega.$$

The cross section for the photoeffect is equal to the number of transitions per unit time divided by the photon beam intensity, which is just the velocity of light c, since we have normalized the problem to one photon per unit volume. Consequently, we get the order-of-magnitude result

$$\sigma \sim \frac{4\pi p}{(2\pi)^2 c}|V_{0p}|^2.$$

We shall now show that the exponent $e^{i\underset{\sim}{k}\cdot\underset{\sim}{r}}$ occurring in (1.23) may be replaced by unity. The wave number $k = \omega/c$ is of order I/c, where I is the ionization potential of the atom; $I \sim Z^2$ for the inner electron shells. The radius of these inner shells is of order $1/Z$, and so $kr \sim (Z^2/c)(1/Z) \sim Z/c \ll 1$ and $e^{i\underset{\sim}{k}\cdot\underset{\sim}{r}} \approx 1$.

The replacement is even better justified for the outer shells, where we have $kr \sim 1/c \approx 1/137$.

Since $(\varphi_0 \underset{\sim}{p} \varphi_p) = (\varphi_0 \dot{\underset{\sim}{r}} \varphi_p) = i(\epsilon_0 - \epsilon_p)(\varphi_0 \underset{\sim}{r} \varphi_p) = i\omega(\varphi_0 \underset{\sim}{r} \varphi_p)$, the photoeffect cross section is of order of magnitude

$$\sigma \sim \frac{p}{c}\omega^2 |(\varphi_0 r \varphi_p)|^2 \frac{1}{\omega} \sim \frac{p\omega}{c} |(\varphi_0 r \varphi_p)|^2.$$

We shall consider K-shell photo-ionization in two limiting cases: when the energy of the ejected electron is much less than the ionization potential, and when it is much greater. In the first case the photon energy is nearly equal to the ionization potential $I \sim Z^2$, that is

$$\sigma \sim \frac{pZ^2}{c} |(\varphi_0 r \varphi_p)|^2.$$

The cross section contains a factor $p \sim \sqrt{E}$; if the matrix element r_{op} did not depend on energy (which is not true in this case), we should have $\sigma \sim \sqrt{E}$. This is a general result for a reaction cross section near a threshold (cf. p.243 below). The $\sqrt{E}$ dependence comes from the density of final states:

$$\int \delta(E-\omega)\frac{d^3p}{(2\pi)^3} \sim \int \delta(E-\omega)\, p^2 dp \sim \int \delta(E-\omega)\, p\, dE \sim p.$$

This is not the case here; the photoeffect cross section is independent of energy for small energy of the ejected electron. This is due due to the fact that, because of the smooth decrease of the potential at large distances, the dipole matrix element behaves as $1/\sqrt{p}$ near the reaction threshold. To see this, let us estimate $\underset{\sim}{r}_{op}$ near the threshold. The wave function of a K-electron is $\varphi_0 \sim e^{-Zr}$, and so in the dipole matrix element is is small distance distances $(r \sim 1/Z)$ which are important. To estimate $\underset{\sim}{r}_{op}$, we have to find $\varphi_{\underset{\sim}{p}}(r)$ for $r \sim 1/Z$. Since the operator $\underset{\sim}{r}$ changes the angular momentum by 1, the wave function of the ejected electron corresponds to a state with unit angular momentum and in the limit $r \to \infty$ has the form

$$\varphi_p \to \frac{1}{pr} \cos(pr + \delta_1) \cos\vartheta. \tag{1.24}$$

On the other hand, the radial part of this function, $\varphi_p/\cos\theta$, can be written u_p/r, where u_p obeys the Schrödinger equation

$$u_p'' + P_1^2 u_p = 0,$$

where $P_1^2 = 2(E + Z/r - 2/r^2)$, $E = p^2/2$. The approximate solution to this equation found in section 1.1 (the quasiclassical solution) is given by the expression

$$u_p = \frac{A}{\sqrt{P_1(r)}} \cos\left(\int_{r_1}^{r} P_1(r)\,dr - \frac{\pi}{2}\right). \tag{1.25}$$

Comparing (1.24) with (1.25), we get $A = 1/\sqrt{p}$.

The condition for applicability of the quasiclassical approximation is $d\lambda/dr \ll 1$ (cf. Sections 1.1 and 3.1). Using the fact that $d\lambda/dr \sim 1/\sqrt{Zr}$, we find that the condition for quasiclassical motion is $r > 1/Z$. Thus, the quasiclassical approximation is applicable, at least to an order of magnitude, right down to the K-shell radius.

To find φ_p for $r \sim 1/Z$ we write $\varphi_p \sim p^{-1/2}(Z/r)^{-1/4} r^{-1} \sim (Z/p)^{1/2}$. Thus, the dipole matrix element has the order of magnitude

$$(\varphi_0 r \varphi_p) \sim Z^{3/2} \int_0^\infty e^{-Zr} r \frac{\sqrt{Z}}{\sqrt{p}} r^2 dr \sim \frac{1}{Z^2 \sqrt{p}}.$$

Thus, the p-dependence of $|r_{\underset{\sim}{op}}|^2$ cancels that of the density of states, and the photoeffect cross section is independent of energy at low energies. We notice that this result arises only because of the smooth behaviour of the Coulomb potential; if the potential in which the electron moves had the form of a potential well, there would be a region at the edge of the well where the quasiclassical approximation breaks down, and the above estimate would be invalid. In fact, by using the formulae of chapter 3 (see p.170) it is easy to verify that in this case $r_{\underset{\sim}{op}}$ does not depend on p, and the behaviour of the photoeffect cross section near threshold is determined by the density of states.

Thus, for $E - I \ll I$ we get for the total photoeffect cross section

$$\sigma \sim \frac{pZ^2}{c}\left(\frac{1}{Z^2}\right)^2 \frac{1}{p}\,,$$

or

$$\sigma = \frac{C_1}{cZ^2}\,, \qquad C_1 \sim 1. \tag{1.26}$$

We now consider the opposite limit, when the energy of the ejected electron is much greater than the ionization potential. We estimate the dipole matrix element and the photoeffect cross section. In the continuous spectrum the wave function has the form $\varphi_p \underset{r\to\infty}{=} F e^{i\underset{\sim}{p}\cdot\underset{\sim}{r}}$, where F is a function which tends to unity for $\underset{\sim}{p}\cdot\underset{\sim}{r} \gg 1$. In the matrix element $(\varphi_o \underset{\sim}{p}\, \varphi_{\underset{\sim}{p}})$ we see that it is distances r of the order of $1/p$ which are important. The deviation of F from unity is determined by the quantity $V(r)/E$. For $r \sim 1/p$ we have $V/E \sim \sqrt{I/E}$ and hence can put $F \approx 1$. Taking the polarization of the photon into account introduces only a factor of order 1. Thus we have

$$\sigma \sim \frac{1}{pc} |(\varphi_0 p \varphi_p)|^2 \sim \frac{p}{c} |(\varphi_0 e^{ipr})|^2.$$

Integrating over angle, we obtain

$$(\varphi_0 e^{ipr}) \sim Z^{3/2} \int_0^\infty e^{-Zr} \frac{\sin pr}{pr} r^2 dr = \frac{Z^{3/2}}{2p} \int_{-\infty}^{\infty} r e^{-Z|r|} \sin pr \, dr.$$

To estimate this expression we note that the second derivative of the function $f(r) = re^{-Z|r|}$ has a discontinuity at $r = 0$; therefore, by the argument used in Section 1.1, the high Fourier components of $f(r)$ are of order of magnitude Z/p^3.

If the potential V regarded as a function of r^2 had no singularities on the real axis, the wave function, being a scalar, would have to be an analytic function of r^2, and the high Fourier components of the wave function would fall of exponentially with energy (cf. Section 1.1). The power law dependence of the matrix element on energy in the present case arises from the square-root singularity of the Coulomb wave function (cf. p. 243).

Thus for $p \to \infty$ the matrix element $(\varphi_0, e^{i\mathbf{p}\cdot\mathbf{r}})$ is of order $Z^{5/2}/p^4$, and the photoeffect cross section is of order

$$\sigma \sim \frac{Z^5 p}{cp^8}, \text{ i.e. } \sigma = \frac{C_2 Z^5}{cE^{7/2}}, \quad \text{or} \quad \sigma \sim \frac{2^{7/2} C_2}{cZ^2} \left(\frac{I}{E}\right)^{7/2}. \tag{1.27}$$

Here I is the K-electron ionization potential, which is just $Z^2/2$. Numerical calculation actually shows that $C_2 \sim 10$.

If in (1.27) we put the energy E equal to the ionization potential, I, we should get formula (1.26), at least to within an order of magnitude. Actually, however, numerical calculation shows that formula (1.27) for E = I gives a cross section an order of magnitude larger than (1.26). This shows that the transition from the region $E - I \ll I$ to the region $E \gg I$ takes place over a very extended region.

Problem: Estimate the photoeffect cross section at energies near the threshold and at high energies in the case of a square well.

Lifetimes of Excited Atomic States

Case 1. We shall first estimate the lifetime of the 2p state of a hydrogenic atom. An electron can make a "dipole" transition from the 2p to the 1s state by emitting a photon. The number of transitions per unit time (inverse lifetime) is given by

$$W_{01} = 2\pi \int |V_{01}|^2 \frac{d^3k}{(2\pi)^3} \delta(E_0 - E_1 - \omega),$$

where $V = -(1/c)\underset{\sim}{p} \cdot \underset{\sim}{A}$ is the interaction between the electron and the electromagnetic field. Here the subscripts 0 and 1 correspond to the 1s and 2p states respectively, and ω is the energy of the photon emitted. Since we have

$$d^3k = k^2\, dk\, d\Omega = k^2 \frac{dk}{d\omega} d\Omega\, d\omega,$$

it follows that

$$W_{01} = 2\pi \int |V_{01}|^2 \frac{k^2 \dfrac{dk}{d\omega} d\Omega}{(2\pi)^3} .$$

We estimate the matrix element V_{01} for the transition. Remembering the normalization of the vector potential $\underset{\sim}{A}$ of the electromagnetic field (cf. p. 57), we find

$$V_{01} = -\frac{1}{c}(pA)_{01} \sim \sqrt{\frac{2\pi}{\omega}}\, p_{01}.$$

(We are not interested in the polarization of the photon, which introduces only a numerical factor of order unity.) We can estimate the matrix element of the momentum p:

$$p_{01} = i\omega r_{01} \sim \frac{\omega}{Z}.$$

Hence we get

$$W_{01} \sim \frac{\omega^2}{Z^2}\frac{1}{\omega}k^2\frac{dk}{d\omega} \sim \frac{\omega^3}{Z^2c^3}.$$

Since the transition frequency ω is of order Z^2, we finally find

$$W_{01} \sim \frac{Z^4}{c^3}.$$

Let us make a numerical estimate of the lifetime of a state against dipole transitions. The atomic unit of time is of order

$$\tau_{at} \sim \frac{\hbar}{E_{at}} \sim \frac{10^{-27}\text{ erg sec}}{27\text{ eV} \times 1.6 \times 10^{-12}\text{ erg/eV}} \approx 2 \times 10^{-17}\text{ sec.}$$

Thus, $\tau_{dipole} \sim (c^3/Z^4)\tau_{at}$; for a hydrogen atom we get $\tau_{dip} \sim 10^7 \times 10^{-17} = 10^{-10}$ seconds. Exact calculation gives rather larger values of τ_{dipole}.

Case 2. We next estimate the lifetimes for quadrupole transitions, for example, 3d → 1s. Above we replaced the factor $e^{i\underset{\sim}{k}\cdot\underset{\sim}{r}}$ which occurs in the electromagnetic field operator $\underset{\sim}{A}$ by unity ($\underset{\sim}{k}$ is the wave vector of the photon, so that $kr = (\omega/c)\, r \sim Z/c \ll 1$). To consider quadrupole transitions we must expand $d^{i\underset{\sim}{k}\cdot\underset{\sim}{r}}$ in a power series; the principal contribution comes from the term $i\underset{\sim}{k}\cdot\underset{\sim}{r}$. We get

$$V_{01} = i\sqrt{\frac{2\pi}{\omega}}\,((\eta p)(kr))_{01},$$

where $\underset{\sim}{\eta}$ is the photon polarization. Let us take the z axis along the vector $\underset{\sim}{\eta}$ and the x axis along $\underset{\sim}{k}$ (the photon polarization vector $\underset{\sim}{\eta}$ is perpendicular to the wave vector $\underset{\sim}{k}$). Then we have

$$\begin{aligned} V_{01} &= ik\sqrt{\frac{2\pi}{\omega}}(p_z x)_{01} = \\ &= ik\sqrt{\frac{2\pi}{\omega}}\left[\frac{1}{2}(p_z x - p_x z) + \frac{1}{2}(p_z x + p_x z)\right]_{01} = \\ &= ik\sqrt{\frac{2\pi}{\omega}}\left[-\frac{1}{2}[\boldsymbol{rp}]_y + \frac{1}{2}\frac{d}{dt}(zx)\right]_{01}. \end{aligned}$$

The first term in this expression is just the matrix element of the y component of the orbital angular momentum, so that it corresponds to a magnetic dipole transition. The product zx is a component of the quadrupole moment, so that the second term determines the probability of an electric quadrupole transition. The two terms are of the same order of magnitude.

Let us make a numerical estimate of the lifetime of a state against quadrupole transitions. The difference from the case of a dipole transition lies in the extra factor of $\underset{\sim}{k}\cdot\underset{\sim}{r}$ in the matrix element. Hence we find

$$\tau_{\text{quad}} \sim \tau_{\text{dipole}} \times (kr)^{-2} \sim \tau_{\text{dip}} \frac{c^2}{Z^2}$$

For a hydrogen atom $\tau_{\text{quad}} \sim 10^{-10} . 10^{4} \sim 10^{-6}$ sec.

Problem: Estimate the lifetime of a hydrogenic atom against the emission of two photons.

Solution: $\tau \sim c^6/Z^6$.

Bremsstrahlung

We estimate the effective cross section for emission of gamma radiation when an electron is scattered by a nucleus, on the assumption that the gamma-ray frequency is much less than the electron energy. Then the emission of the photon has little effect on the motion of the electron and the effective cross section is approximately equal to the product of the cross section for scattering of the electron and the probability of emission of a photon.

Let us then estimate the probability of emission of a photon when the electron motion is assumed given. Schrödinger's equation for the photon in the energy representation has the form

$$i\dot{C}_n = \sum_{n'} V_{nn'} C_{n'} e^{i(\omega_n - \omega_{n'})t}, \tag{1.28}$$

where V describes the interaction of the electron with the electromagnetic field and C_n is the amplitude of the state containing n photons with the given wave vector and polarization. In the ground state $C_o = 1$ and all other C_n are zero. In first-order perturbation theory there is a nonzero transition amplitude from the ground state to the state with one photon described by C_1. From (1.28) we get

$$i\dot{C}_1 = V_{10} e^{i\omega t},$$

whence

$$C_1 = -i \int_{-\infty}^{\infty} V_{10} e^{i\omega t} dt = +i \int_{-\infty}^{\infty} \left(\frac{1}{c} vA\right)_{10} e^{i\omega t} dt.$$

Here we have used the form of interaction with the radiation field appropriate to spinless particles; for particles with spin the formula is more complicated, but the order of magnitude of the interaction can still be estimated from the above expression.

Since the radiation field does not affect the scattering of

the electron, we can take the electron velocity $\underset{\sim}{v}$ to be a given function of time. Consequently, since

$$A \sim \sqrt{\frac{2\pi c^2}{\omega}}\, \eta e^{-ikr} \approx \sqrt{\frac{2\pi c^2}{\omega}}\, \eta e^{-i\omega \frac{vn}{c} t},$$

we get

$$C_1 = i\sqrt{\frac{2\pi}{\omega}} \int_{-\infty}^{\infty} (v\eta)\, e^{i\omega t - i\omega \frac{vn}{c} t} dt, \tag{1.29}$$

where $\underset{\sim}{\eta}$ is the unit photon polarization vector and $\underset{\sim}{n}$ the unit vector $\underset{\sim}{k}/k$.

Since the duration of the collision of the electron with the nucleus is much shorter than the time interval

$$t \sim \frac{1}{\omega\left(1 - \frac{vn}{c}\right)},$$

important in this integral, we can assume that at time $t = 0$ the particle velocity $\underset{\sim}{v}$ changes discontinuously from $\underset{\sim}{v}_1$ to $\underset{\sim}{v}_2$. Then we get from (1.29)

$$C_1 = i\sqrt{\frac{2\pi}{\omega}} \left\{ (v\eta) \frac{e^{i\omega\left(t - \frac{vn}{c} t\right)}}{i\omega\left(1 - \frac{vn}{c}\right)} \Bigg|_{-\infty}^{0} + (v\eta) \frac{e^{i\omega\left(t - \frac{vn}{c} t\right)}}{i\omega\left(1 - \frac{vn}{c}\right)} \Bigg|_{0}^{\infty} \right\} =$$
$$= \sqrt{\frac{2\pi}{\omega^3}}\, \eta \left(\frac{v_1}{1 - \frac{v_1 n}{c}} - \frac{v_2}{1 - \frac{v_2 n}{c}} \right). \tag{1.30}$$

If the angle of deflection of the electron (θ_e) and of the photon (θ_{ph}) is small, we get from (1.30)

$$C_1^2 \sim \frac{1}{\omega^3} \frac{(\Delta v)^2}{(1 - v/c)^2}; \tag{1.31}$$

where Δv is the change in the electron velocity. Since for small θ_{ph} and θ_e the denominators in (1.30) have the form

$$1-\frac{v}{c}+\frac{\theta_{ph}^2}{2}, \quad 1-\frac{v}{c}+\frac{(\theta_{ph}+\theta_e)^2}{2}$$

the only angles θ_{ph}, θ_e of importance are those which satisfy

$$\theta_{ph}, \theta_e \lesssim \sqrt{1-\frac{v}{c}} \tag{1.32}$$

We now estimate the probability of emission of a photon in the frequency interval $[\omega, \omega + d\omega]$ for a given change of the electron velocity $\Delta \underset{\sim}{v}$. To do this we must multiply (1.31) by the density of final states for the photon $d\underset{\sim}{k}/(2\pi)^3$:

$$dW \sim \frac{1}{\omega^3} \cdot \frac{(\Delta v)^2}{(1-v/c)^2} k^2 \, dk \, d\Omega_{ph} \sim \frac{(\Delta v)^2}{c^3(1-v/c)^2} \cdot \frac{d\omega}{\omega} \theta_{ph} \, d\theta_{ph} \tag{1.33}$$

The final cross section for bremsstrahlung is given by the product of the probability dW for emission of a photon and the Coulomb scattering cross section for relativistic electrons:

$$d\sigma_{Br} \sim \frac{(\Delta v)^2 \theta_{ph}^2}{c^3(1-v/c)^2} \cdot \frac{d\omega}{\omega} \cdot \frac{Z^2}{E^2} \frac{d\Omega_e}{\sin^4(\theta_e/2)}$$

or for small scattering angles $(\sin\theta_e \approx \theta_e \approx q/p)$

$$d\sigma_{Br} \sim \frac{q^2(1-v/c)}{c^3(1-v/c)} \cdot \frac{d\omega}{\omega} \frac{Z^2}{p^2c^2} \frac{\theta_e d\theta_e}{\theta_e^4} \sim \frac{Z^2}{c^5} \cdot \frac{d\omega}{\omega} \cdot \frac{dq}{q} \tag{1.34}$$

Let us find the probability of emission of a photon with arbitrary change of the electron momentum. To do this we must integrate (1.34) over q. We shall assume the electron to be ultrarelativistic $(1-v/c \ll 1)$. Since $Ze^2/\hbar c \ll 1$, the minimum deflection angle will be determined by the diffraction of the electron by the atom (size $a \sim Z^{-1/3}$), and the maximum angle by condition (1.32) (cf. p. 53). Integration of (1.34) over q then gives

$$\int \frac{d\sigma_{Br}}{dq}\, dq \sim \frac{Z^2}{c^5} \cdot \frac{d\omega}{\omega}\, \ell n\left(\frac{q_{max}}{q_{min}}\right) \sim \frac{Z^2}{c^5}\, \frac{d\omega}{\omega}\, \ell n\left(\frac{c}{Z^{1/3}}\right) \tag{1.35}$$

We can estimate the total cross section for bremsstrahlung radiation of any frequency. Integrating (1.35) over ω, we get

$$\sigma_{Br} \sim \frac{Z^2}{c^5}\, \ell n\, \frac{c}{Z^{1/3}}\, \ell n\, \frac{\omega_{max}}{\omega_{min}} \quad . \tag{1.36}$$

The maximum photon frequency is of the order of the electron energy

$$\omega_{max} \sim \frac{mc^2}{\sqrt{1 - v^2/c^2}} \; .$$

If we let $\omega_{min} \to 0$ expression (1.36) will diverge (cf. p. 71). However, the energy loss rate of the electron due to bremsstrahlung will be finite: $-dE/dx = \int n\omega\, d\sigma_{Br}$, where n is the number of nuclei per unit volume. Thus we finally get

$$-\frac{dE}{dx} \sim \frac{Z^2 n}{c^5}\, E \ln \frac{c}{Z^{1/3}} \; .$$

Pair Creation

We remind the reader that for given momentum p the relativistically invariant equation of motion of the electron - the Dirac equation - has eigenfunctions corresponding to negative as well as positive values of the energy: $E = \pm\sqrt{m^2c^4 + p^2c^2}$. To explain why the electron does not fall into a negative-energy state by emitting a photon, Dirac postulated that in the vacuum all the negative-energy states are already filled with electrons, so that transitions into them are forbidden by the Pauli principle. An unfilled negative-energy state is equivalent to the existence of a positron in the vacuum.

From this point of view the mechanism of pair formation is analogous to that of the photoeffect: a photon excites an electron from a negative-energy state to a positive-energy one. In free space this process is impossible, as is immediately obvious from the laws of conservation of energy and momentum. However, it becomes possible in the field of a nucleus, since the electron can transfer its surplus momentum to the nucleus. The cross section for pair formation in the field of a nucleus can be estimated in the same way as that for bremsstrahlung: we first find the probability of pair creation for a given value of momentum transfer to the nucleus, then multiply this probability by the scattering cross section for the electron-nuclear scattering, and finally integrate over all values of the momentum transfer.

Actually, we can get the cross section for pair creation simply by regarding it as the process inverse to bremsstrahlung. Strictly speaking, the process truly inverse to bremsstrahlung is the absorption of a photon by an electron making a transition from one positive-energy state to another. However, in the ultrarelativistic case (to which we confine ourselves for simplicity) the eigenfunctions corresponding to negative energy (i. e. to a positron) are only slightly different from those corresponding to positive energy (the correction due to the Coulomb field of the nucleus is small). This means that it is a good approximation to regard pair formation as the process inverse to bremsstrahlung.

The ratio of the cross section for direct and inverse processes is just the ratio of the corresponding densities of final states. In the case of bremsstrahlung the final state contains an electron and a photon, while in the case of pair formation it contains a positron and an electron. However, at high energies the density of final states for the photon is identical to that for an electron of the same energy, and so the pair-creation cross section is of the same order of magnitude as the bremsstrahlung cross section integrated over photon energies of the order of the electron energy. Thus we get

$$\sigma_{\text{pair}} \sim \frac{Z^2}{c^5} \ln\left(\frac{c}{Z^{1/3}}\right)$$

The energy loss of a photon per unit path length due to pair formation is therefore of order

$$-\frac{dE}{dx} \sim \frac{nZ^2}{c^5} E \ln \frac{c}{Z^{1/3}} ,$$

that is, of the same order of magnitude as the energy loss rate of an ultrarelativistic electron due to bremsstrahlung.

Creation of Soft Photons in the Scattering of Charged Particles (The "Infrared Catastrophe")

Consider the problem of creation of soft photons in the scattering of a particle by a nucleus. We shall assume that the particle is nonrelativistic. The Hamiltonian of the system has the form

$$H = T + V - \frac{1}{c} pA + H_{\gamma}. \tag{1.37}$$

where T is the kinetic energy of the particle, V the scattering potential, $-(1/c)\mathbf{p}.\mathbf{A}$ the interaction of the particle with the photon field, and H_{γ} the Hamiltonian of the photon field (cf. above)

Let us assume that the energy of the photons under consideration is much less than that of the particle. Then the motion of the particle can be taken as given and we need consider only that part of the Hamiltonian which contains operators which act on the photon wave functions. The photon Hamiltonian changes over the time of the collision (which will be assumed small) from $H_o = H_{\gamma} - (1/c)\mathbf{p}_o.\mathbf{A}$ before the collision to $H_1 = H_{\gamma} - (1/c)\mathbf{p}_1.\mathbf{A}$ afterwards. Let us go over to a system of coordinates in which the particle was at rest before the scattering. Then we have

$$H_0 = H_{\gamma}, \quad H_1 = H_{\gamma} - \frac{1}{c} qA, \tag{1.38}$$

where $\mathbf{q} = \mathbf{p}_1 - \mathbf{p}_o$ is the change in the momentum of the particle due to scattering.

The perturbation $-(1/c)\mathbf{q}.\mathbf{A}$ is large for small frequencies, since $A_{\mathbf{k}\omega} \sim 1/\sqrt{\omega}$. Therefore, instead of using perturbation theory, we shall use only the suddenness of the change of the photon Hamiltonian. We introduce the eigenfunctions χ_n of the Hamiltonian H_o and the eigenfunctions χ_n' of the Hamiltonian H_1, and assume that before the collision the photon system was in the state χ_o. Because of the suddenness of the collision the wave function of the system, χ_o, is practically unchanged, so that the transition amplitude into the state χ_n' is given by $(\chi_n' | \chi_o)$ (cf. p.102).

We represent the Hamiltonian of the electromagnetic field, which is given by

$$H_\gamma = \int \left(\frac{1}{8\pi c^2} \dot{A}^2 + \frac{1}{8\pi} (\mathrm{rot}\, A)^2 \right) dr, \tag{1.39}$$

in the form of a sum of Hamiltonians for the various field oscillators. To do this we write the vector potential $\underset{\sim}{A}$ in the form

$$A = \sum_{k\lambda} \sqrt{2\pi c^2}\, \eta_{k\lambda} (q_{k\lambda} \exp (ikr - i\omega_{k\lambda} t) + \\ + \overset{*}{q}_{k\lambda} \exp (-ikr + i\omega_{k\lambda} t)). \tag{1.40}$$

where $\underset{\sim}{k}$, λ, and $\omega_{\underset{\sim}{k}\lambda}$ are respectively the wave vector, polarization, and energy of the photons, and $\underset{\sim}{\eta}_{\underset{\sim}{k}\lambda}$ is the unit polarization vector. Omitting the indices $\underset{\sim}{k}$, λ in what follows and writing

$$q = Q + \frac{i}{\omega} P, \quad q^* = Q - \frac{i}{\omega} P,$$

and substituting in (1.39), we get

$$H_\gamma = \sum \frac{1}{2} (P^2 + \omega^2 Q^2). \tag{1.41}$$

For soft photons the exponential $e^{i\underset{\sim}{k}\cdot\underset{\sim}{r}}$ may be replaced by unity. We shall take the time factors $\exp(\pm i\omega t)$ to be included in q, q^*. Then we have

$$A = \sum \sqrt{2\pi c^2}\, \eta \cdot 2Q.$$

and so the Hamiltonian H_1 for the photons has the form

$$H_1 = \sum \left\{ \frac{1}{2} (P^2 + \omega^2 Q^2) - 2\sqrt{2\pi}\, (\eta \dot{q})\, Q \right\}. \tag{1.42}$$

We can represent this in the form of a sum of oscillator Hamiltonians with shifted coordinate:

$$H_1 = \sum \left\{ \frac{1}{2} P^2 + \frac{1}{2} \omega^2 (Q - \delta)^2 - \frac{1}{2} \omega^2 \delta^2 \right\}, \tag{1.43}$$

where $\delta = (2\sqrt{2\pi/\omega^2})\,\underset{\sim}{\eta}\cdot\underset{\sim}{q}$. Thus the wave functions after collision $\chi'_n(Q)$ have the form of oscillator wave functions, but with shifted coordinate:

$$\chi'_n(Q) = \chi_n(Q - \delta).$$

Thus the transition amplitude into the state χ'_n has the form

$$C_n = \int \chi_n (Q - \delta)\, \chi_0 (Q)\, dQ. \tag{1.44}$$

We first consider two limiting cases: when δ is much larger than the amplitude of the zero-point vibrations, and when it is much smaller. If δ is small, then it follows from (1.44) that

$$C_n \approx -\delta \int \frac{\partial \chi_n}{\partial Q}\, \chi_0 dQ = i\delta (P)_{0n}, \tag{1.45}$$

where P is the momentum operator. The latter has nonzero matrix elements only for transitions which change the number of photons by unity. Since the zero-point energy is $\omega/2$, it follows from (1.41) that $|(P)_{01}|^2 = \omega/2$. Hence the probability of emission of a single photon is $W_1 = \frac{1}{2}\omega\delta^2 = (4\pi/\omega^3)(\underset{\sim}{\eta}\cdot\underset{\sim}{q})^2$. This formula agrees with the result obtained by ordinary perturbation theory.

Let us now estimate the matrix element $C_n = \int \chi_n (Q-\delta)\, \chi_0 (Q)\, dQ$ when the shift δ is much greater than the vibrational amplitude of the oscillator. For this purpose we use the quasiclassical expression for the functions χ_n (cf. Sections 1.1 and 3.1):

$$\chi_n(Q) = \frac{a_n}{\sqrt[4]{E_n - {}^1/_2 Q^2}} \cos\left(\int \sqrt{E_n - {}^1/_2 Q^2}\, dQ - \frac{\pi}{4}\right).$$

For simplicity we have taken the oscillator frequency ω equal to unity; the results are easily transposed to the case $\omega \neq 1$. The normalization factor a_n is of order unity, since

$$1 = \int \chi_n^2\, dQ \sim \frac{a_n^2}{\sqrt{E_n}} l_n \sim \frac{a_n^2}{\sqrt{E_n}} \frac{\sqrt{E_n}}{\omega}.$$

Thus we have

$$C_n \sim \int \frac{\exp(-Q^2/2)}{\sqrt[4]{E_n - {}^1/_2 (Q-\delta)^2}} \times$$
$$\times \cos\left(\int \sqrt{E_n - {}^1/_2 (Q-\delta)^2}\, dQ - \frac{\pi}{4}\right) dQ. \tag{1.46}$$

Since by hypothesis the quantity δ is much larger than the zero-point vibrational amplitude, we can neglect the term $Q^2/2$ under the square root sign in (1.46). Let us write $E_n - \delta^2/2 \equiv \Delta E$ and change the variable of integration Q to $\Delta E/\delta + Q \equiv x$. Then we get from (1.46)

$$C_n \sim$$
$$\sim \int_0^\infty \frac{\cos(\delta^{1/2} x^{3/2})}{\sqrt[4]{\delta x}} \exp\left(-\frac{x^2}{2} + 2x\frac{\Delta E}{\delta}\right) dx \exp\left[-\frac{1}{2}\left(\frac{\Delta E}{\delta}\right)^2\right]. \tag{1.47}$$

The values of x significant in this integral correspond to the argument of the cosine being of order unity, that is, have the order of magnitude $x \sim \delta^{-1/3}$. Thus,

$$C_n \sim \frac{1}{\sqrt[4]{\delta}} \sqrt[12]{\delta}\, \delta^{-1/3} \exp\left[-\frac{1}{2}\left(\frac{\Delta E}{\delta}\right)^2\right],$$
$$|C_n|^2 \sim \frac{1}{\delta} \exp\left[-\left(\frac{\Delta E}{\delta}\right)^2\right].$$

Thus the probability distribution for emission of photons has the form of a Gaussian curve with a width $\Delta E \sim \delta$ and a maximum $w_{max} \sim 1/\delta$ near the value $E = \delta^2/2$. It is easy to see that this value is just that of the energy of a classical oscillator whose origin of coordinates is suddenly shifted by an amount δ.

We now proceed to calculate C_n for arbitrary δ. We first calculate the probability that no photon of the type in question is created, that is

$$w_0 = \left| \int \chi_0(Q - \delta)\, \chi_0(Q)\, dQ \right|^2.$$

Since the ground-state oscillator wave function has the form

$$\chi_0(Q) = \sqrt[4]{\frac{\omega}{\pi}} \exp\left(-\frac{1}{2}\omega Q^2\right),$$

it follows that

$$w_0 = \left| \sqrt{\frac{\omega}{\pi}} \int_{-\infty}^{\infty} \exp\left(-\omega Q^2 - \frac{1}{2}\omega\delta^2 + \omega Q\delta\right) dQ \right|^2 = \exp\left(-\frac{1}{4}\omega\delta^2\right) = \exp\left[-\frac{4\pi}{\omega^3}(q\eta)^2\right]. \tag{1.48}$$

We now put back the subscripts $\underset{\sim}{k}$, λ: then we find that

$$w_0^{k\lambda} = \exp\left[-\frac{4\pi}{\omega_{k\lambda}^3}(q\eta_{k\lambda})^2\right]$$

is the probability that no photon $(\underset{\sim}{k}, \lambda)$ will be radiated.

The probability that no photon in any state will be radiated at all is given by

$$W_0 = \prod_{k\lambda} w_0^{k\lambda} = \exp\left[-\sum_{k\lambda} \frac{4\pi}{\omega_{k\lambda}^3} (q\eta_{k\lambda})^2\right].$$

To calculate W_o we must find the sum

$$\sum_{k\lambda} \frac{4\pi}{\omega_{k\lambda}^3} (q\eta_{k\lambda})^2 = \sum_{\lambda} 4\pi \, (q\eta_{k\lambda})^2 \int \frac{\omega^2 d\omega d\Omega}{(2\pi)^3 c^3 \omega^3}. \tag{1.49}$$

We see that this expression is logarithmically divergent at the low-frequency end. Thus $W_o = 0$, that is, any deceleration, however small, must always be accompanied by radiation of soft photons.

Let us now calculate the matrix element $C_n = \int \chi_n(Q-\delta)\chi_o(Q)dQ$ for arbitrary n. The oscillator wave functions $\chi_n(Q)$ have the form

$$\chi_n(Q) = \sqrt[4]{\frac{\omega}{\pi}} \frac{1}{\sqrt{2^n n!}} \exp\left(-\frac{1}{2}\omega Q^2\right) H_n(\sqrt{\omega} Q),$$

where the H_n are Hermite polynomials. Thus,

$$C_n = \frac{\sqrt{\omega}}{\sqrt{\pi 2^n n!}} \int \exp\left[-\frac{1}{2}\omega Q^2 - \frac{1}{2}\omega (Q-\delta)^2\right] H_n(\sqrt{\omega} Q) dQ.$$

This integral may be easily calculated by using the generating function for the Hermite polynomials:

$$\exp(-t^2 + 2t\xi) = \sum_{n=0}^{\infty} \frac{t^n}{n!} H_n(\xi).$$

The result is

$$C_n = \frac{\alpha^n}{\sqrt{n!}} \exp\left(-\frac{1}{2}\alpha^2\right), \tag{1.50}$$

where

$$\alpha^2 = \frac{1}{2}\omega\delta^2 = \frac{4\pi}{\omega^3}(q\eta)^2.$$

The probability of simultaneous emission of $n_{\underset{\sim}{k}\lambda}$ photons for each $\underset{\sim}{k}, \lambda$ is equal to the product of squares of expressions of the type (1.50), that is,

$$W = \exp\left[-\sum_{k\lambda}\alpha_{k\lambda}^2\right]\prod_{k\lambda}\frac{(\alpha_{k\lambda})^{2n_{k\lambda}}}{(n_{k\lambda})!}.$$

Let us find the average number of photons radiated in deceleration

$$\bar{n}_{k\lambda} = \sum_{n_{k\lambda}} n_{k\lambda} w_n^{k\lambda} = \exp\left[-\alpha_{k\lambda}^2\right] \sum_{n_{k\lambda}} \frac{(\alpha_{k\lambda})^{2n_{k\lambda}}}{(n_{k\lambda})!} n_{k\lambda} =$$
$$= \alpha_{k\lambda}^2 = \frac{4\pi}{\omega_{k\lambda}^3}(q\eta_{k\lambda})^2.$$

Thus we get the following formula for the probability of emission of n photons of type k, λ:

$$w_n^{k\lambda} = \frac{(\bar{n}_{k\lambda})^{n_{k\lambda}}}{(n_{k\lambda})!}\exp(-\bar{n}_{k\lambda}), \tag{1.51}$$

which is just a Poisson distribution.

The probability of emitting a single photon of the type in question is evidently from (1.51)

$$w_1^{k\lambda} = \bar{n}_{k\lambda}\exp(-\bar{n}_{k\lambda}).$$

For $\bar{n}_{\underset{\sim}{k}\lambda} \ll 1$ we get $w_1^{k\lambda} \approx \bar{n}_{k\lambda}$, which agrees with the result obtained by ordinary perturbation theory.

We see from (1.49) that the cross section for purely elastic scattering is strictly zero. Experimentally, elastic scattering means any scattering in which the energy of the radiated photons is less than some quantity E_1 determined by the accuracy of measurement. The observed elastic cross section is given by

$$\sigma_{\text{exp}} = \sigma_0 W_0'$$

where σ_0 is the elastic scattering cross section calculated neglecting radiation, and W_0' is the probability that no photon with energy greater than E_1 be emitted. As we saw above,

$$W_0' = \exp\left[-\int_{E_1}^{E} \frac{4\pi}{\omega^3}(q\eta_{k\lambda})^2 \frac{\omega^2 d\omega}{c^3(2\pi)^3}\right] = $$

$$= \exp\left[-4\pi \frac{2}{3} \frac{1}{(2\pi)^3} \frac{q^2}{c^3} \ln\frac{E}{E_1}\right]. \qquad (1.52)$$

Here we took the energy E of the particle as the upper limit, although actually the assumption of the "softness" of the photons is violated at energies less than this. However, if $\ln(E/E_1)$ is sufficiently large this introduces only a small error.

The cross section for radiation of an arbitrary number of photons of any type is

$$\sigma = \sigma_0 \sum_n W_n = \sigma_0.$$

If we had used perturbation theory, we would have found the results

$$w_1^{k\lambda} = \frac{4\pi}{\omega_{k\lambda}^3}(q\eta_{k\lambda})^2,$$

$$W_1 = \sum_{k\lambda} w_1^{k\lambda} = \frac{4\pi \cdot 2q^2}{(2\pi)^3 c^3}\int_0^E \frac{d\omega}{\omega},$$

that is, an expression which diverges at the lower limit. This divergence, which is due to the use of perturbation theory in a situation where it is illegitimate, is called the infrared divergence or "infrared catastrophe."

The Lamb Shift

The interaction of an atomic electron with the zero-point vibrations of the electromagnetic field will make it fluctuate around its equilibrium orbit. As a result, the electronic charge is smeared out in space and the interaction with the attractive Coulomb field of the nucleus is decreased. Thus the energy levels are pushed up by the interaction with the zero-point vibrations; this phenomenon is known as the Lamb shift. We shall estimate its order of magnitude.

Let δ be the deflection of the electron from its equilibrium position due to interaction with the zero-point vibrations. Then the change of the nuclear potential is given by

$$H' = V(r + \delta) - V(r) = \frac{\partial V}{\partial r}\delta + \frac{1}{2}\frac{\partial^2 V}{\partial x_i \partial x_j}\delta_i\delta_j$$
$$(i, j = x, y, z).$$

where r is the coordinate of the electron measured with respect to the nucleus. We must average this expression over the fluctuations of the electron. Then we have

$$\overline{\delta} = 0, \quad \overline{\delta_x^2} = \overline{\delta_y^2} = \overline{\delta_z^2} = \frac{1}{3}\overline{\delta^2}.$$

Consequently,

$$H' = \frac{1}{6}\overline{\delta^2}\nabla^2 V.$$

Thus to find the magnitude of the Lamb shift we must find the mean-square amplitude $\overline{\delta^2}$ of the fluctuations of the electron.

For purposes of estimation we may write the equation of motion of the electron in the field of the zero-point vibrations $\mathscr{E}$ in the form

$$\ddot{\delta} + \omega_0^2\delta = -\mathscr{E},$$

where ω_0 is the orbital frequency of the electron. We expand the electron fluctuation δ and the field of the zero-point vibrations $\mathscr{E}$ in Fourier series:

$$\delta = \sum_{k\lambda} \delta_{k\lambda} e^{-i\omega_{k\lambda}t}, \quad \mathscr{E} = \sum_{k\lambda} \mathscr{E}_{k\lambda} e^{-i\omega_{k\lambda}t}.$$

Here we have assume that $k\delta \ll 1$; this will be verified below. The Fourier-transformed equation of motion of the electron takes the form

$$(-\omega_0^2 + \omega_{k\lambda}^2)\,\delta_{k\lambda} = \mathscr{E}_{k\lambda}. \quad \text{i.e.}$$

$$\delta_{k\lambda} = \mathscr{E}_{k\lambda}\left(\frac{1}{\omega_{k\lambda} - \omega_0} + \frac{1}{\omega_{k\lambda} + \omega_0}\right)\frac{1}{2\omega_{k\lambda}}.$$

The first term on the right-hand side corresponds to absorption of a photon, the second to emission. In a quantum-mechanical calculation this formula must be modified as follows (cf. p. 57)

$$\delta_{k\lambda} = \mathscr{E}_{k\lambda}\left(\frac{\sqrt{n_{k\lambda}}}{\omega_{k\lambda} - \omega_0} + \frac{\sqrt{n_{k\lambda}+1}}{\omega_{k\lambda} + \omega_0}\right)\frac{1}{2\omega_{k\lambda}},$$

where $n_{k\lambda}$ is the number of photons present. In our present case $n_{k\lambda}$ is zero, and so

$$\delta_{k\lambda} = \frac{1}{2\omega_{k\lambda}(\omega_{k\lambda} + \omega_0)}\mathscr{E}_{k\lambda}.$$

The mean-square fluctuation amplitude $\overline{\delta^2}$ is given by

$$\overline{\delta^2} = \sum_{k\lambda} \overline{\delta_{k\lambda}^2} = \sum_{k\lambda} \frac{\mathscr{E}_{k\lambda}^2}{4\omega_{k\lambda}^2 \, (\omega_{k\lambda} + \omega_0)^2} \, .$$

Let us find the quantity $\mathscr{E}_{k\lambda}^2$. The energy of the zero-point vibrations of type $(\underset{\sim}{k}, \lambda)$ is given by

$$E_{k\lambda} = \int \frac{\mathscr{E}_{k\lambda}^2}{4\pi} \, dV = \frac{\mathscr{E}_{k\lambda}^2}{4\pi} = \frac{1}{2} \, \omega_{k\lambda}$$

(the normalization volume is taken equal to unity).   Hence $\mathscr{E}_{\underset{\sim}{k\lambda}}^2 = 2\pi\omega_{\underset{\sim}{k\lambda}}$, and so

$$\overline{\delta^2} = \frac{\pi}{2} \sum_{k\lambda} \frac{1}{\omega_{k\lambda} \, (\omega_{k\lambda} + \omega_0)^2} \, .$$

We replace the sum over $\underset{\sim}{k}$ by an integration:

$$\sum_{k\lambda} = 2 \int \frac{4\pi k^2 dk}{(2\pi)^3} = \frac{1}{\pi^2} \int \frac{\omega^2 d\omega}{c^3}$$

(where the factor 2 comes from a sum over the two possible photon polarizations).   Then we get

$$\overline{\delta^2} = \frac{1}{2\pi c^3} \int \frac{\omega d\omega}{(\omega + \omega_0)^2} \sim \frac{1}{c^3} \ln \frac{\omega_{max}}{\omega_0} \, ,$$

where ω_{max} is determined by the condition that the motion of the electron in the field of the zero-point vibrations should be non-relativistic (if the electron acquires a relativistic momentum in the course of its fluctuations, then the equation of motion should contain the "relativistic" (effective) mass and $\delta_{k\lambda}$ will be decreased accordingly).   Thus the integral is cut off at the upper end by the condition $k \ll c$, $\omega \ll c^2$.   This means that the quantity $k\delta$, which we neglected above, is of order $k\delta \sim c^{-1/2} \ll 1$.

Since $\sqrt{\overline{\delta^2}}$ depends on ω_{min} and ω_{max} only logarithmically, it does not matter that these quantities are none too precisely defined.

The shift in the electron energy level in first-order perturbation theory is given by the diagonal matrix element of the perturbation H' in the electron state n, that is

$$\delta E = \frac{1}{6}\overline{\delta^2}\int \Psi_n \nabla^2 V \Psi_n dr = \frac{1}{6}\overline{\delta^2}\int \Psi_n 4\pi\rho\Psi_n dr,$$

where ρ is the nuclear charge density. Since the nuclear charge is concentrated in a region small compared to the radius of the electron orbits, we can replace the electron wave function by its value at the origin. The integral $\int \rho\, dr$ is simply the nuclear charge Z, so that we get

$$\delta E = \frac{\pi Z}{6}\overline{\delta^2}\,|\Psi_n(0)|^2 \sim \frac{Z}{c^3}\,|\Psi_n(0)|^2 \ln\frac{c^2}{\omega_0},$$

that is, the Lamb shift is determined by the value of the electron wave function at the origin. Thus the shift is appreciable only for s electrons, since only for them is there no centrifugal barrier, so that the electron can get to the origin. The shift of the levels with nonzero orbital angular momentum will be much smaller.

The Lamb shift thus leads to splitting of the $2s_{1/2}$ and $2p_{1/2}$ levels in hydrogenic atoms. According to the above discussion, the $2p_{1/2}$ level is shifted much less than the $2s_{1/2}$ one. Since $|\Psi_{2s}(0)|^2 = Z^3/8\pi$, we have

$$\delta E = C_1 \frac{Z^4}{c^3}\ln\frac{c^2}{Z^2}\,.$$

(An exact calculation gives $C_1 = 1/6\pi$).

Asymptotic Character of Series in Quantum Electrodynamics

In quantum electrodynamics one uses perturbation theory in the interaction of charged particles with the field: the perturbation series contains the small dimensionless parameter $e^2/\hbar c \approx 1/137$. We shall show that these series are "asymptotic", that is, that they diverge as the number of terms increases. There is in fact an optimum number of terms which best describes any given process; below we shall estimate this number.

We first of all verify that the physical quantitites of interest, regarded as functions of e^2, have a singularity at $e^2 = 0$ and consequently cannot be expanded in a power series in e^2. As an example we consider the energy of the vacuum (Dyson, 1952). If there were no singularity at $e^2 = 0$, the energy of the vacuum would not depend on the way in which e^2 tends to zero. Suppose the value of this energy, as found by perturbation theory, tends to E_o. To what limit does the energy of the vacuum tend when e^2 tends to zero from below ? Consider the creation in the vacuum of N electron-positron pairs under the assumption $e^2 < 0$. For $e^2 < 0$ the electrons attract one another (so of course do the positrons). Let us find out at what value of N the process of pair formation becomes energetically favourable. Suppose the cluster of electrons and the cluster of positrons have separated to a distance sufficiently large that the repulsion between them can be neglected; then we can estimate the energy of each cluster by the Thomas-Fermi method. We find that the energy gain is of order $E \sim N^{7/3}$ (cf. p. 41). The loss of energy is $2mc^2$ for the formation of each electron-positron pair, and hence $2mc^2N$ for the total. Hence if $N^{7/3} > c^2 N$, that is if $N > c^{3/2}$, the creation of electron-positron pairs is energetically advantageous.

Actually, for $N \sim c^{3/2}$, the potential φ occurring in the Thomas-Fermi equations is of order c^2, so that it is necessary to ask what result is obtained in the relativistic case. For simplicity we consider the ultrarelativistic limit. In this case $p_o \sim \varphi/c$, so that (cf. p. 40) $\varphi/\ell^2 \sim (\varphi/c)^3$, and hence $\varphi\ell \sim c^{3/2}$. Since $N \sim (\varphi/\ell^2)\ell^3$, we get $N \sim \varphi\ell \sim c^{3/2}$, that is, the same estimate as in the nonrelativistic case. The energy of the system is then $E \sim Nc^2 - N\varphi \sim c^{7/2} - c^3/\ell$, so that it can be decreased indefinitely by decrease of ℓ. Thus we find that for negative e^2 the electromagnetic vacuum is unstable. Since it is stable for positive e^2, this shows that there must be a singularity at $e^2 = 0$.

The creation of N electron-positron pairs corresponds to the Nth order of perturbation theory. The value of N found above, $N \sim c^{3/2}$, indicates to within an order of magnitude the order of perturbation theory at which the singularity at $e^2 = 0$ will begin to make itself felt. Taking terms of order higher than this into account would not improve the results, but rather the opposite.

Since the asymptotic character of the perturbation series makes itself felt only at very high terms, the question is of purely theoretical interest.

CHAPTER 2

VARIOUS TYPES OF PERTURBATION THEORY

Dimensional and model approximations are only the first stage in the approach to the solution of a physical problem; to carry the problem through to the end we must find a numerical solution. However, it is only rarely that it is possible to find exact solutions to the problems of theoretical physics. For instance, the simplest problem in quantum mechanics - the problem of finding the wave function for the one-dimensional Schrödinger equation - can be solved exactly only for a few forms of potential. We therefore have to be content ourselves either with approximate analytic methods or with a numerical solution; and to check the numerical methods and understand their significance we must in any case find an approximate analytic solution in various limiting cases.

In some cases the nature of the solution is determined by the analytic properties or symmetry properties of the quantities under investigation; examples of this type will be discussed in Chapters 4-6. In the present chapter we will discuss various kinds of perturbation theory, where the solution can be found in the form of a power series in some small parameter.

The simplest case of perturbation theory is the case of a weak external field or a weak interaction between particles. Then the "small parameter" is the ratio of the (external field) energy or interaction potential to a typical value of the energy of free motion. As an example of this type of perturbation theory, which exploits the smallness of some extra term in the Hamiltonian, we consider below the problem of scattering.

A second type of perturbation theory involves perturbation of the boundary conditions. Suppose, for instance, we know the solution of our problem with boundary conditions imposed on a spherical surface, and wish to know the solution with the same boundary conditions, but now imposed on a nearly but not exactly spherical surface. By performing the appropriate coordinate transformation we can reduce the boundary conditions to conditions on a spherical surface, but then the coordinate transformation leads to a correction to the Hamiltonian, that is, the problem reduces to perturbation theory for a weak deformation of the Hamiltonian.

In the case where the perturbation V is not small, but lasts only for some small time τ, then the problem contains the small parameter $\xi \sim V\tau$. This type of perturbation theory will be illustrated by the problems of ionization in β-decay, ionization in nuclear collisions and the Mössbauer effect. Another good example of the "sudden perturbation" method is the problem of the creation of soft photons in electron scattering (the so-called "infrared catastrophe") which was discussed in the last chapter.

The opposite limiting case is when the perturbation

changes over times much greater than the characteristic periods of motion of the system in question (the so-called adiabatic approximation). In this case we have to solve the auxiliary problem of the motion of the system for a fixed perturbation. The "small parameter" in this case is the quantity $\xi \sim 1/\omega\tau$, where ω is a typical frequency of the system, and τ is the time over which the perturbation changes substantially. The classical example of an adiabatic perturbation is the theory of a molecule: here we first solve the auxiliary problem of motion of the electrons for fixed nuclear positions, and then find the solution including the nuclear motion on the assumption that the ratio of molecular to electron frequencies is small. A second example of the adiabatic approximation which we shall consider is the problem of the scattering of a proton on a hydrogen atom.

One other effective approximate method can be developed when the energy levels of the system include some which are close together. In this case, when the perturbation is small compared to the distance in energy to the remaining levels, one can do a calculation which treats the adjacent levels exactly. This method will be illustrated by the problem of the motion of a particle in a periodic potential, the Stark effect and the problem of the change of the lifetime of the $2S_{1/2}$ state of the hydrogen atom in an electric field. In the last two cases the "adjacent level" is the $2P_{1/2}$ level.

Perturbation theory has played an important role in the development of theoretical physics. It can be used to get an idea of the general character of the solution even in cases where the expansion parameter is not small but is of order unity. However, an exception must be made for those phenomena which vanish for small values of the expansion parameter. Thus, for example, if we were to consider the interaction potential as a small quantity, we could never in any order of perturbation theory obtain a function describing a bound state of the particle; this is particularly obvious where the bound state appears only at a finite depth of the potential well.

1. PERTURBATION THEORY IN THE CONTINUOUS SPECTRUM

We shall consider the scattering problem in the case when a small perturbing potential is added to the scattering potential; the aim is to find the change in the scattering amplitude in perturbation theory. As an example we investigate the scattering of charged particles by atomic nuclei and calculate the correction to the Coulomb scattering amplitude due to the finite size of the nucleus.

Suppose the Hamiltonian of the system is of the form

$$H = H_0 + H',$$

where H' is a small perturbation, and the solution of the problem with Hamiltonian H_0 is known:

$$H_0\varphi_p = E_p\varphi_p.$$

In the case of the scattering problem (continuous spectrum) the boundary conditions imposed on $\varphi_{\underset{\sim}{p}}$ may be of either of two types:

$$\varphi_p^{\pm} \to e^{ipr} + \frac{f_0^{\pm}}{r} e^{\pm ipr}.$$

Only one of the two functions $\varphi^{\pm}$ is physically meaningful, namely $\varphi_{\underset{\sim}{p}}^{+}$, which corresponds to an outgoing spherical wave; however it is useful to keep the $\varphi_{\underset{\sim}{p}}^{-}$ too as an aid to calculation. Either of the systems of functions $\varphi_{\underset{\sim}{p}}^{+}$ or $\varphi_{\underset{\sim}{p}}^{-}$ forms a complete orthogonal set, that is

$$\int \varphi_p^{+*}\varphi_{p'}^{+}dr = \int \varphi_p^{-*}\varphi_{p'}^{-}dr = (2\pi)^3\delta\,(p - p').$$

The Schrödinger differential equation with the boundary conditions appropriate to the scattering problem is equivalent to the integral equation

$$\Psi_p = \varphi_p^{+} + \int \frac{dp'}{(2\pi)^3}\,\frac{(\varphi_{p'}^{-}\,|\,H'\,|\,\Psi_p)}{E_p - E_{p'}}\,\varphi_{p'}^{-}. \qquad (2.1)$$

as can be easily verified by operating on (2.1) with the operator $H_o - E_{\underset{\sim}{p}}$. We could have expanded $\Psi_{\underset{\sim}{p}}$ in the functions $\varphi_{\underset{\sim}{p}}^{+}$, but we shall see below that the $\varphi_{\underset{\sim}{p}}^{-}$ are more convenient.

The integrand of (2.1) has a singularity for $|\underset{\sim}{p}| = |\underset{\sim}{p}'|$. We see that the way in which we go round the singularity determines the asymptotic form of $\Psi_{\underset{\sim}{p}}$, that is sets the boundary conditions.

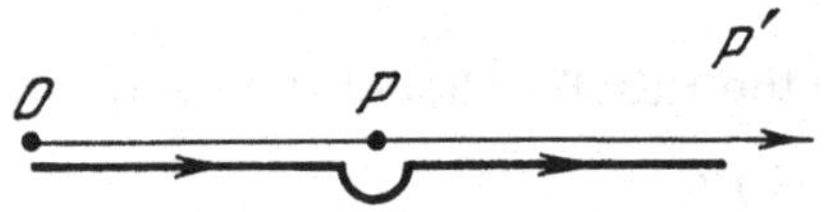

Fig. 17

We shall now show that it is the way indicated in Fig. 17 (rather than a passage above the singularity or a principal value integral) which gives the correct boundary conditions.

Let us proceed to calculate the integral in expression (2.1) for $\Psi_{\underset{\sim}{p}}$ in the limit $r \to \infty$. In this limit the function $\varphi_{\underset{\sim}{p}'}^{-}$ has the form

$$\varphi_{p'}^{-} \xrightarrow[r\to\infty]{} e^{ip'rx} + \frac{f_0^{-}}{r} e^{-ip'r},$$

where

$$x = \cos \angle (\boldsymbol{p}'\boldsymbol{r}).$$

The first term in this expression leads to an integral in (2.1) of the form

$$\int_{-1}^{1} F(x)\, e^{ip'rx}\, dx.$$

Integrating by parts, we get

$$\int_{-1}^{1} F(x)\, e^{ip'rx}\, dx \underset{r\to\infty}{\approx} \frac{F(1)\, e^{ip'r} - F(-1)\, e^{-ip'r}}{ip'r}.$$

We now show that the part of this expression which contains $e^{-ip'r}$ vanishes in the limit $r \to \infty$ when we integrate over dp'. We have to estimate the integral

$$\frac{1}{r} \int_{\Omega} \frac{dp'}{E_p - E_{p'}} e^{-ip'r} \Phi(p'). \tag{2.2}$$

where the integration contour is shown in Fig. 17. We deform the integration contour into $C_1 + C_2$ (see Fig. 18), taking δ small enough so that there are no singularities between the new and the old contours; this ensures that the value of the integral (2.2) is unchanged. In the integral over C_2 we have $p' = \xi - i\delta$, $\delta > 0$,

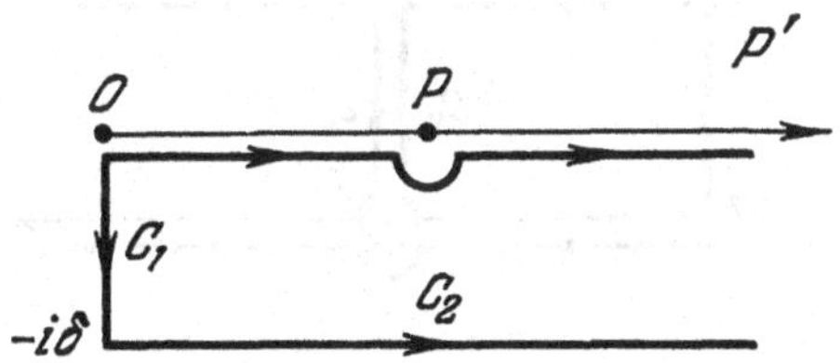

Fig. 18

and hence $e^{-ip'r} = e^{-i\xi r} e^{-\delta r}$. The exponential $e^{-\delta r}$ can be taken out from under the integral sign in (2.2), so that the whole integral over C_2 is exponentially small in the limit $r \to \infty$. For the contour C_1 we have $p' = -it$, so that it contributes the integral

$$\frac{1}{r}\int_0^{\delta}\frac{e^{-rt}}{E_p - E_{it}}\Phi(t)\,dt. \tag{2.3}$$

In the limit $r\to\infty$ the region of t important in the integrand of (2.3) is of the order of the region $[0, 1/r]$ and so E_{it} is small in this region, $E_{it} \ll E_p$. Thus the integral (2.3) is of order

$$\frac{1}{r}\int_0^{\delta} e^{-rt}\,dt \sim \frac{1}{r^2},$$

that is, it decreases as an inverse power of r.

The same line of reasoning shows that the part of the integral in (2.1) which contains the term $(f_0^-/r)\exp{-ip'r}$ from the asymptotic expression for $\varphi^-_{\underset{\sim}{p}'}$, also decreases as an inverse power of r. It is this fact which makes the $\varphi^-_{\underset{\sim}{p}}$ a more convenient set of basis functions than the $\varphi^+_{\underset{\sim}{p}}$.

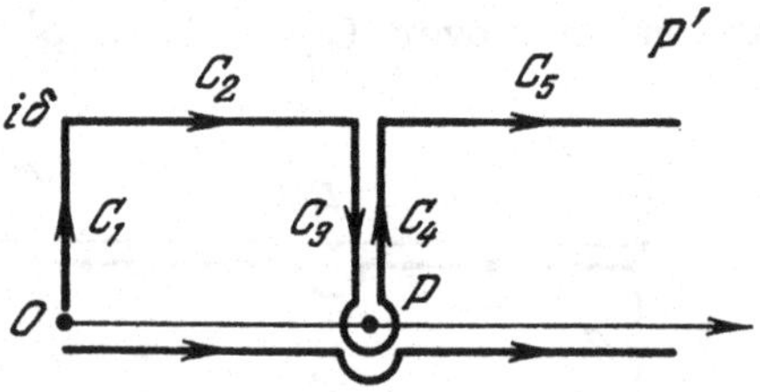

Fig. 19

Thus the only term left in (2.1) is the integral containing $e^{ip'r}$. To calculate it we choose the contour indicated in Fig. 19. Just as in the case of the integral containing $e^{-ip'r}$, the contribution from $C_2 + C_5$ will be exponentially small and that from C_1 will decrease as a power of r as $r\to\infty$. The contributions from

C_3 and C_4 cancel one another. Thus the whole integral reduces to the residue at the point $p' = p$:

$$\Psi_p = \varphi_p^+ + 2\pi i \frac{2\pi p'^2}{(2\pi)^3} \cdot \frac{(\varphi_{p'}^- | H' | \Psi_p)}{dE_p/dp} \frac{e^{ip'r}}{ip'r},$$

where $\underset{\sim}{p}' = p\underset{\sim}{r}/r$. Thus Ψ_p has the correct asymptotic form:

$$\Psi_p = e^{ipr} + \frac{(f_0 + f)}{r} e^{ipr},$$

where

$$f = -\frac{p}{2\pi v} (\varphi_{p'}^- | H' | \Psi_p^+) \Big|_{p'=p\frac{r}{r}}. \qquad (2.4)$$

In the case of nonrelativistic particles we can replace p/v by m.

In the case of free motion $\varphi_{\underset{\sim}{p}}^+$ and $\varphi_{\underset{\sim}{p}}^-$ are plane waves, $f_0 = 0$, and we get the scattering amplitude in the so-called Born approximation:

$$f_B = -\frac{p}{2\pi v} \int H'(r') e^{iqr'} dr',$$

where

$$q = p - p\frac{r}{r}$$

is the momentum transfer in scattering: $|\underset{\sim}{q}| = 2p \sin\theta/2$, where θ is the scattering angle.

To obtain successive iterations it is convenient to reformulate the integral equation (2.1) for the wave function as a corresponding integral equation for the scattering amplitude f. Using (2.1) and (2.4), we get

$$f(p', p) = f_1(p', p) - \frac{2\pi v}{p} \int \frac{g(p', p'') f(p'', p)}{E_p - E_{p''}} \frac{dp''}{(2\pi)^3}, \qquad (2.5)$$

where

$$f_1(p', p) = \frac{-p}{2\pi v}\int \varphi_{p'}^{-*} H' \varphi_p^{+} dr,$$

$$g(p', p) = \frac{-p}{2\pi v}\int \varphi_{p'}^{-*} H' \varphi_p^{-} dr.$$

In this expression $\underset{\sim}{p}$ and $\underset{\sim}{p}'$ are free parameters. The cross section is determined by the quantity $|f(\underset{\sim}{p}, \underset{\sim}{p}')|^2$ for $|\underset{\sim}{p}'| = |\underset{\sim}{p}|$.

Let us find the criterion of applicability of perturbation theory in the continuous spectrum. The basic principle involved in finding the criterion of validity of any approximation method is that the physical quantity calculated in the approximation should be large in comparison to the correction to it from the next order of approximation.

In the first order of perturbation theory we have

$$f = f_1 = -\frac{p}{2\pi v}\int \varphi_{p'}^{-*} H' \varphi_p^{+} dr'.$$

In the second order of perturbation theory we get from (2.1)

$$f_2 = -\frac{p}{2\pi v}\int \varphi_{p'}^{-*} H' \Psi_p^{(1)} dr' \sim m\int \varphi_{p'}^{-} H' \frac{f_1}{r'} e^{ipr'} dr'.$$

The condition $f_2/f_1 \ll 1$ serves as the criterion of applicability of perturbation theory.

Suppose a is the order of magnitude of the region of space in which the perturbing potential $H'(r)$ is appreciably different from zero. Consider first the case $pa \lesssim 1$. Then the function $\varphi_{\underset{\sim}{p}}^{\pm}(a)$ is of order of magnitude unity. Consequently the amplitude f_1 is of order $H'(a)\, a^3$ and is only weakly dependent on the scattering angle. Thus $f_2/f_1 \sim H' a^2$ and the criterion for the applicability of perturbation theory has the form

$$H'a^2 \ll 1.$$

Secondly consider the case $pa \gg 1$. (This case will be considered in more detail in Section 3.1.) In this high-energy limit the scattering is strongly anisotropic, being concentrated in the forward direction. We can see this from the fact that in expression (2.4) the quantity

$$\bar{\varphi}^{*}_{p'}\varphi^{+}_{p} \sim e^{iqr} \sim e^{ipa \sin\frac{\theta}{2}}$$

is fast oscillating for $pa \gg 1$ for all scattering angles θ except those for which $\theta \lesssim 1/pa \ll 1$. For scattering angles $\theta \lesssim 1/pa$ the quantity $\bar{\varphi}^{*}_{\underset{\sim}{p}'}(a)\varphi^{+}_{\underset{\sim}{p}}(a)$ is of order of magnitude unity and the scattering amplitude can be estimated to be

$$f_1(\theta) \underset{\theta \lesssim 1/pa}{\sim} H'(a)\,a^3.$$

We see that the criterion of applicability of perturbation theory for these scattering angles is the same as that at small momenta.

For angles $\theta \gg 1/pa$ the amplitude $f(\theta)$ is much reduced by the oscillatory behaviour of the wave functions $\varphi^{+}_{\underset{\sim}{p}}$.

Scattering of Charged Particles by the Atomic Nucleus

As an example we consider the scattering of charged particles by an atomic nucleus. We assume that the particles interact only with the electric field of the nucleus (as do e.g. electrons). We shall assume that the nuclear charge is distributed uniformly over the volume of the nucleus. Then the potential energy of the particle will be

$$V = \begin{cases} -\frac{Z}{R}\left(\frac{3}{2}-\frac{1}{2}\frac{r^2}{R^2}\right), & 0<r<R, \\ -\frac{Z}{r}, & r>R. \end{cases}$$

where Z is the charge of the nucleus and R its radius. Then the perturbation will have the form

$$H' = \begin{cases} -\frac{Z}{R}\left(\frac{3}{2}-\frac{1}{2}\frac{r^2}{R^2}\right)+\frac{Z}{r}, & 0<r<R, \\ 0, & r>R. \end{cases}$$

The correction to the scattering amplitude to first order in H' is

$$f_1 = -\frac{1}{2\pi}\int \varphi_{p'}^{-*} H' \varphi_p^+ \, d\mathbf{r} \approx -\frac{1}{2\pi}\varphi_{p'}^{-*}(0)\,\varphi_p^+(0)\int H'\,d\mathbf{r}.$$

The above estimate is legitimate because H' changes over a distance of order R, while $\varphi_{\mathbf{p}}$ changes over a distance of order $1/p$. For not-too-high electron velocities we have $pR \ll 1$ ($R \lesssim 10^{-4}$ in atomic units). Thus in the expression for f the functions $\varphi^-_{\mathbf{p}'}$ and $\varphi^+_{\mathbf{p}}$ may be taken out from under the integral sign and evaluated at the point $\mathbf{r} = 0$.

Calculation gives

$$\int H'\,d\mathbf{r} = \frac{2\pi Z R^2}{5}.$$

and hence we find

$$f_1 = -\frac{ZR^2}{5}\varphi_{p'}^{-*}(0)\,\varphi_p^+(0).$$

The nonrelativistic Coulomb wave functions $\varphi^{\pm}_{\mathbf{p}}(0)$ are given,[1] for a repulsive potential, by

---

1 See L. D. Landau and E. M. Lifshitz, Quantum Mechanics (Pergamon Press, London, 1959), pp. 121-2.

$$\varphi_p^{\pm}(0) = \frac{1}{\sqrt{(2\pi)^3}} e^{-\pi/2p}\, \Gamma(1 \pm i/p),$$

and for an attractive potential by

$$\varphi_p^{\pm}(0) = \frac{1}{\sqrt{(2\pi)^3}} e^{\pi/2p}\, \Gamma(1 \mp i/p),$$

where $\Gamma(1 \pm i/p)$ is the gamma function. One is usually interested in the scattering of electrons with momentum much greater than that of the atomic electrons; in that case $\Gamma(1 \pm i/p) \approx 1$.

We thus finally get

$$f_1 = -\frac{1}{80\pi^3} \frac{ZR^2}{a_0} e^{\pm 2\pi/p}, \tag{2.6}$$

where the minus sign in the exponential refers to a repulsive potential and the plus sign to an attractive one, and a_0 is the Bohr radius.

The differential cross section $d\sigma/d\Omega$ is given by

$$d\sigma/d\Omega = | f_0 + f_1 |^2,$$

where f_0 is the Rutherford scattering amplitude. Since f_1 is independent of the scattering angle while f_0 decreases with increasing angle, it is important to take the finite dimensions of the nucleus into account when considering large-angle scattering.

The criterion for the applicability of perturbation theory in the case considered $(pR \ll 1)$ is $H'R^2 \ll 1$ or $ZR \ll 1$, which is fulfilled for all real nuclei.

2. PERTURBATION OF THE BOUNDARY CONDITIONS

In this section we consider the case in which a perturbation is applied to the boundary conditions. By means of a coordinate transformation we can transform the boundary conditions back to their unperturbed form; however, this changes the expression for the Hamiltonian of the system. This change of the Hamiltonian is nothing but a perturbing potential, whose effect can be discussed by ordinary perturbation theory. We shall illustrate the theory by calculating the one-particle energy levels of a deformed nucleus.

Suppose we know the solution of the stationary (or non-stationary) problem described by the Hamiltonian H(x) and a boundary condition on some surface S_0:

$$\alpha\Psi + \beta\Psi' |_{S_0} = 0$$

(where Ψ' is the component of the derivative of Ψ normal to S_0). We wish to find the solution of the problem with the same Hamiltonian H(x) but with the boundary condition

$$\alpha\Psi + \beta\Psi' |_S = 0,$$

where the surface S is close to S_0 (it is just this closeness which is the "small parameter" in the problem). To solve the problem we must find a coordinate transformation $x_i = f_i(x'_j)$ such that $S(x) = S_0(x')$, that is, the equation for the new surface in the old variables is the same as that for the old one in the new variables. In other words, if the equation of the surface S_0 is $\Phi_0(x_j) = 0$, and that of S is $\Phi(x_j) = 0$, then we must have

$$\Phi\,(f_i\,(x'_j)) = \Phi_0\,(x'_j).$$

The Hamiltonian $H(x_i) = H(f_i(x'_j))$ is changed when expressed in the new variables. We can write it in the form

$$H(x'_j) + H'(x'_j),$$

where $H'(x'_j) = H(f_i(x'_j)) - H(x'_j)$ is simply a perturbation. From now on we can go straight on to apply ordinary perturbation theory.

The Energy Levels of a Deformed Nucleus

Suppose we know the stationary energy levels in a spherical potential well, and wish to know them in an ellipsoidal well (the walls are taken to be of infinite height). The equation for the surface S_0 has the form

$$\Phi_0(x_j) = 0: \quad \sum_{j=1}^{3} x_j^2 - R^2 = 0,$$

while the equation for the surface S is

$$\Phi(x_j) = 0: \quad \sum_{j=1}^{3} \frac{x_j^2}{a_j^2} - 1 = 0.$$

We introduce new variables

$$x_j = \frac{a_j x'_j}{R}.$$

Then $\Phi(f_i(x'_j)) = \Phi_0(x'_j)$. The kinetic energy operator is changed by this transformation:

$$T(x_i) = -\frac{1}{2M}\sum_{i=1}^{3}\frac{\partial^2}{\partial x_i^2} = -\frac{R^2}{2M}\sum_{i=1}^{3}\frac{1}{a_i^2}\frac{\partial^2}{\partial x_i'^2}.$$

Consequently the perturbation has the form

$$H'(x'_i) = -\frac{1}{2M}\sum_{i=1}^{3}\left(\frac{R^2}{a_i^2} - 1\right)\frac{\partial^2}{\partial x_i'^2}.$$

This will be small provided all the a_i are close to R.

Consider an ellipsoidal deformation of the nucleus, assuming the ellipsoid to be biaxial (with semiaxes a and b). The quantity usually called the deformation parameter is

$$\beta = 2\frac{a-b}{a+b}.$$

Suppose that the volume of the nucleus remains unchanged by the deformation, that is, $ab^2 = R^3$. If we write

$$a = R(1+\delta),\; b = R(1-\delta_1),$$

where $\delta, \delta_1 \ll 1$, then to a first approximation the relation $ab^2 = R^3$ gives $\delta - 2\delta_1 = 0$, that is,

$$\beta = 2\frac{R(1+2\delta_1) - R(1-\delta_1)}{2R} \approx 3\delta_1,$$

or

$$a \approx R\left(1+\frac{2}{3}\beta\right), \qquad b \approx R\left(1-\frac{1}{3}\beta\right).$$

Then the perturbation H' has the following form:

$$H' \approx -\frac{\beta}{3M}\left(\nabla^2 - 3\frac{\partial^2}{\partial z^2}\right).$$

Suppose the energy of a particle before deformation of the nucleus was $\epsilon^0_{n\ell j}$ (it cannot depend on the magnetic quantum number m in a spherically symmetric potential). Taking the deformation into account in first-order perturbation theory leads to the expression

$$\varepsilon_{nljm} = \varepsilon^0_{nlj} + (\varphi^{(0)}_{nljm} \,|\, H' \,|\, \varphi^{(0)}_{nljm}).$$

We then obtain by explicit calculation

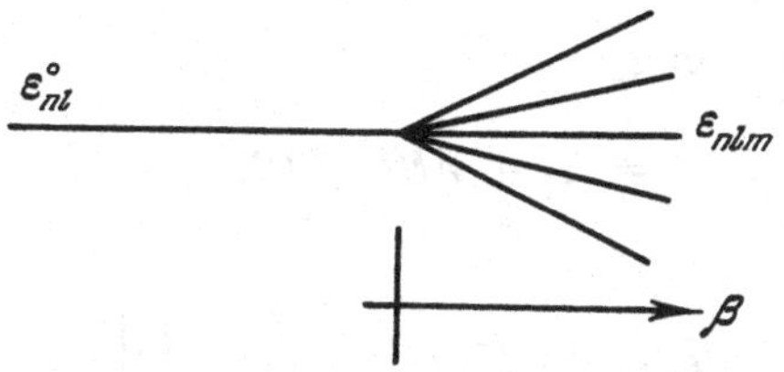

Fig. 20

$$\varepsilon_{nljm} = \varepsilon^0_{nlj}\left[1 + \beta\left(\frac{m^2}{j\,(j+1)} - \frac{1}{3}\right)\right]. \tag{2.7}$$

The dependence of $\epsilon_{n\ell jm}$ on β is illustrated schematically in Fig. 20. We note that

$$\frac{1}{2j+1}\sum_m \varepsilon_{n\ell jm} = \varepsilon^0_{n\ell j}$$

that is, the "centre of gravity" of the multiplet is not shifted by the nuclear deformation.

In the above calculation it has been assumed that the deformation is small enough that the spin-orbit coupling is not broken; then the levels are determined by the total angular momentum j. (cf. problem 2).

Problems

1. Calculate $\epsilon_{n\ell m}$ in the absence of spin-orbit coupling ($\Psi_{n\ell m} = R_{n\ell}\, Y_{\ell m}$).

Solution $\varepsilon_{nlm} = \varepsilon^0_{nl}\left\{1 + \beta\left[\frac{m^2 - \frac{1}{4}}{\left(l - \frac{1}{2}\right)\left(l + \frac{3}{2}\right)} - \frac{1}{3}\right]\right\}.$

2. Calculate ϵ_n for arbitrary relative magnitude

of the spin-orbit interaction and the deformation energy:

$$H = H_0 + ASl + H'.$$

Solution

$$\varepsilon_{nljm}^{+,-} = \varepsilon_{nl}^{+,-} + \beta\varepsilon_{nl}^{0}\left[\frac{m_j^2}{L} - \frac{1}{3}\right] \pm$$
$$\pm\left\{\sqrt{\frac{A^2}{4}\left(l+\frac{1}{2}\right)^2 + \beta\varepsilon_{nl}^{0}\frac{m_j}{L}\left[\beta\varepsilon_{nl}^{0}\frac{m_j}{L} - Am_j\right]} - \frac{A}{2}\left(l+\frac{1}{2}\right)\right\},$$

where

$$\varepsilon_{nl}^{+,-} = \varepsilon_{nl}^{0} + \frac{A}{2}\left[-\frac{1}{2} \mp \left(l+\frac{1}{2}\right)\right],$$

and the sign $\pm$ corresponds to $j = \ell \pm \frac{1}{2}$

(For the sake of conciseness of notation we have introduced in the solution the notation $L = (\ell - \frac{1}{2})(\ell + 2/3)$).

3. SUDDEN PERTURBATIONS

In the examples considered considered so far the "small parameter" of the problem has been the relative change of the system Hamiltonian. We now consider the case when the small parameter is the time of action of the perturbation, while the perturbation itself need not be weak.

One example of the action of such a perturbation is ionization due to β decay. In this case the velocity of the electron emitted in β decay is much greater than the velocity of the atomic electrons. Hence from the point of view of the atomic electrons the change in the nuclear charge occurs suddenly. Another example is ionization due to the collision of a fast particle with the nucleus. In this case the sudden recoil of the nucleus is also a "sudden event" from the point of view of the atomic electrons.

We consider a system whose wave function obeys Schrödinger's equation $i(\partial\Psi/\partial t) = H\Psi$ with the Hamiltonian H given as follows:

$$H = \begin{cases} H_1(r), & t < 0, \\ H_2(r), & t > \tau. \end{cases}$$

We assume that the system Hamiltonian changes sharply over the short time interval τ. Let us introduce the complete set of eigenfunctions of the Hamiltonian H_2:

$$H_2\varphi_n^{(2)} = \varepsilon_n^{(2)}\varphi_n^{(2)}$$

and expand the solution of the problem, $\Psi(t)$, in terms of the functions $\varphi_n^{(2)}$:

$$\Psi(t) = \sum_n a_n(t)\,\varphi_n^{(2)} e^{-i\varepsilon_n^{(2)}t}. \tag{2.8}$$

We write the Hamiltonian H in the form $H = H_2 + (H - H_2) = H_2 + V$. Then we have

$$i\frac{da_n}{dt} = \sum_m V_{nm}(t)\,e^{-i\left[\varepsilon_m^{(2)}-\varepsilon_n^{(2)}\right]t} a_m(t). \tag{2.9}$$

To see this, note that if we temporarily use the notation $b_n(t) = a_n(t)\exp(-i\epsilon_n^{(2)}t)$, then the expansion $\Psi(t) = \Sigma_n b_n(t)\varphi_n^{(2)}$ simply represents the transition from the coordinate representation to the energy representation with respect to the Hamiltonian H_2.

Schrödinger's equation in the new representation has the form

$$i\frac{\partial b_n}{\partial t} = \sum_m [\varepsilon_n^{(2)}\delta_{nm} + V_{nm}(t)]\, b_m,$$

whence Eq. (2.9) follows immediately.

Integrating expression (2.9), we get

$$a_n(t) - a_n(0) = -i\sum_m \int_0^t V_{nm} a_m(t)\, e^{-i[\varepsilon_m^{(2)} - \varepsilon_n^{(2)}]t}\, dt. \tag{2.10}$$

The matrix element V_{nm} tends to zero for $t \gg \tau$. Let us assume that $(\epsilon_n^{(2)} - \epsilon_m^{(2)})\tau \ll 1$. Then the exponential in the integrand is approximately unity. It then follows that if a technique of successive approximations is to be useful, we require the condition $V_\tau \ll 1$ (i.e., the perturbation need not be small, but its period of action must be short).

To zeroth approximation in $V\tau$ we get from (2.10) $a_n^{(0)}(t) = a_n(0)$. Replacing $a_m(t)$ on the right-hand side of (2.10) by $a_m(0)$ to the first approximation and taking account of (2.8), we get

$$a_n^{(1)}(t) = -i\sum_m \int_0^t (\varphi_n^{(2)}\,|\,V\,|\,\varphi_n^{(2)})\, a_m(0)\, dt = \tag{2.11}$$

$$= -i\int_0^t (\varphi_n^{(2)}\,|\,V\,|\,\Psi(0))\, dt.$$

We assume that the initial state $\Psi(0)$ was one of the eigenfunctions of H_1, that is,

$$\Psi(0) = \varphi_{n_0}^{(1)} = \sum a_n(0)\, \varphi_n^{(2)}.$$

so that $a_n(0) = (\varphi_{n_0}^{(1)}|\varphi_n^{(2)})$. Thus in the zeroth approximation the transition probability into the state $\varphi_n^{(2)}$ is

$$W_{nn_0} = |(\varphi_{n_0}^{(1)} | \varphi_n^{(2)})|^2. \tag{2.12}$$

In the Hilbert space of $\Psi(t)$ spanned by the eigenfunctions (basis vectors) φ_n, a sudden perturbation corresponds to a large change of the basis $(\varphi_n^{(1)} \to \varphi_n^{(2)})$ accompanied by only a small change of the wave function $(\Psi(t) \approx \Psi(0))$.

Ionization of Atoms in β-decay

Calculation of the probability of ionization of an atom in β decay is much simplified if we use the fact that the velocity of the electron emitted in β-decay is much greater than the velocities of the atomic electrons. We shall show that ionization is due to the change in the nuclear charge; the direct interaction of the emitted electron with the atomic ones may be neglected.

We first estimate the probability of an electronic transition induced by this direct interaction. According to perturbation theory it is given by

$$W = \frac{1}{\hbar^2} \left| \int_0^\infty V_{nn_0} e^{i\omega_{nn_0} t} \, dt \right|^2. \tag{2.13}$$

Here V_{nn_0} is the interaction matrix element and ω_{nn_0} the frequency of the transition. To estimate this expression, we note that ω_{nn_0} is of the order of the atomic frequencies, while the time of passage of the β decay electron is much less than the atomic periods; thus $\omega_{nn_0}\tau \ll 1$, so that we can replace the exponential in (2.13) by unity. The interaction V of the β-decay electron with the atomic electrons is of order of magnitude e^2/a, where a is a quantity of the order of the atomic dimensions.

Since in β decay the electron is emitted from the atom with a speed of the order of the velocity of light c, the time of passage τ is of order a/c. Thus the probability W is of order of magnitude

$$W \sim \frac{1}{\hbar^2} V^2 \tau^2 \sim \left(\frac{e^2}{\hbar c}\right)^2 \ll 1. \tag{2.14}$$

Next we calculate the probability of ionization due to the change of the nuclear charge. As we have just seen, the wave function of the atomic electrons does not change much over the time of passage of the β decay electron. Thus the probability of ionization of an electron in a state Ψ_0^Z in the potential of the nucleus with charge Z is given according to (2.12) by

$$W_{0,E} = |(\Psi_0^Z | \Psi_E^{Z+1})|^2. \tag{2.15}$$

where Ψ_E^{Z+1} is the wave function of the ejected atomic electron with energy $E > 0$ in the potential of the nucleus with charge $Z + 1$. The total probability of ionization due to change of the nuclear charge is

$$W_0 = \int_0^\infty W_{0,E}\, dE. \tag{2.16}$$

Let us estimate (2.16). To do this we expand the function $f(Z_1, Z_2) = (\Psi_0^{Z_1} | \Psi_E^{Z_2})$ in a series in the difference $Z_1 - Z_2$:

$$f(Z_1, Z_2) \approx (Z_1 - Z_2) \left.\frac{\partial f}{\partial Z_1}\right|_{Z_1 = Z_2}.$$

The derivative is

$$\left.\frac{\partial f}{\partial Z_1}\right|_{Z_1 = Z_2} = \left(\Psi_E^{Z_1} \middle| \frac{\partial \Psi_0^{Z_1}}{\partial Z_1}\right) \sim \frac{1}{Z_1}.$$

Therefore we have

$$(\Psi_0^{Z_1} | \Psi_E^{Z_2}) \sim \frac{Z_1 - Z_2}{Z_1} = \frac{1}{Z_1},$$

or

$$W_0 \sim \frac{1}{Z^2}. \tag{2.17}$$

Thus the effect of the direct interaction between the β decay and atomic electrons is of relative order of magnitude

$$\left(\frac{Ze^2}{\hbar c}\right)^2 \ll 1.$$

From (2.15) we can also draw some conclusions about the selection rules for the process in question. In the case of ionization of the inner shells the self-consistent field in which the electron moves can be taken to be spherically symmetric, so that $\Psi = R_{n\ell}(r)\, Y_{\ell m}(\theta, \varphi)$. Substituting this form of Ψ in (2.15) we get the selection rules

$$l_1 = l_2, \; m_1 = m_2.$$

For the outer shells the effective charge Z felt by the atomic electrons is of order 1. Hence, according to (2.17), the ionization probability W_k is of order 1. This is confirmed by experiments on the accumulation of positive ions in $\beta^{\pm}$ decay.

In the case of the K and L shells the calculation of the ionization probability can be carried through in full. In these cases the spatial regions of importance are close to the nucleus, where the functions Ψ_o^Z and Ψ_E^{Z+1} can be taken to be hydrogenic (the error thus introduced is of order $1/Z$).

The ionization probability falls off fast when the energy of the ejected electron is much larger than the ionization potential, since then the function Ψ_E^{Z+1} oscillates many times over the radius of the K shell, so that the integral (2.15) is small. Thus the ejected electrons have energies in a range from zero up to an energy of the order of the ionization potential.

Let us find the ionization probability for a K electron when the energy of the ejected atomic electron is much larger than the ionization potential. For large energies the radial wave function R_E^{Z+1} must be the same as the radial wave function of a free particle:

$$R_E^{Z+1} \underset{E\to\infty}{\sim} \frac{1}{r}\sqrt{\frac{2}{\pi k}}\sin kr.$$

where $k = \sqrt{2E}$. We have normalized the function R_E to an energy delta function:

$$\int_0^\infty R_E R_{E'} r^2\, dr = \delta(E - E').$$

Hence the probability W_{1E} is given by

$$W_{1E} = \left| \int_0^\infty R_1^Z R_E^{Z+1} r^2\, dr \right|^2 \underset{E\to\infty}{\approx} \frac{2}{\pi k}\left| \int_0^\infty R_1^Z \sin(kr)\, r\, dr \right|^2.$$

For a K electron we have

$$R_1^Z = 2Z^{3/2} e^{-Zr}$$

Thus we have to calculate the integral $I = \int_0^\infty re^{-Zr} \sin kr\, dr$ in the limit $k \to \infty$. Integrating by parts, we get $I \approx 2Z/k^3$. This integral was calculated in section 1.1, p.25 , and on p. 62.

Thus for $E \gg Z^2$ the ionization probability for a K electron is approximately given by

$$W_{1E} \approx \frac{8Z^3}{\pi k}\left(\frac{2Z}{k^3}\right)^2 = \frac{2\sqrt{2}}{\pi}\frac{Z^5}{E^{7/2}}.$$

Ionization of an Atom in Nuclear Reactions

In nuclear collisions involving large energy transfer there must occur ionization of the recoil atoms. If the velocity acquired by the nucleus is not too large, then it can carry its electrons off with it, and ionization takes place only in the outer, weakly bound shells. For large velocities, on the other hand, the nucleus recoils right out of its electronic shells instead of carrying them with it.

We shall calculate the probability of ionization when a neutron collides with the nucleus (Migdal 1939). The duration of the neutron-nucleus collision is of order $\tau \sim R/v$, where R is the nuclear radius and v the neutron velocity. This time is much less than the electronic periods $\tau_{e\ell}$, so that the electron wave function is practically unchanged over the duration of the collision. If we denote the velocity acquired by the nucleus by v_n, the displacement ℓ undergone by the nucleus over the duration of the collision time is of order of magnitude $\ell \sim v_n \tau \sim (v_n/v)R \sim (M/M_n)R < R$ (where M is the neutron mass and M_n that of the nucleus), so that ℓ is much less than the dimensions of the electronic shells. Thus the nucleus is effectively undisplaced over the duration of the collision.

We go over to the coordinate system in which the nucleus is at rest after the collision. Then the wave function of the initial state takes the form

$$\Psi_0 \rightarrow \exp(i\mathbf{v}_n \cdot \sum_i \mathbf{r}_i)\, \Psi_0(\mathbf{r}_1, \mathbf{r}_2, \ldots)$$

(This expression may be most simply obtained by expanding the wave function Ψ_0 in plane waves and shifting the momentum of each by an amount $\underset{\sim}{v}_n$.) Let $\Psi_1(r_1, r_2, \ldots)$ denote the final state of the atom. Then, according to (2.12), the probability of excitation is

$$W = |(\Psi_1|\exp(i\mathbf{v}_n \cdot \sum_i \mathbf{r}_i)|\Psi_0)|^2 \tag{2.18}$$

The criterion of applicability of this formula is $\tau \ll \tau_{e\ell}$; we can equally well write it in the form $(R/v) \ll (a/v_{e\ell})$ or $v_n \gg (R/a)v_{e\ell}$, where a is a quantity of the order of the dimensions of the shell in question.

From expression (2.18) we can get a formula for the transition probability of a single fixed electron. Since the interaction between the electrons is weak compared to the nuclear potential for $Z \gg 1$, we have

$$W = W_1 W_2$$

where W_1 is the transition probability for the electron in question, and W_2 the probability of excitation of all the other electrons. Summing over all final states we get $\sum_1 W_1 = 1$, $\sum_2 W_2 = 1$. Thus the probability of transition of the electron in question from the state ψ_0 to the state ψ_n is given by

$$W = W_1 \sum_2 W_2 = |\int \psi_n^* \exp(i\mathbf{v}_n \cdot \mathbf{r})\, \psi_0 \, d\mathbf{r}|^2 \tag{2.19}$$

If the nuclear velocity is much less than the electron velocity (but still $v_n \gg v_{e\ell} R/a$), we can expand the exponential in (2.19) in a power series. The zeroth term in the expansion vanishes because of the orthogonality of the functions ψ_0 and ψ_n. Suppose the velocity $\underset{\sim}{v}_n$ is chosen to be along the z axis. Then we have

$$W \cong v_n^2 |(\psi_n | z | \psi_0)|^2 \tag{2.20}$$

In the opposite case when $v_n \gg v_{el}$, the exponential $\exp(i\underset{\sim}{v}_n \cdot \sum_i \underset{\sim}{r}_i)$ in (2.18) [or (2.19)] is fast oscillating. It follows that the transition probability W is appreciable only when $\Psi_1 \sim \exp(-i\underset{\sim}{v}_n \cdot \sum_i \underset{\sim}{r}_i)$, that is, when the electrons move in the new system with velocity $-\underset{\sim}{v}_n$. In the laboratory system of coordinates this means that the electron shell is not carried off by the nucleus after the collision.

If the nuclear velocity v_n is greater than the velocity of the outer electrons but less than that of the inner ones, that is, $1 < v_n < Z$, then the inner shells are carried off with the nucleus but the outer ones are not. The charge of the resulting ion is of the order of magnitude of the number of electrons in the outer shells. This line of argument gives us an estimate of the charge of the fragments produced in nuclear fission: the charge is determined by the number of electrons with velocities less than the velocity of the fragments (cf. p.114).

For the hydrogen atom and for the inner shells of other atoms the calculations can be carried out in full; the probability of ionization and of excitation for these cases is given in the problems at the end of this section.

Let us estimate the total probability of excitation or ionization W for $v_n \ll 1$. From (2.20) we find

$$W = C_1 v_n^2$$

where C_1 is a number of order unity. It follows from the solution of Problem 4 (see p.115) that the probability that a hydrogen atom is left in its ground state is

$$W_{00} = \frac{1}{(1 + \frac{1}{4}v_n^2)^4}$$

from which we conclude that for a hydrogen atom $C_1 = 1$.

Transfer of Energy When a Photon Is Emitted by a Nucleus in a Molecule (Mössbauer Effect)

Suppose that a photon of energy $\hbar\omega$ is emitted from a nucleus situated in a molecule. If $\hbar\omega$ is small, the impulse is weak and the molecule absorbs the recoil momentum as a whole. If $\hbar\omega$ is large, the nucleus is ejected from the molecule and excitation of the system takes place. The same considerations apply to metals: for small $\hbar\omega$ the recoil momentum is taken up by the whole crystal lattice.

Let us estimate the duration τ of the process of ejection of the photon from the nucleus: τ is the time of transit of a wave packet of dimension $\sim\lambdabar = c/\omega$ across the nucleus, that is, $\tau \sim \lambdabar/c \sim 1/\omega$. (This estimate is invalid for $\lambdabar < R$, where R is the nuclear radius, that is for $\omega > c/R \sim 137.10^4.27$ eV ~ 40 MeV.) The time τ is quite negligible compared with the characteristic periods of vibration and rotation of the molecule. Consequently the wave function of the molecule cannot change over the time of flight of the photon. The problem is therefore reminiscent of the problem of ionization in nuclear reactions. At the moment immediately after the emission of the photon the molecular wave function has the form

$$\Psi' = \Psi(\boldsymbol{r}_1, \boldsymbol{r}_2, \ldots) e^{iM\boldsymbol{v}\boldsymbol{r}_1},$$

where $\underset{\sim}{v}$ is the recoil velocity, $\underset{\sim}{r}_1$ is the position vector of the nucleus which emitted the photon, M is the mass of this nucleus, and Ψ is the wave function describing the ground state of the molecule. The probability that the molecule remains unexcited is given by

$$W_0 = |(\Psi_0 | e^{iMvr_1} | \Psi_0)|^2. \tag{2.21}$$

If $Mvr_1 \ll 1$, then $W_0 \sim 1$, that is, the recoil is taken up by the molecule as a whole and there is no excitation.

Let us estimate the photon frequency ω at which excitation of the vibrational degrees of freedom of the molecule begins to take place. We will see that the vibrational amplitude is of order of magnitude $M^{-1/4}$; hence in (2.21) we have $r_1 \sim M^{-1/4}$. Thus in order that vibrations should not be excited we must have $Mv.M^{-1/4} \ll 1$. Since the recoil momentum of the nucleus $p_n = Mv$ must be equal to the photon momentum $\hbar\omega/c$, this condition means that $\hbar\omega/c \ll M^{1/4}$, or $\omega \ll 137\ M^{1/4}$ in atomic units; in eV,

$$\omega \ll 137(100.1840)^{1/4}\ 27.\ \text{eV} \sim 70\ \text{keV}$$

(we have taken a nucleus with $A \sim 100$). The criterion of applicability of (2.21) is

$$\omega \gg \omega_{vib} \sim M^{-1/2} \sim 27.(100.1840)^{-1/2} \sim 0.06\ \text{eV}$$

where ω_{vib} is the vibrational frequency of the molecule.

We next estimate the photon frequency at which excitation of the rotational degrees of freedom of the molecule begins. For this purpose we consider a plane rotator. The wave function is $\psi_m = \exp(im\varphi)$. According to (2.21) the probability of a transition from the state ψ_o to the state ψ_m is given by

$$w_{m0} = \left|\int \psi_m e^{iMvr}\psi_0 \, dr\right|^2 \sim \left|\int_0^{2\pi} e^{im\varphi + iMva\cos\varphi} d\varphi\right|^2, \tag{2.22}$$

where a is a quantity of the order of the molecular dimensions. Thus in order that a given rotational level should not be excited the orbital angular momentum imparted to the molecule by the photon, which is of order Mva, must be much less than the rotator angular momentum m. In this case we get from (2.22) $W_{mo_{m\neq o}} \ll 1$. For the lowest excited state (m = 1) we get

$$Mva \ll 1 \quad \text{or} \quad \frac{\omega}{c} a \ll 1,$$

whence $\omega \ll c = 137.27$ eV $\sim$ 4 keV.

Problems

1. Verify that nuclear recoil is not an important factor in the ionization of an atom in β decay.

2. Use the Thomas-Fermi method to estimate the charge of fragments produced in nuclear fission, by assuming it to be equal to the number of electrons with velocity less than the velocity of the fragments.

Solution

$Z_{frag} \sim 8$

3. Estimate the probability of ionization of an atom due to the magnetic interaction of an

incident neutron with the atomic electrons.

Solution

$$\frac{W_{magn}}{W} \leqslant \left|\frac{V_{magn}\tau}{v}\right|^2 \sim \left|\frac{\mu\mu_e}{a^3}\cdot\frac{R}{v^2}\right|^2 \ll \left|\frac{1}{Mc^2}\cdot\frac{1}{R}\right|^2 \sim 10^{-7}$$

4. Determine the probability that in the collision of a neutron with the nucleus of a hydrogen atom the atomic electron is left in its ground state.

Solution

$$W_{11} = \frac{1}{(1 + \frac{1}{4}v_n^2)^4}$$

4. ADIABATIC PERTURBATIONS

In this section we shall consider the case in which a perturbation acting on a quantum mechanical system varies slowly as a function of time. In this case the system is able to adjust itself to the slow change of the parameters. Such a perturbation is called adiabatic.

In the case of an adiabatic perturbation it is convenient to look for the solution of Schrödinger's equation in the form of a superposition of stationary eigenfunctions calculated for arbitrary but fixed values of the parameters. We assume that the Hamiltonian $H(x, \xi)$ depends on the slowly varying parameter ξ. Let us introduce the eigenfunctions $\varphi_m(x, \xi)$ where the value of ξ is fixed:

$$H(x, \xi)\, \varphi_m(x, \xi) = \varepsilon_m(\xi)\, \varphi_m(x, \xi). \tag{2.23}$$

We shall look for a solution of Schrödinger's equation

$$i\frac{\partial \Psi}{\partial t} = H\Psi \tag{2.24}$$

in the form

$$\Psi = \sum_m a_m(t)\, \varphi_m(x, \xi) \exp\left\{-i\int_{-\infty}^{t} \varepsilon_m(\xi)\, dt\right\}. \tag{2.25}$$

Substituting (2.25) in (2.24), we find

$$\frac{da_n}{dt} + \sum_m \left(\varphi_n \middle| \frac{\partial \varphi_m}{\partial \xi}\right) \dot{\xi} a_m \exp\left\{-i\int_{-\infty}^{t} (\varepsilon_m - \varepsilon_n)\, dt\right\} = 0. \tag{2.26}$$

Equation (2.26) is convenient if we want to find the solution by successive approximations for small $\dot{\xi}$ ($\equiv d\xi/dt$). In fact we have

$$a_n^{(0)}(t) = a_n(-\infty),$$

$$a_n^{(1)}(t) = -\int_{-\infty}^{t} dt' \sum_m \left(\varphi_n \middle| \frac{\partial \varphi_m}{\partial \xi}\right) \dot{\xi} a_m(0) \exp\left\{-i\int_{-\infty}^{t'} (\varepsilon_m - \varepsilon_n)\, dt'\right\} \tag{2.27}$$

and so on.

Let us express $(\varphi_n | \partial\varphi_m/\partial\xi)$ in terms of the matrix element of the Hamiltonian H. Differentiating the relation (2.23) with respect to the parameter ξ, we find

$$\frac{\partial H}{\partial \xi} \varphi_m + H \frac{\partial \varphi_m}{\partial \xi} = \frac{d\varepsilon_m}{\partial \xi} \varphi_m + \varepsilon_m \frac{d\varphi_m}{d\xi}.$$

Multiplying this equation from the left by φ_n and integrating over x, we get

$$\left(\varphi_n \middle| \frac{\partial \varphi_m}{\partial \xi}\right)_{n \neq m} = \frac{\left(\varphi_n \middle| \frac{\partial H}{\partial \xi} \middle| \varphi_m\right)}{\varepsilon_m - \varepsilon_n}, \quad \left(\varphi_n \middle| \frac{\partial H}{\partial \xi} \middle| \varphi_n\right) = \frac{d\varepsilon_n}{d\xi}. \tag{2.28}$$

We shall assume that the eigenstates φ_n are not degenerate. Then the wave functions φ_n can be chosen to be real and hence for the diagonal matrix element we have

$$\left(\varphi_n \middle| \frac{\partial \varphi_n}{\partial \xi}\right) = \frac{1}{2} \frac{\partial}{\partial \xi} \int \varphi_n^2 \, d\boldsymbol{r} = 0.$$

Substituting (2.28) in (2.27), we finally get

$$a_n^{(1)}(t) = -\int_{-\infty}^{t} dt' \sum_m{}' \frac{\left(\varphi_n \middle| \frac{\partial H}{\partial \xi} \middle| \varphi_m\right) \dot{\xi}}{\varepsilon_m - \varepsilon_n} \times \tag{2.29}$$

$$\times\, a_m(-\infty) \exp\left\{-i \int_{-\infty}^{t'} (\varepsilon_m - \varepsilon_n)\, dt'\right\}.$$

where the prime on the sum over m indicates that the diagonal term is to be omitted.

The criterion of applicability of adiabatic perturbation theory is that the change of the Hamiltonian over a time of the order of the reciprocal eigenfrequencies of the system should be small compared to the energies corresponding to these frequencies. This criterion of applicability is easily obtained if we estimate the ratio of the correction $a_n^{(2)}(t)$ to the amplitude $a_n^{(1)}(t)$: using (2.29), we find

$$\frac{a_n^{(2)}}{a_n^{(1)}} \sim \frac{\partial H}{\partial t} \frac{1}{(\varepsilon_m - \varepsilon_n)^2}. \tag{2.30}$$

As is clear from (2.30), the case in which for certain values of the parameters ξ the energy levels cross needs special treatment. In this case we should look for the wave function in the form of a superposition of states corresponding to these intersecting levels.

We shall show that if the reciprocal eigenfrequencies of the system ω_{mn}^{-1} are much less than the time τ over which the system Hamiltonian changes appreciably, then as a rule the excitation of the system is exponentially small. This follows from the fact that expression (2.29) for the transition amplitude $a_n^{(1)}(t)$ reduces to the Fourier components of the function

$$\frac{\left(\varphi_n \left| \frac{\partial H}{\partial \xi} \right| \varphi_m\right) \dot{\xi}}{\varepsilon_n - \varepsilon_m}.$$

In Section 1.1 we saw that these Fourier components are exponentially small [$\sim \exp(-\omega_{mn}\tau)$] for $\omega_{mn}\tau \gg 1$, provided the functions themselves and their derivatives do not have any singularities on the real axis. The application of perturbation theory to exponentially small expressions requires special care, since a small change in the exponent can give as large an effect as the next power of $\dot{\xi}$. Formula (2.29) gives the magnitude of the correction within a factor of order unity (Dykhne, 1961). As we will show below (p.243), the Coulomb wave functions do have a singularity at $r^2 = 0$. Therefore the probability of ionization by a slow particle passing by, for instance, decreases according to a power law rather than exponentially (cf. next section).

Problem: Use relation (2.28) to calculate $\overline{r^{-2}}$ for centrally symmetric motion.

Solution:

$$\overline{\frac{1}{r^2}} = \frac{1}{l+\frac{1}{2}}\left(\frac{\partial E}{\partial l}\right)_{n_r}.$$

Ionization of an Atom by the Passage of a Slow Heavy Particle

Let us estimate how the probability of ionization of an atom by the passage past it of a slow, heavy charged particle depends on the velocity of this particle. For simplicity we consider the case of a hydrogen atom. We assume that the velocity v of the passing particle is much less than the velocity of the atomic electron, that is, that $v \ll 1$.

The interaction energy of the passing particle with the atomic electron has the form $-|\mathbf{r} - \mathbf{r}_1|^{-1}$. The matrix element for transition of the electron from the state φ_0 to the state $\varphi_{\mathbf{k}}$, with simultaneous transition of the heavy particle from state $\Psi_{\mathbf{p}}$ to $\Psi_{\mathbf{p}'}$, is given in first-order perturbation theory by

$$M^{0k}_{pp'} = -\left(\Psi_{p'}\varphi_k \frac{1}{|r-r_1|}\varphi_0\Psi_p\right).$$

The criterion for the applicability of perturbation theory will be indicated below, when we estimate this matrix element.

We now show that in the expression for $M^{ok}_{\mathbf{pp}'}$, we may replace $\Psi_{\mathbf{p}}$ and $\Psi_{\mathbf{p}'}$ by plane waves. This can be seen by writing $\Psi_{\mathbf{p}}$ in the form $\Psi_{\mathbf{p}} \sim e^{i\mathbf{p}.\mathbf{r}}\chi$, where $\chi = \exp\{iM\int \frac{V}{p^2}\mathbf{p}.d\mathbf{r}\}$ (cf. section 3.2), V is the Coulomb interaction of the passing particle with the nucleus, and M is the reduced mass of the particle and nucleus. The phase $M\int \frac{V}{p^2} \underset{\sim}{p} \cdot d\underset{\sim}{r}$ is of order $MVa/p \sim 1/v$, where a is the impact parameter. Since $v \ll 1$, the phase of the function χ is not small. However, we shall now see that the phase of the product $\chi\chi'$ is small. In fact, we have:

$$\chi\chi' = \exp\left[iM\int \frac{V}{p}\frac{q\,dr}{p}\right] = \exp\left[iM\int \frac{Vq_{\parallel}}{p^2}\,dr\right].$$

where $q_{\parallel}$ is the momentum transferred along the direction of motion of the particle. Let us estimate this quantity. The energy transferred is given by

$$E_{p'} - E_p = \Delta\left(\frac{1}{2}Mv^2\right) = v\Delta p = vq_{\parallel}.$$

On the other hand, $E_{\underset{\sim}{p}'} - E_{\underset{\sim}{p}} = \epsilon_{\underset{\sim}{k}} - \epsilon_o \sim 1$. Consequently we have $q_{\parallel} \sim v^{-1}$ and so

$$\chi\chi' \simeq \exp\left[iM\frac{Vq_{\parallel}a}{p^2}\right] \simeq \exp\left[i\frac{Mq_{\parallel}}{p^2}\right] \simeq \exp\left[\frac{i}{Mv^3}\right].$$

Under the condition $Mv^3 \gg 1$ we therefore have $\chi\chi' \approx 1$ and the replacement of $\Psi_{\underset{\sim}{p}} \Psi_{\underset{\sim}{p}'}$ by plane waves is justified.

We now estimate $M^{o\underset{\sim}{k}}_{\underset{\sim}{p}\underset{\sim}{p}'}$. Since we have

$$\int e^{-iqr}\frac{1}{|r - r_1|}\,dr = e^{-iqr_1}\frac{4\pi}{q^2},$$

it follows that

$$M^{ok}_{pp'} \sim \frac{1}{q^2}(\varphi_0 e^{-iqr}\varphi_k).$$

The differential cross-section for ionization is obtained by dividing $|M^{o\underset{\sim}{k}}_{\underset{\sim}{p}\underset{\sim}{p}'}|^2$ by the current of incident particles (which is proportional to v), multiplying by a delta-function corresponding to the law of conservation of energy, and summing over the possible positive-energy states of the electron and the possible values of $\underset{\sim}{p}'$:

$$\sigma \sim \frac{1}{v}\int \sum_k \frac{1}{q^4} |(\varphi_0 e^{-iqr}\varphi_k)|^2 \delta(E_p - E_{p'} - \varepsilon_k + \varepsilon_0)\, dp'.$$

We can write $d\underset{\sim}{p}' = d\underset{\sim}{q} = 2\pi q_\perp \, dq_\parallel \, dq_\perp$. The integration over the longitudinal component of momentum transfer $dq_\parallel$ is cancelled by the delta-function:

$$\int \delta(E_p - E_{p'} - \varepsilon_k + \varepsilon_0)\, dq_\parallel = \int \delta(q_\parallel v - \varepsilon_k + \varepsilon_0)\, dq_\parallel = 1/v,$$

where $q_\parallel = (\epsilon_k - \epsilon_o)/v$. Consequently,

$$\sigma \sim \frac{1}{v^2}\int \sum_k \frac{|(\varphi_0 e^{-iqr}\varphi_k)|^2}{\left(q_\perp^2 + \left(\frac{\varepsilon_k - \varepsilon_0}{v}\right)^2\right)^2} q_\perp \, dq_\perp.$$

It is obvious from this expression that the important values of $q_\perp$ are of order $1/v$.

Let us now estimate the matrix element $(\varphi_o e^{-i\underset{\sim}{q}.\underset{\sim}{r}} \varphi_{\underset{\sim}{k}})$. It is equal to $\int e^{-\alpha r - i\underset{\sim}{q}.\underset{\sim}{r}} d\underset{\sim}{r}$, where α is a number of order unity. This sort of high Fourier component was estimated in connection with the discussion of the dipole photoeffect, where we obtained the result (cf. p. 62)

$$(\varphi_0 e^{-iqr}\varphi_k) \sim \frac{1}{q^4}.$$

Substituting the value just obtained for the matrix element into the formula for σ, we get

$$\sigma \sim \frac{1}{v^2}\int \frac{q_\perp \, dq_\perp}{q^4 q^8}.$$

Since $q_\parallel \sim q_\perp \sim 1/v$, we find $\sigma \sim v^{10}/v^2 \sim v^8$. Thus, the cross section for ionization of an atom by the passage of a slow, heavy particle is proportional to the eighth power of the particle velocity.

One might ask why it is not exponentially small, as usually

happens for an adiabatic process ? This is connected with the fact that the Coulomb wave functions $e^{-\alpha r} = e^{-\alpha\sqrt{x^2+y^2+z^2}}$ have a singularity with respect to each of the variables if the other variables are held fixed. This singularity can touch the real axis, resulting in a power-law rather than exponential decrease of the integral in question. (cf. also p.243).

Let us verify that the criterion for the applicability of perturbation theory is satisfied. Since $\epsilon_k - \epsilon_o \sim 1$ and we have

$$M^{0k}_{pp'} \sim \frac{1}{q^2}(\varphi_0 e^{-iqr}\varphi_k) \sim \frac{1}{q^2}\frac{1}{q^4} \sim v^6 \ll 1,$$

it follows that

$$M^{0k}_{pp'}/(\varepsilon_k - \varepsilon_0) \ll 1,$$

which is just what is required for the applicability of perturbation theory.

Finally, let us summarize the conditions under which the results obtained above are valid. They are

$$1 \gg v^3 \gg 1/M.$$

from which it follows in particular that the energy of the passing particle must be much greater than typical atomic energies.

Capture of an Atomic Electron by a Proton (Charge Exchange)

We now apply adiabatic perturbation theory to the problem of charge exchange induced by the passage of a slow proton past a hydrogen atom. We shall assume that the proton velocity is much less than that of the atomic electron, so that the adiabatic approximation is applicable (Firsov 1955). Moreover we assume that the energy of the incident proton is much greater than a typical atomic energy, so that the motion of the proton may be assumed given.

The Hamiltonian of the system has the form

$$H = T_e - \frac{1}{\left|r - \frac{1}{2}R\right|} - \frac{1}{\left|r + \frac{1}{2}R\right|} + \frac{1}{R},$$

where R is the distance between the protons, which we shall take as a given function of time, r is the electron coordinate, and T_e its kinetic energy. The Hamiltonian H is symmetric with respect to interchange of the incident and target protons, which is equivalent to the substitution $r \to -r$. Consequently we can introduce eigenfunctions of H which are symmetric or antisymmetric with respect to this substitution: $\varphi_n^s(r, R)$ and $\varphi_n^a(r, R)$. Let us as assume that initially the electron was attached to the proton with coordinate $R/2$; then the initial state of the electron is of the form

$$\varphi = \varphi^0_{n_0}(r - {}^1\!/_2\, R),$$

where $\varphi^0_{n_0}$ is the Coulomb wave function for the state n_0.

After scattering the hydrogen atom is left with high probability in the state $\varphi^0_{n_0}$ (excitation would require a Fourier component of the perturbation with frequency equal to the excitation energy, $\omega_{nn_0} = \epsilon_n - \epsilon_{n_0}$; for slow passage, when the time of passage is long compared to the reciprocal frequency $\omega^{-1}_{nn_0}$, the Fourier component with frequency ω_{nn_0} is very small and excitation cannot occur). However, the electron may be transferred to the other proton; this phenomenon is called charge exchange. In this case the adiabatic condition (2.30) is not fulfilled. Indeed, when the protons are a large distance apart, the electron wave function is $\varphi^0_{n_0}(r - \frac{1}{2}R)$ if it is attached to one proton and $\varphi^0_{n_0}(r + \frac{1}{2}R)$ if it is attached to the other. From these two wave functions we can form a symmetric and an antisymmetric combination, $\varphi^s_{n_0}$ and $\varphi^a_{n_0}$, both of which correspond to exactly the same energy; consequently we are dealing here with a case of level crossing.

We should therefore look for a solution of Schrödinger's equation

$$i\frac{\partial \Psi}{\partial t} = H\Psi \tag{2.31}$$

in the form of a superposition of these two states $\varphi^s_{n_0}$ and $\varphi^a_{n_0}$, and only then apply adiabatic perturbation theory. Thus we write

$$\Psi = C^s(t)\,\varphi^s_{n_0}(r, R)\exp\left(-i\int_{-\infty}^{t}\varepsilon^s dt\right) + \\ + C^a(t)\,\varphi^a_{n_0}(r, R)\exp\left(-i\int_{-\infty}^{t}\varepsilon^a dt\right). \tag{2.32}$$

We substitute (2.32) in (2.31), multiply by $\varphi^s_{n_0}$ and integrate over the coordinate $\underset{\sim}{r}$; this gives

$$\left\{\frac{dC^s}{dt} + C^s\frac{dR}{dt}\left(\varphi^s_{n_0}\left|\frac{\partial\varphi^s_{n_0}}{\partial R}\right.\right)\right\}\exp\left(-i\int_{-\infty}^{t}\varepsilon^s dt\right) = \\ = -C^a\frac{dR}{dt}\left(\varphi^s_{n_0}\left|\frac{\partial\varphi^a_{n_0}}{\partial R}\right.\right)\exp\left(-i\int_{-\infty}^{t}\varepsilon^a dt\right). \tag{2.33}$$

When the nucleons are far apart we have

$$\varphi^{s,a}_n(r, R)\underset{R\to\infty}{\rightarrow}\frac{1}{\sqrt{2}}\left[\varphi^0_n\left(r - \frac{1}{2}R\right) \pm \varphi^0_n\left(r + \frac{1}{2}R\right)\right]. \tag{2.34}$$

We see from (2.33) with (2.34) that in the zeroth approximation

$$C^s = C^s(t = -\infty) = \frac{1}{\sqrt{2}}, \quad C^a = C^a(t = -\infty) = \frac{1}{\sqrt{2}}$$

(this corresponds to the electron being initially attached to the proton with coordinate $\underset{\sim}{R}/2$). With these values of C^s and C^a we get from (2.32)

$$\Psi|_{t\to\infty} = \frac{1}{2}\Big[\exp\Big(-i\int_{-\infty}^{\infty} \varepsilon^s dt\Big) + \\ + \exp\Big(-i\int_{-\infty}^{\infty} \varepsilon^a dt\Big)\Big]\varphi^0_{n_0}\Big(\boldsymbol{r} - \frac{1}{2}\boldsymbol{R}\Big) + \\ + \frac{1}{2}\Big[\exp\Big(-i\int_{-\infty}^{\infty} \varepsilon^s dt\Big) - \exp\Big(-i\int_{-\infty}^{\infty} \varepsilon^a dt\Big)\Big]\varphi^0_{n_0}\Big(\boldsymbol{r} + \frac{1}{2}\boldsymbol{R}\Big). \tag{2.35}$$

The first term in this expression corresponds to ordinary scattering, the second to charge exchange. The squared modulus of the coefficient of $\varphi^0_{n_0}(\mathbf{r} + \frac{1}{2}\mathbf{R})$ gives the probability of charge exchange:

$$W(\rho) = \sin^2\Big(\frac{1}{2}\int_{-\infty}^{\infty} (\varepsilon^a - \varepsilon^s)\, dt\Big) = \sin^2\Big(\int_{a}^{\infty} (\varepsilon^a - \varepsilon^s)\, dt\Big), \tag{2.36}$$

where ρ is the impact parameter.

We suppose the quantities $\epsilon^a(R)$ and $\epsilon^s(R)$ known from the solution of the problem for an electron in the field of two fixed nuclei. Assuming the trajectory of the incident particle to be a straight line, we replace integration over t by integration over R; since $R^2 = \rho^2 + v^2t^2$, we have

$$dt = \frac{1}{v}\frac{R\,dR}{\sqrt{R^2 - \rho^2}}.$$

Introducing the notation $\epsilon^a(R) - \epsilon^s(R) \equiv \Phi(R)$, we get from (2.36)

$$W(\rho) = \sin^2\Big(\frac{1}{v}\int_{\rho}^{\infty} \Phi(R)\frac{R\,dR}{\sqrt{R^2 - \rho^2}}\Big). \tag{2.37}$$

Consider an impact parameter $\rho \sim 1$. Then the values of the distance between the particles which are important in the interaction will also be of order 1, and hence the integral in (2.37) will be of order 1. Since we have supposed $v \ll 1$ (see above) the expression for $W(\rho)$ is very sensitive to small changes in ρ. If we average $W(\rho)$ over an interval of values of ρ small compared with ρ, we get

$$\overline{W}=\frac{1}{2},$$

that is, charge exchange takes place in 50% of all cases.

Now consider the probability of charge exchange when $\rho \gg 1$. It can be shown[2] that for large distances between the protons the difference in energy between the symmetric and antisymmetric states is

$$\varepsilon^a(R)-\varepsilon^s(R)\underset{R\to\infty}{\approx}\frac{4}{e}Re^{-R}.$$

Thus for $\rho \gg 1$ we have

$$F(\rho)\equiv\frac{1}{v}\int_\rho^\infty \Phi(R)\frac{R\,dR}{\sqrt{R^2-\rho^2}}=$$

$$=\frac{4}{ev}\int_\rho^\infty \frac{R^2e^{-R}}{\sqrt{R^2-\rho^2}}dR=\frac{4\rho^2e^{-\rho}}{ev}\int_0^\infty \frac{(z+1)^2\,dz}{\sqrt{z(z+2)}}e^{-\rho z}\approx$$

$$\approx\frac{4\rho^2}{ev}e^{-\rho}\frac{1}{\sqrt{2}}\int_0^\infty z^{-1/2}e^{-\rho z}dz=\frac{\pi\sqrt{2}}{ev}\rho^{3/2}e^{-\rho}.$$

We see that $F(\rho) \ll 1$ for $\rho \gg \rho_1 \sim \ln(1/v)$. Therefore we get

$$W=\sin^2 F(\rho)\approx F^2(\rho)=\frac{2\pi^2}{e^2v^2}\rho^3e^{-2\rho}.$$

Thus for $\rho \ll \rho_1$ we have $F(\rho) \gg 1$, while for $\rho \gg \rho_1$ we get $F(\rho) \ll 1$.

We can now estimate the total charge exchange cross section

---

[2] L. D. Landau and E. M. Lifshitz, Quantum Mechanics (Pergamon Press, London, 1959), p.292.

$$\sigma = \int_0^\infty \sin^2 F(\rho)\cdot 2\pi\rho\, d\rho.$$

To an order of magnitude we find

$$\sigma = \int_0^{\rho_1} \frac{1}{2} 2\pi\rho\, d\rho = \frac{1}{2}\pi\rho_1^2 \sim \frac{1}{2}\pi \ln^2 \frac{1}{v}. \tag{2.38}$$

Thus for small velocities v the charge exchange cross section is much larger than the geometrical cross section.

5. FAST AND SLOW SUBSYSTEMS

Consider a system composed of two subsystems, one of which is characterized by large and one by small frequencies. The interaction between the subsystems is not assumed to be small. Let ξ denote the set of coordinates of the slow subsystem. The Hamiltonian of the whole system has the form

$$H = H_1(x, \xi) + H_2(\xi),$$

where H_1 includes both the Hamiltonian of the fast subsystem and its interaction with the slow subsystem.

In the last section we assumed that the motion of the slow subsystem was classical - we fixed the trajectory $\xi = \xi(t)$ and introduced the velocity $\dot{\xi}$ as a c number. In the following we shall not make this assumption.

We introduce the system of eigenfunctions of the Hamiltonian H_1

$$H_1(x, \xi)\,\varphi_n(x, \xi) = \varepsilon_n(\xi)\,\varphi_n(x, \xi),$$

as was done above. We seek the solution of the whole problem

$$H\Psi_\lambda = E_\lambda \Psi_\lambda \tag{2.39}$$

in the form

$$\Psi_\lambda = \sum_{n'} u_{n'\lambda}(\xi)\, \varphi_{n'}(x, \xi). \tag{2.40}$$

Substituting (2.40) in (2.39), we get

$$(H_1 + H_2) \sum_{n'} u_{n'\lambda}(\xi)\, \varphi_{n'}(x, \xi) = E_\lambda \sum_{n'} u_{n'\lambda}(\xi)\, \varphi_{n'}(x, \xi).$$

We multiply this equation from the left by $\varphi_n^*(x, \xi)$ and integrate over x; this gives

$$(\varepsilon_n - E_\lambda)\, u_{n\lambda} + \int \varphi_n^* H_2 \sum_{n'} u_{n'\lambda} \varphi_{n'}\, dx = 0.$$

Here for simplicity we assumed that H_1 commutes with ξ. Moreover, since

$$H_2 u_{n'\lambda} \varphi_{n'} = \varphi_{n'} H_2 u_{n'\lambda} + [H_2, \varphi_{n'}] u_{n'\lambda},$$

we have

$$(\varepsilon_n - E_\lambda)\, u_{n\lambda} + H_2 u_{n\lambda} = -\sum_{n'} \int \varphi_n^* [H_2, \varphi_{n'}]\, dx \cdot u_{n'\lambda}. \tag{2.41}$$

This relation is analogous to the corresponding expression in the preceding section; the only difference is that $i\dot{\xi}(\partial\varphi_n/\partial\xi) = i(d\varphi_n/dt)$ is replaced by the commutator $[H_2, \varphi_n]$.

In relation (2.41) we can transfer the term with $n' = n$ to the left:

$$\left\{H_2 - \left(E_\lambda - \varepsilon_n - \int \varphi_n^* [H_2, \varphi_n]\, dx\right)\right\} u_{n\lambda} = \\ = -\sum_{n' \neq n} \int \varphi_n^* [H_2, \varphi_{n'}]\, dx \cdot u_{n'\lambda}. \tag{2.42}$$

This equation is in a form suitable for solution by iteration. In the zeroth approximation we get

$$\{H_2(\xi) - [E_\lambda - \varepsilon_n(\xi)]\}\, u_{n\lambda}^{(0)}(\xi) = 0.$$

We see that the energy of the fast subsystem plays the role of a potential energy in the effective Hamiltonian of the slow subsystem.

Vibrational Energy Levels of a Molecule

Let us estimate the vibrational energy of a molecule. For simplicity we shall write expressions explicitly for an ionic molecule of hydrogen, although the results are easily taken over to more complex molecules. The Hamiltonian of the molecular hydrogen ion has the form (in atomic units)

$$H = -\frac{1}{2}\nabla_r^2 - \frac{1}{\left|r - \frac{1}{2}R\right|} - \frac{1}{\left|r + \frac{1}{2}R\right|} + \frac{1}{R} - \frac{1}{M}\nabla_R^2.$$

Here M is the proton mass, $\underset{\sim}{r}$ the coordinate of the electron, $\pm(1/2)\underset{\sim}{R}$ the coordinates of the two protons (Fig. 21). The slow subsystem is formed by the nuclei, the fast one by the electron.

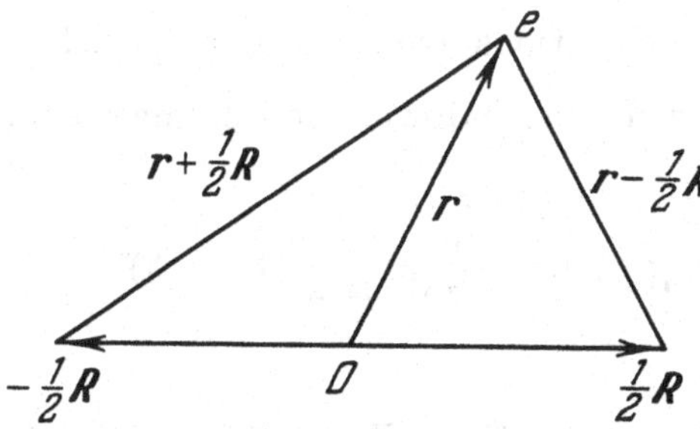

Fig. 21

In the notation of the theory developed above, we take for $H_1(x, \xi)$ the Hamiltonian

$$H_1(r, R) = -\frac{1}{2}\nabla_r^2 - \frac{1}{\left|r - \frac{1}{2}R\right|} - \frac{1}{\left|r + \frac{1}{2}R\right|} + \frac{1}{R}$$

and for $H_2(\xi)$ the Hamiltonian

$$H_2(R) = -\frac{1}{M}\nabla_R^2$$

(Of course, we could count the term $1/R$ in H_2 if we wished.) We introduce functions $\varphi_n(\underset{\sim}{r}, \underset{\sim}{R})$ which satisfy the equation

$$H_1\varphi_n(r, R) = \varepsilon_n(R)\varphi_n(r, R).$$

They correspond to the electronic eigenfunctions for a fixed distance between the nuclei. The solution of Schrodinger's equation with the Hamiltonian H has the form

$$\Psi_\lambda = \sum_{n\nu} C_\lambda^{n\nu}\chi_{n\nu}(R)\varphi_n(r, R), \tag{2.43}$$

where the functions $\chi_{n\nu}$ satisfy the equation

$$\left\{-\frac{1}{M}\nabla_R^2 + \varepsilon_n(R)\right\}\chi_{n\nu}(R) = \omega_{n\nu}\chi_{n\nu}(R). \tag{2.44}$$

The nuclei of the molecule will vibrate about some mean position, and these vibrations are small in amplitude (see below). Therefore we can write the following approximate expression for $\epsilon_n(R)$:

$$\varepsilon_n(R) \approx \varepsilon_n(R_0) + \frac{1}{2}\left(\frac{d^2\varepsilon_n}{dR^2}\right)_{R=R_0}\cdot(R - R_0)^2.$$

Substituting this form of $\epsilon_n(R)$ in Eq. (2.44), we get Schrodinger's equation for a three-dimensional harmonic oscillator. The vibrational energy levels of the molecule are therefore given by

$$\omega_{n\nu} = \varepsilon_n(R_0) + \left(\nu + \frac{1}{2}\right)\omega_0,$$

where ω_0 is the classical oscillator frequency and ν is an integer.

Let us find the dependence of ω_0 on the mass of the molecule. From the relation

$$\frac{1}{2} M\omega_0^2 (R - R_0)^2 = \frac{1}{2}\left(\frac{d^2\varepsilon_n}{dR^2}\right)_{R=R_0} \cdot (R - R_0)^2$$

we find that

$$\omega_0 = \sqrt{\frac{1}{M}\left(\frac{d^2\varepsilon_n}{dR^2}\right)_{R=R_0}}.$$

Since in atomic units the electronic energies ϵ_n are of order 1 and the dimensions of the molecule R_0 are also of order 1, we have $d^2\epsilon_n/dR^2 \sim 1$ and so

$$\omega_0 \sim \frac{1}{\sqrt{M}}.$$

Since the frequencies of the electronic transitions ω_{el} are of order $\omega_{el} \gtrsim 1$, we see that $\omega_0 \ll \omega_{el}$.

Let us next estimate the amplitude of the nuclear oscillation. In the ground state of the oscillator the mean values of the kinetic and potential energies are equal and each equal to half the total energy: $\overline{T} = \overline{U} = E_0/2$; that is

$$\frac{1}{2}\left(\frac{d^2\varepsilon_n}{dR^2}\right)_{R=R_0} \cdot \overline{(R - R_0)^2} = \frac{1}{4}\omega_0.$$

Since $\omega_0 \sim 1/\sqrt{M}$, the root-mean-square amplitude of the deviation of the nuclei from their equilibrium position is of order

$$\sqrt{\overline{(R - R_0)^2}} \sim \frac{1}{\sqrt[4]{M}} \ll 1.$$

We shall show that it is legitimate to calculate the coefficients $C_\lambda^{n\nu}$ in the expression (2.43) for Ψ_λ, which are defined by (2.42), by iteration. To show this we must estimate the right-hand side of (2.42). Writing the commutator out explicitly, we get terms of the following forms:

$$\left(\frac{1}{M}\chi_{n\nu}\varphi_n \left| \frac{d\varphi_{n'}}{dR}\frac{d\chi_{n'\nu'}}{dR}\right.\right), \qquad \left(\frac{1}{M}\chi_{n\nu}\varphi_n \left| \frac{d^2\varphi_{n'}}{dR^2}\chi_{n'\nu'}\right.\right),$$

(There are no terms of the form $[(1/M)\chi_{n\nu}\varphi_n|\varphi_{n'}(d^2\chi_{n'\nu'}/dR^2)]$.) We can estimate the derivatives of the function $\chi_{n\nu}$:

$$\frac{d\chi_{n\nu}}{dR} \sim \frac{\chi_{n\nu}}{\sqrt{\overline{(R-R_0)^2}}} \sim \sqrt[4]{M}\chi_{n\nu}.$$

Thus the larger of the two types of terms indicated above will be the first, which is of order $M^{-3/4}$. Thus we have justified the solution of this problem by iteration: the expansion parameter is of order $M^{-3/4}\omega_0^{-1} \sim M^{-1/4}$.

Besides vibrational levels, a molecule may have also rotational levels. The frequencies of these levels are inversely proportional to the moment of inertia of the molecule, that is,

$$\omega_{rot} \sim 1/M$$

Thus the electronic, vibrational, and rotational frequencies have the following ratio:

$$\omega_{rot} : \omega_{vib} : \omega_{e\ell} = 1 : M^{1/2} : M$$

Excitation of Nuclear Dipole Levels by a Fast Particle

We consider the problem of the excitation of an atomic nucleus by the passage near it of a fast charged particle. In this case the fast subsystem (the incident particle) has a continuous spectrum and the slow subsystem (the nucleus) a discrete one. We shall find the solution under the assumption that the time of passage of the particle is small compared to the important reciprocal frequencies of the system.

First of all we obtain general formulas enabling us to apply the adiabatic approximation to a scattering problem. We introduce the eigenfunctions $\varphi_p(\underset{\sim}{r}, \xi_i)$ of the Hamiltonian of the fast subsystem $H_1 = T_r + V(\underset{\sim}{r}, \xi_i)$, where T_r is the kinetic energy of the incident particle and $V(\underset{\sim}{r}, \xi_i)$ its interaction with the slow subsystem; ξ_i is the coordinate (or set of coordinates) of the slow and $\underset{\sim}{r}$ that of the fast subsystem. Thus the functions φ_p satisfy the equation

$$H_1\varphi_p(r, \xi_i) = \varepsilon_p\varphi_p(r, \xi_i). \tag{2.45}$$

The energy of the particle is just $\epsilon_p = p^2/2M$ (where M is its mass) and thus does not depend on ξ_i, in distinction to the case of a discrete spectrum.

We shall seek a solution (2.45) which has the asymptotic form

$$\varphi_p \underset{r\to\infty}{\longrightarrow} e^{ipr} + \frac{f(\vartheta, \xi_i)}{r} e^{ipr}. \tag{2.46}$$

Since ϵ_p is independent of ξ_i, the equation for the wave function of the slow subsystem (the nucleus) will be the same as it was without ϵ_p (the only effect being to shift the eigenvalues by a constant amount ϵ_p):

$$H_2\chi_n(\xi_i) = \omega_n\chi_n(\xi_i).$$

We look for a solution of the problem in the form

$$\Psi_{pn} = \sum_n \int \frac{dp'}{(2\pi)^3} C_{pn}^{p'n'} \varphi_{p'}\chi_{n'},$$

where the coefficients $C_{\underset{\sim}{p}n}^{\underset{\sim}{p}'n'}$ satisfy the equation [cf. (2.42)]

$$(\omega_n - E_{p_0 n_0}) C_{pn}^{p_0 n_0} = -\sum_{n'} \int \frac{dp'}{(2\pi)^3} (\varphi_p \chi_n [H_2, \varphi_{p'}] \chi_{n'}) C_{pn}^{p'n'}.$$

This equation may be solved by iteration. Suppose initially (at $t = -\infty$) the system was in the state $(\underset{\sim}{p}_0, n_0 = 0)$. Then in the zeroth approximation

$$\Psi_{p_0 0}^{(0)} = C_0 \varphi_{p_0} \chi_0.$$

According to (2.46) we have

$$\Psi_{p_0 0}^{(0)} \underset{r \to \infty}{\to} C_0 \left(e^{ip_0 r} \chi_0(\xi_i) + \frac{e^{ip_0 r}}{r} f(\vartheta, \xi_i) \chi_0(\xi_i) \right) =$$

$$= C_0 \left(e^{ip_0 r} \chi_0(\xi_i) + \frac{e^{ip_0 r}}{r} \sum_n (\chi_n f \chi_0) \chi_n \right). \qquad (2.47)$$

The exact solution must have the asymptotic form

$$\Psi \underset{r \to \infty}{\to} e^{ip_0 r} \chi_0(\xi_i) + \frac{1}{r} \sum_n f_n(\vartheta) e^{ip_n r} \chi_n(\xi_i).$$

where the momentum p_n is determined by the law of conservation of energy:

$$\varepsilon_{p_n} + \omega_n = \varepsilon_{p_0} + \omega_0.$$

Neglecting the change in the energy of the particle, we get

$$\Psi \underset{r \to \infty}{\to} e^{ip_0 r} \chi_0(\xi_i) + \frac{e^{ip_0 r}}{r} \sum_n f_n(\vartheta) \chi_n(\xi_i). \qquad (2.48)$$

We compare this expression with the asymptotic form of the approximate solution $\Psi_{p_0 0}^{(0)}$. From a comparison of (2.47) and (2.48) we find find C_0 and the amplitudes for elastic and inelastic scattering:

$$C_0 = 1,$$
$$f_0(\vartheta) = (\chi_0(\xi_i) \,|\, f(\vartheta, \xi_i) \,|\, \chi_0(\xi_i)),$$
$$f_n(\vartheta) = (\chi_n(\xi_i) \,|\, f(\vartheta, \xi_i) \,|\, \chi_0(\xi_i)). \qquad (2.49)$$

Let us use these results to calculate the cross section for excitation of nuclear dipole levels by a fast charged particle. The Coulomb interaction of the incident particle with the nucleons of the nucleus has the form (where Z_1 is the charge of the particle and Z_2 that of the nucleus)

$$V = Z_1 \sum_i \frac{1}{|\mathbf{r} - \mathbf{r}_i|} = \frac{Z_1 Z_2}{r} + Z_1 \frac{\mathbf{d}\mathbf{r}}{r^3} + \dots,$$

where $\mathbf{d}$ is the nuclear electric dipole moment. This potential can be represented in the form of a Coulomb potential with a shifted origin:

$$V \approx \frac{Z_1 Z_2}{\left| \mathbf{r} - \frac{1}{Z_2} \mathbf{d} \right|}.$$

For this to be valid it is necessary that the quantity $a = |\mathbf{d}/Z_2|$ should be small in comparison with the values of r of interest, that is, in comparison with the impact parameter ρ: $a \ll \rho$. Then the wave functions will be Coulomb wave functions with shifted argument:

$$\varphi_{\mathbf{p}}(\mathbf{r}, \xi_i) = \varphi_{\mathbf{p}}^{Q}\left(\mathbf{r} - \frac{1}{Z_2}\mathbf{d}\right) \gamma(\xi_i).$$

We choose $\gamma(\xi_i)$ so that $\varphi_{\mathbf{p}}$ should have the asymptotic form (2.46):

$$\varphi_{\mathbf{p}}(\mathbf{r}, \xi_i) \underset{r \to \infty}{\longrightarrow} \left\{ \exp\left[i\mathbf{p}\left(\mathbf{r} - \frac{1}{Z_2}\mathbf{d}\right)\right] + \right.$$
$$\left. + \frac{f^{Q}(\vartheta)}{r} \exp\left[ip \left|\mathbf{r} - \frac{1}{Z_2}\mathbf{d}\right|\right]\right\} \gamma(\xi_i),$$

where f^Q is the Coulomb scattering amplitude. If this is to agree with (2.46), we must take $\gamma(\xi_i) = \exp(i\mathbf{p}.\mathbf{d}/Z_2)$. Then

$$\varphi_p(r, \xi_i) \underset{r\to\infty}{\longrightarrow} e^{ipr} + \frac{e^{ipr}}{r} f^Q(\vartheta) \exp\left[ip\left|r - \frac{1}{Z_2} d\right| + ipd\frac{1}{Z_2} - ipr\right].$$

Since

$$\left|r - \frac{1}{Z_2} d\right| \approx r - \frac{1}{Z_2}\frac{dr}{r}$$

we get

$$\varphi_p(r, \xi_i) \underset{r\to\infty}{\longrightarrow} e^{ipr} + \frac{e^{ipr}}{r} f^Q(\vartheta) \exp\left[iqd\frac{1}{Z_2}\right],$$

where $\underset{\sim}{q} = \underset{\sim}{p} - (p/\underset{\sim}{r})r$ is a vector characterizing the change of the momentum of the incident particle due to the collision.

Thus according to (2.49) we have

$$f_n(\vartheta) = f^Q(\vartheta)\left(\chi_n\left|\exp\left(iqd\frac{1}{Z_2}\right)\right|\chi_0\right). \tag{2.50}$$

where $f^Q(\theta)$ is the Coulomb scattering amplitude. We note that the sum of cross sections for all possible excitations is

$$\sum_n |f_n(\vartheta)|^2 = \sum_n \left(\chi_0\left|\exp\left(-iqd\frac{1}{Z_2}\right)\right|\chi_n\right) \times \times \left(\chi_n\left|\exp\left(iqd\frac{1}{Z_2}\right)\right|\chi_0\right) |f^Q(\vartheta)|^2 = |f^Q(\vartheta)|^2,$$

that is, it is just the differential cross section for Coulomb scattering.

If the deflection angle θ is small (i. e. the momentum transfer is small) then we get from (2.50) the perturbation formula

$$f_n(\vartheta) \approx f^Q(\vartheta)\left(\chi_n\left|\frac{i}{Z_2} qd\right|\chi_0\right), \tag{2.51}$$

that is, the amplitude of excitation is proportional to the dipole matrix element.

Scattering of a Proton by a Hydrogen Atom (Charge Exchange)

Consider the scattering of a proton by a hydrogen atom, when the speed of the proton is much less than that of the atomic electron. In contrast to the case just discussed, here it is the fast subsystem which has a discrete spectrum and the slow subsystem which has a continuous one. The Hamiltonian of the system has the form

$$H = T_r - \frac{1}{\left| \mathbf{r} - \frac{1}{2}\mathbf{R} \right|} - \frac{1}{\left| \mathbf{r} + \frac{1}{2}\mathbf{R} \right|} + \frac{1}{R} - \frac{1}{M}\nabla_{\mathbf{R}}^2 .$$

In Section 2.4 we considered $\mathbf{R}$ to be a given function of time; in this section we shall consider the incident proton and the hydrogen atom as two interacting quantum mechanical systems.

The Hamiltonian H is symmetric with respect to interchange of the incident and target particles ($\mathbf{r} \to -\mathbf{r}$). So we can introduce symmetric and antisymmetric functions $\varphi_n^s(\mathbf{r},\mathbf{R})$ and $\varphi_n^a(\mathbf{r},\mathbf{R})$. When the nucleons are far apart, the electron will be localized near one or the other proton, and therefore

$$\varphi_n^{s,a}(\mathbf{r},\mathbf{R}) \underset{R\to\infty}{\longrightarrow} \frac{1}{\sqrt{2}}\left[\varphi_n^0\left(\mathbf{r} - \frac{1}{2}\mathbf{R}\right) \pm \varphi_n^0\left(\mathbf{r} + \frac{1}{2}\mathbf{R}\right)\right],$$

where the functions on the right-hand side are hydrogen-atom wave functions. We assume that when the nuclei approach one another there is no crossing of the levels; in that case we can label them by the states from which they evolved.

Let us suppose that initially the electron was attached to the proton with coordinate $\mathbf{R}/2$, that is, the wave function had the form

$$\varphi = \varphi^0_{n_0}\left(r - \frac{1}{2}R\right).$$

The Hamiltonian of the slow subsystem has the form

$$H_2(R) = -\frac{1}{M}\nabla^2_R,$$

Its eigenfunctions can be found from the equation

$$\left[-\frac{1}{M}\nabla^2_R + \varepsilon^{s,a}_n(R)\right]\chi^{s,a}_{pn}(R) = E_p\chi^{s,a}_{pn}(R)$$

and have the asymptotic form

$$\chi^{s,a}_{pn}(R) \underset{R\to\infty}{\longrightarrow} e^{ipR} + \frac{f^{s,a}_n}{R}e^{ipR}.$$

In the zeroth approximation the solution to the problem has the form (cf. p. 124)

$$\Psi_\lambda = C^{\lambda s}_{np}\varphi^s_n(r, R)\chi^s_{pn}(R) + C^{\lambda a}_{np}\varphi^a_n(r, R)\chi^a_{pn}(R). \tag{2.52}$$

The coefficients $C^{\lambda s}_{\underset{\sim}{np}}$ and $C^{\lambda a}_{\underset{\sim}{np}}$ can be found from the asymptotic form of Ψ_λ:

$$\Psi_\lambda \underset{R\to\infty}{\longrightarrow} \varphi^0_{n_0}\left(r - \frac{1}{2}R\right)e^{ip_0R} + \\ + \left[f_1\varphi^0_{n_0}\left(r - \frac{1}{2}R\right) + f_2\varphi^0_{n_0}\left(r + \frac{1}{2}R\right)\right]\frac{e^{ip_0R}}{R}, \tag{2.53}$$

where f_1 is the elastic scattering amplitude and f_2 the charge exchange amplitude.

To ensure the asymptotic form (2.53) we must take $p = p_0$ and $n = n_0$ in (2.52). Then the wave function (2.52) is equal to

$$\Psi_\lambda = C^s\varphi^s_{n_0}(r, R)\chi^s_{p_0n_0}(R) + C^a\varphi^a_{n_0}(r, R)\chi^a_{p_0n_0}(R)$$

and has the asymptotic form

$$\Psi_\lambda \underset{R\to\infty}{\longrightarrow} e^{ip_0R}(C^s\varphi^s_{n_0} + C^a\varphi^a_{n_0}) + \frac{1}{R}e^{ip_0R}(f_sC^s\varphi^s_{n_0} + f_aC^a\varphi^a_{n_0}). \quad (2.54)$$

The expressions (2.53) and (2.54) must agree, that is, we must require

$$C^s = C^a = \frac{1}{\sqrt{2}}.$$

Then we get from (2.54)

$$\frac{f_s\varphi^s_{n_0} + f_a\varphi^a_{n_0}}{\sqrt{2}} = \frac{1}{2}\Big\{f_s\Big[\varphi^0_{n_0}\Big(\boldsymbol{r} - \frac{1}{2}\boldsymbol{R}\Big) + \varphi^0_{n_0}\Big(\boldsymbol{r} + \frac{1}{2}\boldsymbol{R}\Big)\Big] +$$
$$+ f_a\Big[\varphi^0_{n_0}\Big(\boldsymbol{r} - \frac{1}{2}\boldsymbol{R}\Big) - \varphi^0_{n_0}\Big(\boldsymbol{r} + \frac{1}{2}\boldsymbol{R}\Big)\Big]\Big\} =$$
$$= f_1\varphi^0_{n_0}\Big(\boldsymbol{r} - \frac{1}{2}\boldsymbol{R}\Big) + f_2\varphi^0_{n_0}\Big(\boldsymbol{r} + \frac{1}{2}\boldsymbol{R}\Big),$$

that is,

$$f_1 = \frac{f_s + f_a}{2}, \quad f_2 = \frac{f_s - f_a}{2}. \quad (2.55)$$

Thus to find the charge exchange amplitude f_2 and elastic scattering amplitude f_1 we must first calculate the amplitudes f^s and f^a. To find the latter, we must solve the equation

$$\Big[-\frac{1}{M}\nabla^2_R + \varepsilon^{s,a}_n(R)\Big]\chi^{s,a}_{pn} = E_p\chi^{s,a}_{pn}$$

with the quantities $\epsilon^{s,a}_n(R)$ defined by the solution of the problem of an electron in the field of two fixed protons with separation R. We shall find the solution of this scattering problem in the quasi-classical approximation on p.211.

6. PERTURBATION THEORY FOR ADJACENT LEVELS

Consider a system with two adjacent levels, to which a perturbation is applied (the case of a large number of adjacent levels introduces only algebraic complications). Suppose the perturbation is weak enough so that we can neglect the admixture of states with energies far from the pair in question, while these two adjacent levels are strongly mixed.

We should look for a solution of Schrödinger's equation

$$(H_0 + H')\,\Psi = E\Psi,$$

(where H' is the perturbation) in the form of a superposition of states corresponding to the two adjacent levels:

$$\Psi = C_1\Psi_1 + C_2\Psi_2, \tag{2.56}$$

where the functions $\Psi_{1,2}$ satisfy the unperturbed equation

$$H_0\Psi_{1,2} = \varepsilon_{1,2}\Psi_{1,2}.$$

The coefficients C_1, C_2 in (2.56) are in general of the same order of magnitude.

Let us find the energy levels of the perturbed system. To do so we write Schrödinger's equation in the (C_1, C_2) representation:

$$\left.\begin{aligned}(E - \varepsilon_1)\,C_1 &= H'_{11}C_1 + H'_{12}C_2,\\ (E - \varepsilon_2)\,C_2 &= H'_{21}C_1 + H'_{22}C_2.\end{aligned}\right\} \tag{2.57}$$

This has a nontrivial solution for the coefficients C_1, C_2 if the determinant of the system of equations (2.57) is zero, that is,

$$\begin{vmatrix} E - \varepsilon_1 - H'_{11} & -H'_{12} \\ -H'_{21} & E - \varepsilon_2 - H'_{22} \end{vmatrix} = 0. \tag{2.58}$$

From (2.58) we get

$$E = \frac{\varepsilon_1 + \varepsilon_2 + H'_{11} + H'_{22}}{2} \pm$$
$$\pm \sqrt{\frac{(\varepsilon_1 + \varepsilon_2 + H'_{11} + H'_{22})^2}{4} - (\varepsilon_1 + H'_{11})(\varepsilon_2 + H'_{22}) + |H'_{12}|^2}.$$

Writing $\epsilon_1 + H'_{11} = \tilde{\epsilon}_1$, $\epsilon_2 + H'_{22} = \tilde{\epsilon}_2$, we finally get

$$E = \frac{\tilde{\varepsilon}_1 + \tilde{\varepsilon}_2}{2} \pm \frac{1}{2}\sqrt{(\tilde{\varepsilon}_1 - \tilde{\varepsilon}_2)^2 + 4|H'_{12}|^2}. \tag{2.59}$$

In the case where ordinary perturbation theory is applicable, $|H'_{12}| \ll |\epsilon_1 - \epsilon_2|$, we get from (2.59)

$$E^{(1)} \approx \varepsilon_1 + H'_{11} + \frac{|H'_{12}|^2}{\varepsilon_1 - \varepsilon_2},$$
$$E^{(2)} \approx \varepsilon_2 + H'_{22} - \frac{|H'_{12}|^2}{\varepsilon_1 - \varepsilon_2},$$

as of course we should expect. For $H' \to 0$ the quantity $E^{(1)}$ tends to ϵ_1 and $E^{(2)}$ to ϵ_2. In the opposite limit, $\epsilon_1 = \epsilon_2 = \epsilon$, we get from (2.59)

$$E = \varepsilon + \frac{H'_{11} + H'_{22}}{2} \pm |H'_{12}|. \tag{2.60}$$

We note that for any perturbation which is a function of a parameter ξ such that $H'_{12} \neq 0$, the levels $E^{(1)}$ and $E^{(2)}$, regarded as functions of ξ, do not intersect, since in (2.59) $\sqrt{(\epsilon_1 - \epsilon_2)^2 + 4|H'_{12}|^2} > 0$: the energy levels have the general form represented in Fig. 22.

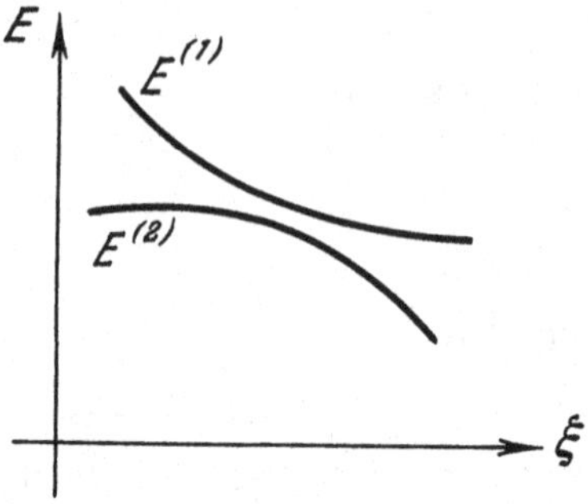

Fig. 22

A Particle in a Periodic Potential

Let us examine how the motion of a free particle is changed by the application of a periodic potential

$$V = V_0\,(e^{ikx} + e^{-ikx}).$$

The wave functions of the unperturbed problem have the form $\Psi_p = e^{ipx}$. The mean value of V in any unperturbed state is zero, and in fact the only nonzero matrix elements are $V_{p,p-k} = V_{p,p+k} = V_0$ (we take the width of the well in which the particle moves equal to unity). In ordinary perturbation theory the shift of the energy levels is given by

$$E_p = \varepsilon_p + \frac{V_0^2}{\varepsilon_p - \varepsilon_{p-k}} + \frac{V_0^2}{\varepsilon_p - \varepsilon_{p+k}},$$

where $\epsilon_p = p^2/2M$. However, this expression is applicable only in the case $V_0 \ll |\epsilon_p - \epsilon_{p-k}|$, that is, when p is not close to $\pm k/2$.

When $p \to k/2$, the energies of states Ψ_p and Ψ_{p-k} approach one another, and we must look for the wave function of the particle in the form

$$\Psi = C_1\Psi_p + C_2\Psi_{p-k}$$

(for $p \to -k/2$ we should have $\Psi = C_1\Psi_p + C_2\Psi_{p+k}$). From (2.59) we get the following expression for the energy eigenvalues:

$$E_p = \frac{\varepsilon_p + \varepsilon_{p-k}}{2} \pm \sqrt{\frac{(\varepsilon_p - \varepsilon_{p-k})^2}{4} + V_0^2}. \tag{2.61}$$

The choice of sign in (2.61) is determined by the condition that for $|\epsilon_p - \epsilon_{p-k}| \gg V_0$ the energy E_p should tend to ϵ_p. Since for $p < k/2$ the quantity $\sqrt{(\epsilon_p - \epsilon_{p-k})^2} = |\epsilon_p - \epsilon_{p-k}| = \epsilon_{p-k} - \epsilon_p$, while for $p > k/2$, $\sqrt{(\epsilon_p - \epsilon_{p-k})^2} = \epsilon_p - \epsilon_{p-k}$, we must choose the minus sign in (2.61) for $p < k/2$ and the plus sign for $p > k/2$. We then get for E_p the discontinuous curve shown in Fig. 23.

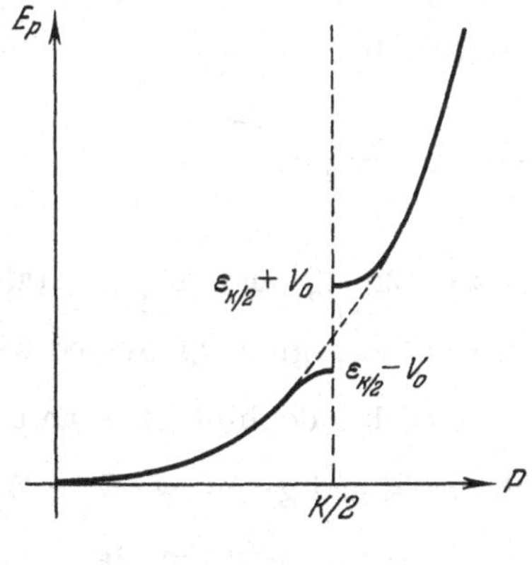

Fig. 23

The magnitude of the discontinuity is $E_{k/2+o} - E_{k/2-o} = 2V_o$.

Thus the application of a periodic potential causes the appearance of a gap in the energy spectrum of a free particle It is this fact, of course, which is responsible for the existence of forbidden regions for the electrons in metals (of course, there one must really consider the three-dimensional case).

It follows from (2.61) that $dE/dp|_{p=k/2} = 0$, that is, the group velocity of a particle on the edge of a zone $(p = k/2)$ is zero. This corresponds to the fact that at the edge of the zone the wave function of the particle is a standing wave rather than a travelling one.

The Stark Effect in the Case of Adjacent Levels

Consider a system with two adjacent levels, to which is applied a uniform electric field $V(x) = -\mathcal{E}x$; let us find how the energy of the system changes in the field. The mean value of V in any state with definite parity is zero (the dipole moment $d_x = \int \Psi_\lambda^*(\underset{\sim}{r})\, x \Psi_\lambda(\underset{\sim}{r})\, d\underset{\sim}{r}$ changes sign when we make the change of integration variable $\underset{\sim}{r} \to -\underset{\sim}{r}$, that is, $d_x = 0$). The energy levels $E_{1,2}$ in the field are given by

$$E_{1,2} = \frac{\varepsilon_1 + \varepsilon_2}{2} \pm \sqrt{\frac{(\varepsilon_1 - \varepsilon_2)^2}{4} + |V_{12}|^2}. \tag{2.62}$$

Consider for instance the $2s_{1/2}$ and $2p_{1/2}$ states of the hydrogen atom, which have identical energies (provided we neglect the Lamb shift). The splitting of such a doublet when an electric field is applied is linear in the field and given by $E_2 - E_1 = 2|V_{12}|$.

We now calculate V_{12}. For the $2s_{1/2}$ state the wave function is $(4\pi)^{-1/2} R_{20}(r)$ and for the $2p_{1/2}$ state it is $(3/4\pi)^{1/2} R_{21}(r) \cos\theta$, where $R_{n\ell}(r)$ are the Coulomb radial wave functions. Thus we have

$$V_{12} = -\mathcal{E}(R_{20}|r|R_{21}) \frac{\sqrt{3}}{4\pi} \int \cos^2\theta \, d\Omega = -\frac{\mathcal{E}}{\sqrt{3}} (R_{20}|r|R_{21}).$$

The matrix elements of the dipole moment for the Coulomb radial wave functions are available from tables; in particular $(R_{20}|r|R_{21}) = 3\sqrt{3}$. Thus we have $|V_{12}| = 3\mathcal{E}$, and we conclude that the splitting of the doublet $(2s_{1/2}, 2p_{1/2})$ in a weak electric field is equal to $6\mathcal{E}$ in atomic units.

Note that the $2p_{3/2}$ level has the same parity as the $2p_{1/2}$ one, so that

$$\int \Psi_{2p_{3/2}} x \Psi_{2p_{1/2}} dr = 0.$$

Thus these levels are not split in our present approximation, that is, up to terms linear in the field. Actually, to calculate their splitting we should have to consider a mixture of all three states, $2s_{1/2}$, $2p_{1/2}$, and $2p_{3/2}$.

For all atoms other than the hydrogen atom, the level splitting is quadratic in the field:

$$E = E_0 + \frac{1}{2}\alpha\mathcal{E}^2.$$

The reason for this is that the energy levels of complex atoms corresponding to different values of the orbital quantum number ℓ are not degenerate, so that there are no states with adjacent energies which can be mixed by a weak electric field.

The Change of the Lifetime of the $2s_{1/2}$ State of the Hydrogen Atom under an Applied Electric Field

The $2s_{1/2}$ state of atomic hydrogen has a very long lifetime: the $2s_{1/2} \to 1s_{1/2}$ transition can take place only with the emission of two photons, and the lifetime of the $2s_{1/2}$ state relative to this transition turns out to be of the order of 1/7 sec. There is in addition the possibility of the dipole transition $2s_{1/2} \to 2p_{1/2}$ (the energies of the two states differ because of the Lamb shift). Let us estimate the probability of this transition. In Section 1.3 we found that the probability of a dipole transition is of order $w \sim \omega^3/c^3$, where ω is the transition frequency. In our case this is the energy difference $\omega = E_{2s_{1/2}} - E_{2p_{1/2}}$, which as we saw (p.83) is of order $1/c^3$. Consequently, $w \sim c^{-12}$. In atomic units $c \approx 137$, and so $w \approx 10^{-24}$; that is, the lifetime of the $2s_{1/2}$ state against the transition $2s_{1/2} \to 2p_{1/2}$ is of order $10^{24} \tau_{at}$, where τ_{at}, the atomic unit of time, is $\sim 10^{-17}$ seconds. Thus the lifetime in question is of order 10^7 seconds, so that the transition $2s_{1/2} \to 1s_{1/2}$ is a great deal more probable than the transition $2s_{1/2} \to 2p_{1/2}$.

When an external field is applied the state of the atom will be a superposition of the states $2s_{1/2}$ and $2p_{1/2}$. But as soon as the atom goes into the state $2p_{1/2}$, it instantly (in a time of order 10^{-10} seconds) makes the dipole transition $2p_{1/2} \to 1s_{1/2}$. We see therefore that the lifetime of the $2s_{1/2}$ state is strongly changed by the application of an electric field.

We therefore put

$$\Psi = C_1 e^{-i\varepsilon_1 t}\psi_1 + C_2 e^{-i\varepsilon_2 t}\psi_2,$$

where the index 1 refers to the state $2s_{1/2}$ and the index 2 to the state $2p_{1/2}$. For the quantities C_1, C_2 we have the equations

$$i\dot{C}_1 = V_{12}C_2 \exp[i(\varepsilon_1 - \varepsilon_2)t],$$
$$i\dot{C}_2 = -i\frac{1}{\tau}C_2 + V_{21}C_1 \exp[-i(\varepsilon_1 - \varepsilon_2)t]. \quad (2.63)$$

where $V = -\mathcal{E}x$. The diagonal matrix elements V_{11}, V_{22} are zero. We have not taken into account the decay of the $2s_{1/2}$ state, which is associated with a time $\sim 1/7$ second.

Even in the absence of an external field C_2 is not constant; because of the interaction with the radiation field we have $C_2^{(0)} \sim \exp -t/\tau$, where τ is of the order of the lifetime of the dipole state, $(\tau \sim 10^{-10}$ seconds). We have taken $C_1^{(0)}$ to be constant because the decay time of the $2s_{1/2}$ state is very long.

Let us write $C \equiv C_2 \exp i\epsilon_{12}t$, where $\epsilon_{12} \equiv \epsilon_1 - \epsilon_2$ is the Lamb shift splitting $(\epsilon_{12} \sim 1/\tau)$. Then we get from (2.63)

$$i\dot{C}_1 = V_{12}C, \; i\dot{C} = V_{21}C_1 - (\varepsilon_{12} + i/\tau)\,C. \tag{2.64}$$

Eliminating C_1, we have

$$\ddot{C} + \left(-i\varepsilon_{12} + \frac{1}{\tau}\right)\dot{C} + |V_{12}|^2 C = 0.$$

We look for the solution of this equation in the form $C = Ae^{i\omega t}$. Then we find

$$\omega^2 - \omega\left(\frac{i}{\tau} + \varepsilon_{12}\right) - |V_{12}|^2 = 0, \tag{2.65}$$

which gives two complex roots for ω.

Initially we have $C_1 = 1$, $C = 0$. This initial condition on C will be automatically fulfilled if we seek C in the form $C = a[\exp i\omega^{(1)}t - \exp i\omega^{(2)}t]$. From (2.64) we get

$$C_1 = \frac{i\dot{C} + (\varepsilon_{12} + i/\tau)\,C}{V_{21}} = \\ = \frac{a}{V_{21}}\left\{\left(\frac{i}{\tau} + \varepsilon_{12} - \omega^{(1)}\right)\exp(i\omega^{(1)}t) + \right. \\ \left. + \left(-\frac{i}{\tau} - \varepsilon_{12} + \omega^{(2)}\right)\exp(i\omega^{(2)}t)\right\}.$$

The quantity a is determined from the condition $C_1(t = 0) = 1$:

$$a = V_{21}(\omega^{(2)} - \omega^{(1)})^{-1}.$$

From Eq. (2.65) we find the frequencies $\omega^{(1)}$, $\omega^{(2)}$:

$$\omega = \frac{1}{2}\left(\varepsilon_{12} + \frac{i}{\tau}\right) \pm \sqrt{\frac{1}{4}\left(\varepsilon_{12} + \frac{i}{\tau}\right)^2 + |V_{12}|^2}. \tag{2.66}$$

Consider first the case $|V_{12}| \ll 1/\tau$. Then we get from (2.66)

$$\omega^{(1)} \approx \varepsilon_{12} + \frac{i}{\tau}, \qquad \omega^{(2)} \approx \frac{V_{12}^2}{\varepsilon_{12}^2 + \frac{1}{\tau^2}}\left(\frac{i}{\tau} - \varepsilon_{12}\right).$$

Thus, in this case the $2s_{1/2}$ state has a very long lifetime, of order $1/|V_{12}|^2\tau$.

Next consider the opposite limit, when the applied field is so large that $|V_{12}| \gg 1/\tau$. Then we get

$$\omega^{(1,2)} \approx \frac{1}{2}\left(\varepsilon_{12} + \frac{i}{\tau}\right) \pm |V_{12}| \pm \frac{1}{8|V_{12}|}\left(\varepsilon_{12} + \frac{i}{\tau}\right)^2$$

and the decay time of the state C_1 is of the same order of magnitude as the lifetime of the dipole state, τ.

Thus the lifetime of the $2s_{1/2}$ state of atomic hydrogen is decreased extremely strongly (in fact by about nine orders of magnitude) by the application of an electric field of order $\mathcal{E} \sim 1/\tau \sim c^{-3} \sim (10^9/137^3)\,\mathrm{V} \sim 500\ \mathrm{V}$.

CHAPTER 3

THE QUASICLASSICAL APPROXIMATION

In this chapter we shall consider an approximate method of solving the Schrödinger equation in the case when the wavelength of the particle is small by comparison with the distance over which the potential changes appreciably. Some very simple examples of such an approximation were considered in Section 1.1 above.

Since the limit of short wavelength corresponds to the limit of classical mechanics, the quasiclassical approximation allows us to explore the relation between classical and quantum mechanics. The relation comes out particularly clearly in the Feynman space-time formulation of quantum mechanics*. According to this approach, the wave function at the point (x, t) is obtained from the wave function $\Psi(x_o, t_o)$ by multiplication by $\exp(i \sum_k S_k)$, where $S_1, S_2 \ldots S_k \ldots$ are the action functions for all possible trajectories connecting the points x_o, t_o and x, t and integration over x.

*R. P. Feynman and A. Hibbs, Quantum Mechanics and Path Integrals, McGraw-Hill, New York 1965.

Since the possible trajectories form a continuum, the exponent contains an integral over all possible trajectories (functional integral).

Suppose there is a classical trajectory connecting the points (x_0, t_0) and (x, t). Since the action function is a minimum along the classical trajectory, the pencil of trajectories adjacent to the classical one gives the minimum phase factor. However, trajectories far from the classical one oscillate strongly and cancel one another. As a result, the only effective contribution to the functional integral comes from the tube of trajectories adjacent to the classical one, and hence we have

$$\Psi(x, t) = \int A(x, x_0)\, e^{iS_0(x,\, t;\, x_0,\, t_0)}\, \Psi(x_0, t_0)\, dx_0,$$

where S_0 is the action calculated along the classical trajectory. The factor A is determined by the effective width of the tube of contributing trajectories.

For a stationary problem we have $S_0(x, t: x_0, t_0) = S_0(x, x_0) - E(t - t_0)$, and hence, since $S_0(x, x_0) = S_0(x, x_1) + S_0(x_1, x_0)$, where x_1 is an arbitrary point on the trajectory, we get

$$\Psi(x) \sim a(x, x_1)\, e^{iS_0(x, x_1)}.$$

which is just the result obtained in the usual formulation of quantum mechanics.

This result can be generalized to the case of trajectories for which the action S is complex; here the role of "classical trajectory" is played by that trajectory which together with its neighbours gives the maximum contribution to the functional integral. The action function for such a trajectory, just as in the case of real S, obeys the Hamilton-Jacobi equation. In particular, for motion under a potential barrier S takes imaginary values. Once this generalization is made, the quasiclassical approximation gives sensible results even in those cases where the conditions of applicability of the approximation would appear not to be fulfilled. We already met a similar case when we obtained the asymptotic expression for the Γ-function: the expression obtained in the limit $x \gg 1$ turned out to be valid with high accuracy even for $x = 1$. In a similar way, the stationary solutions of the Schrödinger equation for the ground and first excited states are quite well described by the quasiclassical approximation, even though to obtain this approximation formally we have to require that the number of nodes of the wave function, n, should be much larger than 1. The reason for this will become clear below: the expansion parameter is not $1/n$, but the quantity $1/\pi^2 n^2$, which is sufficiently small even for $n = 1$.

In some problems (e.g. reflection above a barrier, perturbation of the system by a passing particle) there occur integrals of fast oscillating functions. These integrals can be calculated by the method of steepest descents and are generally determined by how close to the real axis are the closest singularities of the potential. Below we shall clarify the conditions under which these integrals are not exponentially small; these calculations will be used as the occasion for a more detailed investigation of the method of steepest descents. We will examine, in particular, the problem of the escape of a particle from a potential barrier; this problem reduces to the problem of the spreading of a wave packet which describes a particle which at the initial time was inside the barrier.

It will be shown that the probability to find the original system undecayed after a time t decreases exponentially with time. The additional terms arising in the calculation, which do not depend exponentially on time, are determined by the spreading of the packet of slow particles which are formed in the preparation of the system, and have nothing to do with the decay process.

For a better understanding of the quasiclassical approximation in the three-dimensional case the reader should follow through the calculation of the distribution of charge in the atom and of the quasiclassical scattering problem.

1. THE ONE-DIMENSIONAL CASE

We write the Schrödinger equation in the form

$$\varphi'' + k^2(x)\,\varphi = 0, \tag{3.1}$$

where $k^2(x) = 2\,[E - V(x)]$.

Suppose the potential is sufficiently slowly changing in space that the condition

$$kl \gg 1, \tag{3.2}$$

is fulfilled, where ℓ is a characteristic distance over which V(x) changes appreciably. Then the approximate solution of the problem has the form (cf. p. 26)

$$\varphi(x) = \frac{a}{\sqrt{k(x)}} \exp\left[\pm i \int_{x_1}^{x} k\,dx\right]. \tag{3.3}$$

This is the form appropriate to the case $E > V$, when k is real. In the case $E < V$ the solution will contain increasing and decreasing exponentials, that is,

$$\varphi(x) = A \exp\left[\pm \int_{x_1}^{x} |k|\,dx\right], \tag{3.4}$$

where $A = ak^{-1/2}$. The above results are the first-order approximation in an expansion in powers of the quantity $1/(k\ell)^2$. We may obtain also higher approximations. It turns out that the series expansion in the parameter $1/(k\ell)^2$ so obtained is of the asymptotic rather than the ordinary type; we now discuss this.

Problem: Show that for an approximate estimate of the derivative of a quasiclassical function it is sufficient to differentiate only the exponential.

Asymptotic Series

Let us recall what we mean by an asymptotic series. Let

$$s_n = \sum_{\nu=0}^{n} \frac{a_\nu}{z^\nu}$$

be the partial sum of a series. Suppose that for fixed z and $n \to \infty$ the quantity $s_n \to \infty$, but that for fixed n and $z \to \infty$ the sum s_n gives an ever better approximation to some function $f(z)$; that is,

$$\lim_{z \to \infty} z^n [s_n(z) - f(z)] = 0. \tag{3.5}$$

(where $z \to \infty$ in some given interval of arg z). Then we say that s_n is an asymptotic representation of $f(z)$. This is not a series in the usual meaning of the term, since s_n diverges; however, we can derive from it information on the function $f(z)$, since for $z \to \infty$ it forms a good description of this function. In other words, an ordinary convergent series tends to $f(z)$ when $n \to \infty$ for given z, whereas an asymptotic series tends to $f(z)$ when $z \to \infty$ for given n.

Let us construct $s_n(z)$ for a given $f(z)$. The quantity s_0 is just a_0, so that $a_0 = f(\infty)$. The quantity s_1 is equal to a $a_0 + a_1/z$, so

$$\lim_{z \to \infty} z \left(a_0 + \frac{a_1}{z} - f(z) \right) = 0,$$

that is,

$$a_1 = \lim_{z\to\infty} z\,[f(z) - a_0] = \lim_{z\to\infty} z\,[f(z) - f(\infty)].$$

All the required coefficients a_n can be found similarly.

However, it is not possible to invert the process and reconstruct the function $f(z)$ from its asymptotic representation. For instance, suppose $f(z) = e^{-z}$. Then

$$a_0 = 0, \quad a_1 = \lim_{z\to\infty} ze^{-z} = 0, \quad a_2 = a_3 = \ldots = a_n = 0.$$

Hence for the function e^{-z} the whole asymptotic series is equal to zero. It follows that the asymptotic representations of the function $f(z)$ and of $(f(z) + e^{-z})$ are identical.

Let us write $z = k\ell$. It turns out that the quasiclassical solution

$$\varphi_{\text{quasicl}}(x) \approx \sum_{\nu=0}^{n} a_\nu(x) z^{-\nu} \tag{3.6}$$

is an asymptotic series, that is, it diverges as $n\to\infty$. However, in the limit $z\to\infty$ the quantity $a_0(x)$ represents the function $\varphi(x)$ (the exact solution) with arbitrary accuracy.

Suppose furthermore z is fixed. Since for $n\to\infty$ the quasiclassical series diverges, it follows that for any z there is an optimum number of terms $n(z)$ which best represents the function $\varphi(x)$. It can be proved that the optimum number of terms in the quasiclassical approximation is $n\sim k\ell$. We shall in fact only use the zeroth term of this series.

Matching of Quasiclassical Functions

In this section we shall find how to match the quasiclassical wave function φ in the region accessible to classical motion with the wave function in the forbidden region. For definiteness we assume that the potential energy of the particle has the form represented in Fig. 24. (The origin of coordinates is taken to be at the point where V(x) = E.)

The quasiclassical solution in region II (the classically accessible region) has, according to (3.5), the form

$$\varphi = \frac{a_1}{\sqrt{k}} \exp\left(i\int_0^x k\,dx\right) + \frac{a_2}{\sqrt{k}} \exp\left(-i\int_0^x k\,dx\right), \tag{3.7}$$

while in region I (the classically inaccessible region) it is

$$\varphi = \frac{b_1}{\sqrt{|k|}} \exp\left(\int_x^0 |k|\,dx\right) + \frac{b_2}{\sqrt{|k|}} \exp\left(-\int_x^0 |k|\,dx\right). \tag{3.8}$$

If the quasiclassical solution were an analytic function of x in the neighbourhood of x = 0, then to find the relation between a_1, a_2

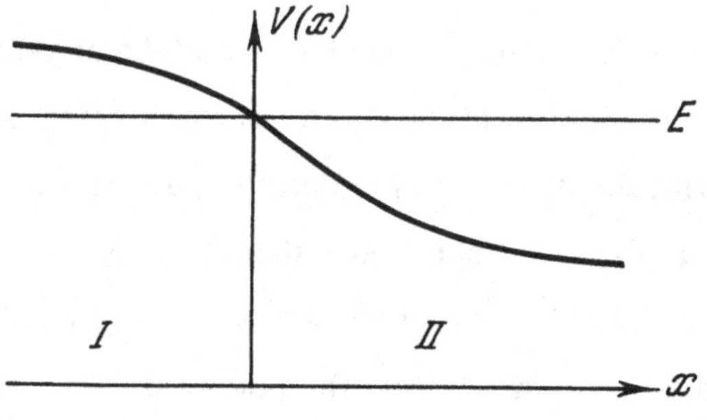

Fig. 24

and b_1, b_2 all we should have to do is to continue the solution analytically from region II to region I (the exponents in (3.7) would simply go over into the exponents in (3.8)). Actually, we could not pass from II to I along the real axis, since in the neighbourhood of $x = 0$ the quasiclassical approximation is inapplicable (we have $k(0) = 0$, so that the condition $k\ell \gg 1$ is not fulfilled). We could overcome this difficulty by going around the point $x = 0$ in the complex plane along an arc of sufficiently large radius (where the quasiclassical approximation should be applicable).

However, as we shall see, this analytic continuation procedure gives the wrong result. It turns out that in such a detour in the complex plane we inevitably cross certain lines (the Stokes lines) where the analytic continuation process breaks down. Such behaviour is characteristic of asymptotic series. We shall show below how to find the analytic continuation in cases like this.

Another possibility consists in looking for the exact solution of the problem near $x = 0$, taking the potential to be a linear function of x in this region. The appropriate solution is the so-called Airy function; we must then join up the exact solution with the quasiclassical ones for $x > 0$ and $x < 0$. Of course, such a procedure is legitimate only if we know that the quasiclassical approximation is valid at distances from the turning point small in comparison with the distance at which the potential departs appreciably from linearity. We shall now show that for large values of the "quasiclassicality" parameter ($k\ell \gg 1$) this condition is indeed fulfilled. To do this, we expand the potential $V(x)$ near $x = 0$ in a series: $V(x) \approx E + V'(0)x$. Then the momentum $k(x)$ is given by $k(x) = \sqrt{-2V'(0)x} \equiv \alpha\sqrt{x}$. The order of magnitude of α is given by $\alpha \sim \sqrt{|V'(0)|} \sim \sqrt{V(0)/\ell} \sim k_0/\sqrt{\ell}$, where $k_0 = \sqrt{2E}$ and ℓ is the distance over which the potential $V(x)$ changes appreciably. Thus in the neighbourhood of the point $x = 0$ we have $k(x) \sim k_0\sqrt{x/\ell}$. In order that the quasicalssical approximation should be applicable, we have to go far enough away from the point $x = 0$ that we should have $d(1/k)/dx \ll 1$ or $\ell^{1/2}/k_0 x^{3/2} \ll 1$. This means we must have $x \gg \ell/(k_0 \ell)^{2/3}$. Since $k_0 \ell \gg 1$, we can always choose a value x_1 such that

$$\frac{l}{(k_0 l)^{2/3}} \ll x_1 \ll l.$$

At distance of the order x_1 the quasiclassical approximation is already valid, while at the same time we can still take the potential to be linear.

For the potential shown in Fig. 24 we find from the properties of the Airy function that the right matching is[1]

$$\frac{1}{\sqrt{k}} \exp\left[i\int_0^x k\,dx - \frac{\pi}{4}\right] \to \frac{1}{i\sqrt{|k|}} \exp\left[\int_x^0 |k|\,dx\right] + \frac{1}{2\sqrt{|k|}} \exp\left[-\int_x^0 |k|\,dx\right],$$

$$\frac{1}{\sqrt{k}} \exp\left[-i\int_0^x k\,dx + \frac{\pi}{4}\right] \to \frac{-1}{i\sqrt{|k|}} \exp\left[\int_x^0 |k|\,dx\right] + \frac{1}{2\sqrt{|k|}} \exp\left[-\int_x^0 |k|\,dx\right]. \tag{3.9}$$

We shall now derive these relations without using the Airy function. We choose the radius ρ of our arc around the point $x = 0$ to be of order x_1 (see Fig. 25). Then we can use the quasiclassical approximation and at the same time take the potential to be linear in x. Since we then have $k(x) = \alpha x^{1/2}$, it follows that

[1] L. I. Schiff, Quantum Mechanics (McGraw-Hill Book Company, New York, 1955).

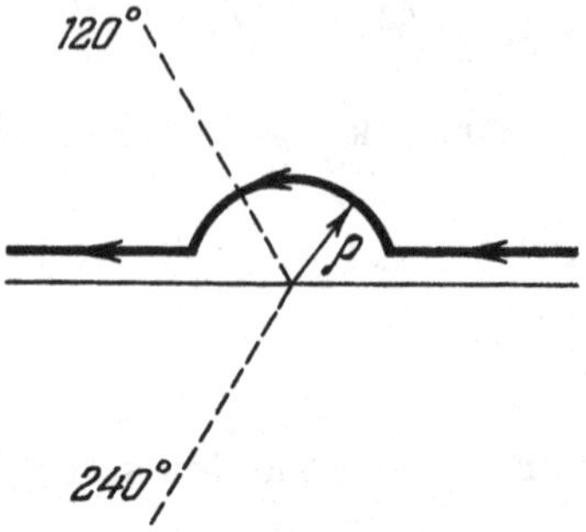

Fig. 25

$\int_0^x k\,dx = (2/3)\alpha x^{3/2}$. In the complex plane we have $x = \rho e^{i\varphi}$ and $\int_0^x k\,dx = (2/3)\alpha\rho^{3/2}\exp[(3/2)i\varphi]$. Hence we get

$$\frac{1}{\sqrt{k}}\exp\left[i\int_0^x k\,dx\right] =$$
$$= \frac{1}{\sqrt{\alpha\rho^{1/2}}}e^{-i\frac{\varphi}{4}}\exp\left[\frac{2}{3}\alpha\rho^{3/2}\left(i\cos\frac{3}{2}\varphi - \sin\frac{3}{2}\varphi\right)\right]. \qquad (3.10)$$

As we go around the contour, so long as $\varphi < 2\pi/3$, $\sin 3/2\,\varphi$ is positive and the expression (3.10) contains a decreasing exponential. However, when $\varphi > 2\pi/3$, the exponential in (3.10) is an increasing one. As soon as an increasing exponential appears we can neglect all terms containing a decreasing one, since the quasi-classical approximation, being an asymptotic representation, is accurate only to a power of the "quasiclassicality" parameter ($\sim 1/k^2\ell^2$) and not to within exponentially small terms. Consequently, the analytic continuation procedure gives us only the coefficient of the increasing exponential. We get in fact

$$\frac{1}{\sqrt{k}}\exp i\left[\int_0^x k\,dx - \frac{\pi}{4}\right] \to \frac{1}{i\sqrt{|k|}}\exp\left[\int_x^0 |k|\,dx\right]. \qquad (3.11)$$

where we have lost a term which is exponentially small compared to the right-hand side of (3.11).

If we go around the point $x = 0$ in the lower half-plane the function (3.11) goes over first into an increasing exponential, and then, after crossing the line $\varphi = -2\pi/3$, into a decreasing one. The exponentially small correction lost in the region $-2\pi/3 < \varphi < 0$ owing to the inexactness of the quasiclassical approximation goes over, after we cross into the region $-\pi < \varphi < -2\pi/3$, into an exponentially large term, which we have therefore managed to lose. This, then, is no way to obtain the correct value of the coefficient of the decreasing exponential either.

How then are we to obtain it ? Suppose that in the classically inaccessible region $(V > E)$ there is only a decreasing exponential:

$$\frac{1}{\sqrt{|k|}} \exp\left[-\int_x^0 |k|\, dx\right]. \tag{3.12}$$

The general solution for $V < E$ has the form

$$\frac{C_1}{\sqrt{k}} \exp\left[i\int_0^x k\, dx - i\varphi_1\right] + \frac{C_2}{\sqrt{k}} \exp\left[-i\int_0^x k\, dx + i\varphi_2\right]. \tag{3.13}$$

Let us find C_1, C_2 and φ_1, φ_2 in this expression. To do this we continue (3.12) analytically into the region $x > 0$. The analytic continuation into the upper half-plane has the form

$$\frac{1}{\sqrt{|k|}} \exp\left[-\int_x^0 |k|\, dx\right] \rightarrow$$

$$\rightarrow \frac{\exp[i\pi/4]}{\sqrt{\alpha\rho^{1/2}\exp(i\varphi/2)}} \exp\left[\frac{2}{3}\alpha\rho^{3/2}\left(\sin\frac{3}{2}\varphi - i\cos\frac{3}{2}\varphi\right)\right].$$

For $\varphi = 0$ we get the second term in (3.13), with $\varphi_2 = \pi/4$, $C_2 = 1$. If we had carried out the analytic continuation in the lower half-plane, we should have got the first term in (3.13) with the values

$$\varphi_1 = \varphi_2 = \pi/4,$$
$$C_1 = C_2 = 1.$$

In each case one of the two terms is lost because of an exponentially small error in the region where the function is exponentially large. Thus we finally get

$$\frac{1}{\sqrt{|k|}} \exp\left[-\int_x^0 |k|\, dx\right] \to \frac{2}{\sqrt{k}} \cos\left(\int_0^x k dx - \frac{\pi}{4}\right)$$

in agreement with (3.9).

The Quantization Condition

In the classically accessible region ($x_1 < x < x_2$, see Fig. 26) the quasiclassical solution of Schrödinger's equation may be written in either of two ways, one of which is obtained by matching the oscillating solution to the exponentially decreasing one at x_1, the other by doing the same at x_2. The requirement that the two functions so obtained be identical gives us a quantization condition.

The first solution has the form

$$\varphi_1 = \frac{a_1}{\sqrt{k}} \cos\left(\int_{x_1}^x k\, dx - C_1\pi\right),$$

where $C_1\pi$ is the phase resulting from matching to the decreasing exponential for $x < x_1$; in cases where the potential has linear behaviour near $x = x_1$ the result is, as we have just seen,

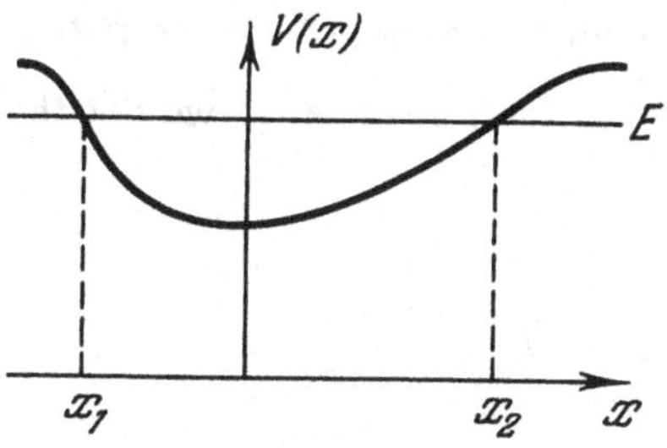

Fig. 26

$C_1 = 1/4$. The second solution has the form

$$\varphi_2 = \frac{a_2}{\sqrt{k}} \cos\left(\int_x^{x_2} k\,dx - C_2\pi\right).$$

Thus the requirement that the two solutions should agree gives

$$\int_{x_1}^{x_2} k dx = (n + C_1 + C_2)\pi \tag{3.14}$$

(with $a_1 = a_2(-1)^n$). It is easy to see that $\cos(\int_{x_1}^{x} k\,dx - C_1\pi)$ then goes through zero n times as x goes from x_1 to x_2, so that n is just the number of nodes of the wave function in this interval.

We can check the quantization condition (3.14) (the so-called Bohr quantization condition) by considering the example of a one-dimensional square well with infinitely high walls. Since the wave function must now go to zero at both walls, we have in this case $C_1 = C_2 = 1/2$. Thus we get

$$\int_{x_1}^{x_2} k_n dx = k_n L = (n+1)\pi, \qquad E_n = \frac{1}{2} k_n^2 = \frac{\pi^2}{2L^2}(n+1)^2, \tag{3.15}$$

where L is the width of the well. The lowest state corresponds to n = 0, that is, the wave function has no nodes.

In the more common case where the potential has a linear behaviour both near x_1 and near x_2, we get the quantization rule

$$\int_{x_1}^{x_2} k\,dx = \left(n + \frac{1}{2}\right)\pi. \tag{3.16}$$

We shall see below (p. 193) that application of the quasiclassical approximation for the Coulomb radial wave functions gives, for $\ell = 0$, $C_1 = 3/4$, $C_2 = 1/4$, and for $\ell \neq 0$, $C_1 = C_2 = 1/4$.

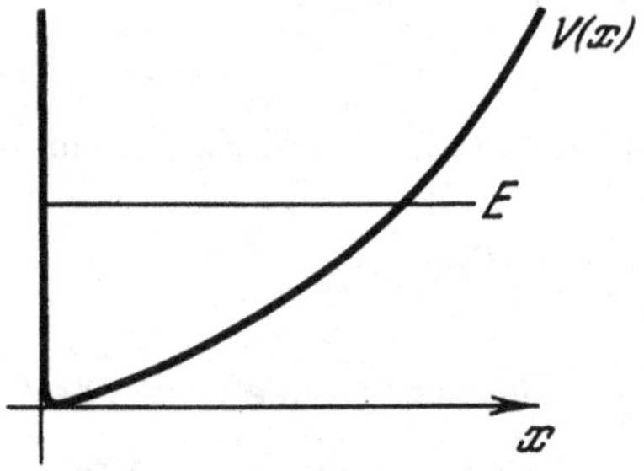

Fig. 27

Problem: Show that for the potential shown in Fig. 27 the quantization condition is

$$\int_{x_1}^{x_2} k\,dx = \left(n + \frac{3}{4}\right)\pi.$$

Accuracy of the Quasiclassical Approximation

The only approximation we made in deriving the formulas used above consisted in the neglect of the term

$$\frac{A''}{k^2A} \sim \frac{A/l^2}{k^2A} \sim \frac{1}{(kl)^2}.$$

To estimate $k\ell$ we observe that

$$\int k\,dr \sim kl \sim n\pi,$$

where n is the number of nodes. Consequently the quasiclassical approximation is valid to an accuracy of order $1/\pi^2 n^2$ (not 1/n). This makes it legitimate to keep the extra term 1/2 which is added to the number of nodes n in the Bohr quantization rule, since this is a correction of relative order 1/n.

This estimate of the relative accuracy refers to all non-exponential expressions. For the wave function itself, which contains the factor $\exp(\int S'\,dx)$, the relative correction is of order $i\int \delta S'\,dx \sim iA''\ell/Ak \sim i/k\ell$. When we calculate the matrix element $(\varphi^*_n, V\varphi_m)$ for functions φ_0 and φ_1 which differ only slightly in their respective numbers of nodes n, m, there is a partial cancellation of the corrections to φ_0 and φ_1, and the error in the calculation of the matrix element will be of order $(k\ell)^{-1}(n-m)/n$.

Normalization of Quasiclassical Functions

The only appreciable contribution to the normalization integral comes from the region $x_1 < x < x_2$ (Fig. 26), since outside this region the wave function decreases exponentially. For the region $x_1 < x < x_2$ the wave function is

$$\varphi_n = \frac{a}{\sqrt{k}} \cos\left(\int_{x_1}^{x} k\,dx - \frac{\pi}{4}\right),$$

and consequently

$$\int |\varphi_n|^2 \, dx = 1 \approx \int_{x_1}^{x_2} \frac{a^2}{k} \cos^2 \Phi \, dx,$$

where

$$\Phi = \int_{x_1}^{x} k \, dx - \frac{\pi}{4} .$$

For $E = E_n$ the wave function φ_n has n nodes in the region $[x_1, x_2]$, since the expression $\left(\int_{x_1}^{x_2} k \, dx\right)$ changes from zero to $(n + 1/2)\pi$, and so $\cos \Phi$ will go through zero n times. If n is large, then the phase Φ is also large. Moreover, we can write $\cos^2 \Phi = \frac{1}{2} + \frac{1}{2} \cos 2\Phi$; the second term then changes sign many times and so gives only a small contribution. Consequently,

$$\frac{a^2}{2} \int_{x_1}^{x_2} \frac{dx}{k} = \frac{a^2}{2} \frac{T}{2} \approx 1,$$

where T is the classical period of the motion. Thus,

$$a^2 = \frac{4}{T} = \frac{2}{\pi} \omega,$$

where ω is the frequency of oscillation in the corresponding classical problem. Therefore we finally have

$$\varphi_n = \sqrt{\frac{2\omega}{\pi k}} \cos \left(\int_{x_1}^{x} k \, dx - \frac{\pi}{4} \right). \tag{3.17}$$

Problem: Estimate the contribution of the classically inaccessible region to the normalization integral.

Solution:

$$\omega \left(\frac{\partial V}{\partial x} \right)_{V=E}^{-2/3} .$$

The Correspondence Principle

By differentiating the quantization rule (3.9) with respect to n we get

$$\pi = \frac{d\varepsilon_n}{dn} \int_{x_1}^{x_2} \frac{dx}{k} = \frac{d\varepsilon_n}{dn} \frac{\pi}{\omega},$$

whence

$$\frac{d\varepsilon_n}{dn} = \omega. \tag{3.18}$$

This relation is called the correspondence principle: at large quantum numbers the energy difference between neighbouring levels is equal to the classical frequency of motion.

Mean Kinetic Energy

Let us express the mean kinetic energy of a particle undergoing quasiclassical motion in terms of its total energy:

$$\overline{T} = \frac{1}{2} \int \varphi_n^* \hat{p}^2 \varphi_n dx = -\frac{1}{2} \int \varphi_n^* \frac{d^2}{dx^2} \varphi_n \, dx \approx \frac{1}{2} \int \varphi_n^* k^2 \varphi_n \, dx.$$

Differentiation of the factor $1/\sqrt{k}$ in the expression

$$\varphi_n = \frac{a}{\sqrt{k}} \cos \left(\int_{x_1}^{x} k \, dx - \frac{\pi}{4} \right)$$

gives only a small contribution to $\overline{T}$. Using the correspondence principle, we get

$$\overline{T} = \frac{a^2}{2} \int_{x_1}^{x_2} k^2 \frac{1}{k} \cos^2 \Phi \, dx = \frac{a^2}{4} \int_{x_1}^{x_2} k \, dx =$$

$$= \frac{a^2}{4} \left(n + \frac{1}{2} \right) \pi = \frac{1}{2} \left(n + \frac{1}{2} \right) \frac{d\varepsilon_n}{dn}. \tag{3.19}$$

In the case of a Coulomb potential the quantization condition for the $\ell = 0$ wave does not contain the term $1/2$ and we get

$$\overline{T} = \frac{1}{2} n \frac{d\varepsilon_n}{dn}.$$

By applying also the virial theorem

$$\overline{T} = \frac{1}{2} \overline{x \frac{dV}{dx}},$$

we can use (3.19) to find the energy eigenvalues in the quasi-classical case. For instance, for a harmonic oscillator $V \sim x^2$ and so $\overline{x \frac{dV}{dx}} = 2\overline{T}$ gives $\overline{T} = \overline{V}$. Therefore,

$$\left(n + \frac{1}{2}\right) \frac{d\varepsilon_n}{dn} = \varepsilon_n.$$

Solution of this equation gives $\epsilon_n = C(n + 1/2)$. The correspondence principle then implies that $C = \omega$, the frequency of classical motion, and so $\epsilon_n = (n + 1/2)\omega$. Again, for the $\ell = 0$ waves in a Coulomb potential we get

$$\overline{T} = -\frac{1}{2} \overline{V} = \frac{1}{2} n \frac{d\varepsilon_n}{dn},$$

and so

$$\varepsilon_n = \overline{T} - 2\overline{T} = -\frac{1}{2} n \frac{d\varepsilon_n}{dn}.$$

the solution of which is $\epsilon_n = C/n^2$. We saw (p. 37) that the correspondence principle implies that $C = -1/2$.

Connection between the Quasiclassical Matrix Elements and the Fourier Components of Classical Motion

Let us find the connection between the matrix element $\int \varphi_n^* U(x) \varphi_{n'} \, dx$ for the one-dimensional problem and the Fourier component of the classical quantity $U[x(t)]$ (for example, we can relate the dipole matrix element

$$\int \varphi_n^* x \varphi_{n'} \, dx$$

to the Fourier component of the classical coordinate $x(t)$).

Suppose n and n' are not very different, that is

$$\frac{|n'-n|}{n} \ll 1, \; n \gg 1, \; n' \gg 1.$$

Then the matrix element $U_{nn'}$ is

$$U_{nn'} = \int \varphi_n^* U \varphi_{n'}^* dx = a_n a_{n'} \int \frac{U}{\sqrt{kk'}} \cos \Phi_n \cos \Phi_{n'} \, dx \approx$$
$$\approx a_n^2 \int U \left[\cos(\Phi_n - \Phi_{n'}) + \cos(\Phi_n + \Phi_{n'})\right] \frac{dx}{2k} \, .$$

Since Φ_n and $\Phi_{n'}$ are large, the second term in the integrand may be neglected, owing to the strongly oscillatory behaviour of the quantity $\cos(\Phi_n + \Phi_{n'})$. Thus,

$$U_{nn'} = \frac{a_n^2}{2} \int_{x_{n1}}^{x_{n2}} U(x) \cos(\Phi_n - \Phi_{n'}) \frac{dx}{v} \, .$$

Since we have

$$\Phi_n - \Phi_{n'} \approx \int_{x_{n1}}^{x} \frac{dx}{k_n} (\varepsilon_n - \varepsilon_{n'}) = \frac{d\varepsilon_n}{dn} (n-n') \int_{x_{n1}}^{x} \frac{dx}{v} = \frac{d\varepsilon_n}{dn} (n-n') \, t,$$

it follows that

$$U_{nn'} = \frac{2}{T} \int_0^{T/2} U[x(t)] \cos \frac{2\pi (n-n') \, t}{T} \, dt.$$

Thus the matrix element of the quantity $U(x)$ in the quasiclassical approximation is equal to the Fourier component of the quantity $U[x(t)]$ in the classical problem.

Criterion for the Applicability of Perturbation Theory to the Calculation of Not Too Small Quantities

When the n-th state of the discrete spectrum is perturbed, the criterion for the applicability of perturbation theory has the form

$$H' \ll E_{nm},$$

where H' is the matrix element of the perturbation and $E_{nm} = E_n - E_m$. Here E_n and E_m denote the closest levels for which the matrix element H'_{nm} is appreciable. Since we have

$$E_{nm} \sim \frac{dE_n}{dn} \sim \frac{E_n}{n},$$

where n is the number of nodes of the wave function, this condition reduces to

$$\frac{H'}{E_n} n \ll 1. \tag{3.20}$$

In the quasiclassical case we have $n \gg 1$ and this condition may not be fulfilled. We shall prove that provided the quantity we are calculating is not too small we can replace (3.20) by the weaker condition

$$\frac{H'}{E_n} \ll 1. \tag{3.21}$$

To prove this assertion, we consider the matrix element $U_{nm} = (\varphi_n|U|\varphi_m)$. Suppose $U(x)$ is a slowly varying function; then U_{nm} will be non-zero only when n and m are close in value (otherwise the integrand will have many nodes). In fact, using the quasiclassical wave functions, we get

$$U_{nm} \approx \frac{a_n a_m}{2} \int U(x) \frac{dx}{\sqrt{k_n k_m}} \{\cos(\Phi_n - \Phi_m) + \cos(\Phi_n + \Phi_m)\}. \tag{3.22}$$

The second term in the integrand oscillates $(n + m)$ times, and its contribution is accordingly negligible.

Suppose now we add to the potential $V(x)$ a small quantity $H'(x)$. Then since $k_n = \sqrt{2[E_n - V(x)]}$, we get

$$\Delta k_n = \sqrt{2[E_n - V(x) - H'(x)]} - \sqrt{2[E_n - V(x)]} \sim \frac{H'}{k_n},$$

and so $\Delta k_n/k_n \sim H'/k_n^2 \sim H'/E_n$, provided that V is not too close to E_n (when the quasiclassical expressions are invalid). Consequantly the condition $\Delta k_n/k_n \ll 1$ is equivalent to $H'/E_n \ll 1$, that is, we can neglect the change of k provided (3.21) is fulfilled. Of course, we cannot do this when we calculate the phase $\Phi_n = \int_{x_{n_1}}^{x} k_n \, dx - \pi/4$, since

$$\delta\Phi_n \approx \int_{x_{n1}}^{x} \delta k_n \, dx \approx \int \frac{H'}{E_n} k_n dx \sim \frac{H'}{E_n} n,$$

where n is the number of nodes; thus $\delta\Phi_n \ll 1$ is equivalent to $H'n/E_n \ll 1$, which just gives us our original criterion for the applicability of perturbation theory. However, we see that when we are calculating quantities which contain the phase difference $\Phi_n - \Phi_m$, where $|n-m| \sim 1$, the change of this difference is of order

$$\delta(\Phi_n - \Phi_m) \sim \frac{H'}{E_n}(n - m) \sim \frac{H'}{E_n}.$$

It follows, then, that if the potential is perturbed by an amount $H'(x)$ the change of the large matrix elements U_{nm} is determined by the condition $H'/E_n \ll 1$, which is what we set out to prove. If $|n-m|$ is not small, the factor $\cos(\Phi_n - \Phi_m)$ is fast oscillating and U_{nm} is small anyway.

Let us see how the quasiclassical wave function $\Psi = Ae^{iS}$ changes when a small perturbation $H'(x)$ is added to the potential $V(x)$. As we saw above, the change of A is of order H'/E, since $A \sim 1/\sqrt{k}$ and $\Delta k/k \sim H'/E$. The quantity S, on the other hand, is of order $\int k\, d\ell$, and hence $\delta S \sim n(H'/E)$, so that δS may even be greater than 1, in spite of the condition $H'/E \ll 1$. Thus the change of Ψ due to the addition of H' is given by

$$\Psi + \delta\Psi \approx \Psi e^{i\delta S} \approx \Psi \exp\left(i\int \frac{H'\,dx}{k}\right).$$

The "perturbation" induced by a change of the shape of a nucleus does not satisfy (3.20), but does satisfy (3.21). Thus, provided the quantities we wish to calculate are not too small, perturbation theory is applicable (see p. 99).

Calculation of Matrix Elements in the Case of Fast Oscillating Functions

Let us suppose we want to find the matrix element

$$I = \int_{-\infty}^{\infty} \Psi_1^*(x)\, f(x)\, \Psi_2(x)\, dx,$$

where $f(x)$ is some slowly varying function which has no singularities near the real axis; and suppose the states 1 and 2 differ strongly in energy, so that the integrand is a fast oscillating function. Matrix elements of this type occur in the problem of excitation of a system by a passing particle. When we considered, above, high Fourier coefficients we saw that such matrix elements are exponentially small. As we will show, in many cases the matrix element is determined by the behaviour of Ψ_1 and Ψ_2 near the singularity of the potential in the complex x-plane. Near this singularity the quasiclassical approximation is inapplicable, and to obtain a quantitative result we must match the exact solution near the singularity to the quasiclassical one far from it, and only then apply the method of steepest descents, which we shall use below. However, this refinement affects only the numerical coefficient in the pre-exponential factor; the argument of the exponential and the order of magnitude of the pre-exponential factor can be found with the aid of the quasiclassical approximation alone. We will meet a similar circumstance below, in the problem of reflection above a barrier.

Thus, we take the functions Ψ_1 and Ψ_2 to be given by the quasiclassical approximation and try to apply the method of steepest descents (Landau, 1932). We shall take the functions Ψ_1 and Ψ_2 to be real and normalized to unit amplitude for $x \to \infty$:

$$\Psi_{1,2} \underset{x>a_{1,2}}{=} \sqrt[4]{\frac{E_{1,2}}{E_{1,2}-V_{1,2}}} \cos\left(\int_{a_{1,2}}^{x} k_{1,2}\,dx - \frac{\pi}{4}\right).$$

The point $x = a_{1,2}$ corresponds to the condition $E_{1,2} = V_{1,2}$. We decompose the function Ψ_1 into two complex conjugate terms $\Psi_1^{+,-}$, corresponding to the asymptotic expressions $\Psi_1^+ \sim e^{iS_1}$, $\Psi_1^- \sim e^{-iS_1}$; then the matrix element I is the sum of two terms I^+ and I^-. If $E_1 > E_2$, then in calculating I^+ one may deform the contour into the upper half-plane, and in calculating I^- into the lower, (since for $x \to i\xi$ the function $\exp iS_1$ becomes $\exp(-\sqrt{2E_1}\,\xi)$, while the increasing term in Ψ_2 contains $\exp(\sqrt{2E_2}\,\xi)$.

We now find the integral I^+. As we saw above (p.158), we have in the upper half-plane

$$\Psi_1^+ = \frac{1}{2i}\sqrt[4]{\frac{E_1}{V_1 - E_1}}\exp\int_z^{a_1}\sqrt{2(V_1 - E_1)}\,dz,$$

$$\Psi_2 = \frac{1}{2}\sqrt[4]{\frac{E_2}{V_2 - E_2}}\exp\left(-\int_z^{a_2}\sqrt{2(V_2 - E_2)}\,dz\right).$$

For real z we take the positive root. The expression for I^+ takes the form

$$I^+ = \frac{\sqrt[4]{E_1E_2}}{4}\int\frac{dz}{\sqrt[4]{(V_1 - E_1)(V_2 - E_2)}}\exp\Phi(z),$$

where

$$\Phi(z) = \int_z^{a_1}\sqrt{2(V_1 - E_1)}\,dz - \int_z^{a_2}\sqrt{2(V_2 - E_2)}\,dz.$$

To apply the method of steepest descents we must find the stationary point (i. e. the point where the exponent has an extremum) closest to the real axis, and then draw through it a curve corresponding to constant imaginary part of the exponent. Along this curve, as may be easily seen from the Cauchy relations, the real part of the exponent decreases in both directions from the stationary point. We must then verify that the integration contour can be deformed so as to lie along the curve of constant imaginary part without crossing any singularities (if this is not the case, then we have to add integrals around the loops enclosing the singularities). After this deformation of the contour the value of the integral is determined by the region of integration near the stationary point, and in the simplest case reduces to a Gaussian integral.

The stationary point is defined by the relation $d\Phi/dz = 0$ or

$$\sqrt{V_1 - E_1} = \sqrt{V_2 - E_2}.$$

Suppose that the two Ψ-functions correspond to motion in the same potential $V_1 = V_2 = V$. Then the stationary point z_0 is also the point where $V(z_0) = \infty$. Because of this the procedure for applying the method of steepest descents must be somewhat modified. Suppose z_0 is a simple pole of V, i.e. $V \underset{z\to z_0}{=} A/(z - z_0)$. Near the stationary point the argument of the exponential has the form

$$\Phi(z) = \Phi(z_0) - \sqrt{2}\int_{z_0}^{z} (\sqrt{V - E_1} - \sqrt{V - E_2})\, dz \simeq$$

$$\simeq \Phi(z_0) + \frac{2}{3\sqrt{2}}\frac{E_1 - E_2}{\sqrt{A}}(z - z_0)^{3/2}.$$

The point $z = z_0$ is a branch point of the function $\Phi(z)$. To make $\Phi(z)$ single-valued we make a cut from $z = z_0$ to $z = i\infty$. For $E_1 > E_2$ the function $\Phi(z)$ has a negative real part and $\exp \Phi$ decreases exponentially for $\operatorname{Im} z \to \infty$. For large z, $\Phi(z) \to i(\sqrt{2E_1} - \sqrt{2E_2})z$ and the beginning and end of the curve $\operatorname{Im}\Phi = \text{const.}$ run parallel to the imaginary axis. At the point $z = z_0$ the curve of constant imaginary part has a kink.

To see this, consider first the case where the residue A is positive (for instance, the case of a Coulomb potential, for which $A = Z$). If we put $E_1 > E_2$, then the curve of constant imaginary part of $\Phi(z)$ is determined, near z_0, by the requirement that the quantity $(z - z_0)^{3/2} = \rho^{3/2} \exp(3i\psi/2)$ be real and negative, whence $\psi = \pm 2\pi/3$. (Note that for positive A we must measure the angles from the line along which $(z - z_0)$ is real and positive, since $\sqrt{V - E}$ is by definition positive for real and positive $V - E$).

A typical behaviour of the curve $\mathrm{Im}[\Phi(z) - \Phi(z_0)] = 0$ is shown in Fig. 28a, where the heavy line represents the cut. When $\sqrt{A}$ is complex, the curve swings round near z_0. Thus, for

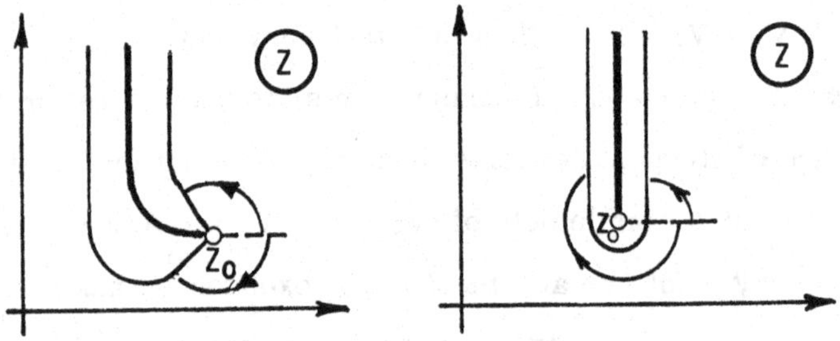

Fig. 28

instance, for the potential $V = V_0/(1 + x^2)$ the residue is $A = V_0/2i$ and it is easy to see that near z_0 the curves swings round the point z_0 by an angle $-\pi/6$.

To clarify the method of steepest descents in its application to the case in question we note the following. The integral along the curve of constant imaginary part is of course equal to the integral over the contour shown in Fig. 28b, since one contour can be deformed into the other without crossing any singularities. However, the integral over the contour of Fig. 28b is not determined only by the neighbourhood of the point z_0 and hence cannot be calculated in general. We can see this as follows: on the right-hand edge of the cut we have $\psi = \pi/2$ and

$$(z - z_0)^{3/2} = \rho^{3/2}\left(\cos\frac{3\pi}{4} + i\sin\frac{3\pi}{4}\right).$$

Since the real part of the exponent is negative, the integral is determined by the neighbourhood of the point z_0. However, on the left-hand edge we have $\psi = -\frac{3}{2}\pi$ and

$$(z - z_0)^{3/2} = \rho^{3/2}\left(\cos\frac{\pi}{4} - i\sin\frac{\pi}{4}\right),$$

so that the real part of the exponent is positive and the integral is determined by the behaviour of the function $\Phi(z)$ far from the point z_0.

The integral we are interested in is determined by the neighbourhood of z_0 and can be written in the form

$$I^+ \cong \frac{\sqrt[4]{E_1E_2}}{4\sqrt{A}} e^{\Phi(z_0)} f(z_0) \int dz\,(z - z_0)^{1/2} \times \\ \times\left[1 + \frac{E_1 + E_2}{4A}(z - z_0)\right]\left[1 + \left(\frac{df}{dz}\frac{1}{f}\right)_{z_0}(z - z_0)\right] \exp\alpha(z - z_0)^{3/2},$$

where

$$\alpha = \frac{2}{3\sqrt{2A}}(E_1 - E_2).$$

Here we used the first two terms in the expansion of the integral near z_0. Integration over the curve of constant imaginary part near z_0 corresponds to the replacement $z - z_0 = \rho e^{i\psi_1}$ and $z - z_0 = \rho e^{i\psi_2}$ before and after the kink respectively, where $\psi_1 = -2\pi/3$, $\psi_2 = 2\pi/3$. It is easily seen that the first term in the square bracket in the integrand drops out (the contributions from the integrals before and after the kink exactly cancel.) Thus we get

$$I^+ = \frac{\sqrt[4]{E_1E_2}}{4\sqrt{A}} e^{\Phi(z_0)} f(z_0)\left[\frac{E_1 + E_2}{4A} + \left(\frac{df}{dz}\frac{1}{f}\right)_{z_0}\right]\int_0^\infty e^{-\alpha\rho^{3/2}}\rho^{3/2}d\rho,$$

or

$$I^+ \sim i\sqrt[4]{E_1E_2}\left[\frac{E_1+E_2}{4A}+\left(\frac{df}{dz}\frac{1}{f}\right)_{z_0}\right]\frac{A^{1/3}}{(E_1-E_2)^{5/3}}f(z_0)\,e^{\Phi(z_0)}.$$

A similar calculation can be carried out for I^-; here the appropriate integration contour lies in the lower half-plane. If $f(x)$ is real, then $I^- = (I^+)^*$ and so $I = 2\,\mathrm{Re}\,I^+$. The criterion of applicability of the expression obtained is the condition that the next terms in the expansion of Φ in $z - z_0$ be small.

We will find $\Phi(z)$ for two cases: for the Coulomb potential $V_Q = Z/|x|$ and for a potential of the form

$$V_R(x) = \frac{V_0}{1+e^{\alpha(x^2-a^2)}}\,.$$

which for $\alpha a^2 \to \infty$ goes over into a square barrier of height V_0 and width $2a$. We first calculate the quantity $\Phi_1 = \int_{z_0}^{a}\sqrt{2(V-E)}\,dz$. Introducing the variable $\xi = E/V$, we write Φ_1 in the form

$$\Phi_1 = \sqrt{2E}\int_0^1 \sqrt{1-\xi}\,\frac{1}{\xi^{3/2}}\,\frac{d\xi}{d\xi/dz}\,.$$

In the case of the Coulomb potential we have $z_0 = 0$, $\xi = (E/Z)z$, $\Phi_1 = \pi Z/\sqrt{2E}$, and so $\Phi(0) = -\pi Z(v_2^{-1} - v_1^{-1})$. In the case of the potential V_R the nearest singularity to the real axis in the upper half-plane is the pole defined by the condition

$$e^{\alpha(z_0^2-a^2)} = e^{\pm i\pi}.$$

Assuming that $\alpha a^2 \gg 1$, we get

$$z_0 = \pm a + \frac{i\pi}{2a\alpha}\,.$$

As we see, the largest contribution comes from the pole at $z_0 = a + \frac{i\pi}{2a\alpha}$. Near this pole the variable ξ is related to z by the equation

$$\xi = \xi_0 (1 + e^{\beta(z-a)}),$$

where $\xi_0 = E/V$, $\beta = 2a\alpha$.

For Φ_1 we get

$$\Phi_1 = \frac{\sqrt{2E}}{\beta} \int_0^1 \sqrt{\frac{1}{\xi} - 1}\, \frac{d\xi}{\xi - \xi_0}.$$

To evaluate this integral we write $\sqrt{\xi^{-1} - 1} = t$ and suppose that $\xi_0 > 1$. We have

$$I = \int_0^1 \sqrt{\frac{1}{\xi} - 1}\, \frac{d\xi}{\xi - \xi_0} = 2 \int_0^\infty \frac{t^2 dt}{(1+t^2)\,(1 - \xi_0(1+t^2))}.$$

We extend the integral from $-\infty$ to ∞ and shift the contour into the upper half-plane. The integral reduces to the sum of the residues at the poles $t_1 = i$ and $t_2 = i\sqrt{1 - \xi_0^{-1}}$. Thus we find

$$I = \pi\left(\sqrt{1 - \frac{1}{\xi_0}} - 1\right),$$

and hence if $\xi_0 > 1$ $(E > V_0)$ we get for Φ_1

$$\Phi_1 = \frac{\pi\sqrt{2}}{\beta}(\sqrt{E - V_0} - \sqrt{E}).$$

We must make the transition to the case $V_0 > E$ by moving around a circle around the point $E = V_0$ in the upper half-plane, that is, by replacing $E - V_0$ by $|E - V_0|e^{i\pi}$. Thus for $E < V_0$ we find

$$\Phi_1 = -\frac{\pi\sqrt{2}}{\beta}\sqrt{E} + \frac{i\pi\sqrt{2}}{\beta}\sqrt{V_0 - E}.$$

When the two states Ψ_1 and Ψ_2 both have energies less than the height of the barrier (E_1, $E_2 < V_0$), then $\Phi(z_0)$ is given by

$$\Phi(z_0) = = -\frac{\pi\sqrt{2}}{\beta}(\sqrt{E_1}-\sqrt{E_2})+\frac{i\pi\sqrt{2}}{\beta}(\sqrt{V_0-E_1}-\sqrt{V_0-E_2}).$$

It is significant that in contrast to the Coulomb case, the exponential smallness of the matrix element is determined not by the penetrability of the whole barrier, but by the penetrability of a narrow transition layer of width $\sim 1/\beta$. In the limit $\beta\to\infty$ the singularity approaches the real axis and the matrix element contains only terms falling off according to a power law (not exponentially).

Barrier Penetration

It is possible to formulate the problem of the penetration of a potential barrier by a particle in two different ways. One corresponds to the case in which a steady current of particles is incident on the potential barrier, and it is required to find the amplitudes of the transmitted and reflected waves. A physical example of this case is provided by the field emission of electrons from a metal under the influence of an electric field; here we get a stationary current of electrons transmitted through the potential barrier (cf. the problem posed below).

The second formulation of the problem corresponds to the case when at the initial time a particle is known to be inside a potential well and it is required to find its state after a time t, and in particular to find the probability of observing it to be still in its original state at this later time. A physical example is the problem of α-decay; as a result of a nuclear reaction is produced a "parent" nucleus, which subsequently decays into a "daughter" nucleus and an α-particle.*

First let us consider the stationary formulation of the problem. Assume that the particles are incident from the left. Then the solution for $x \to \pm\infty$ has the form

$$\Psi_{x\to-\infty} = e^{ikx} + ae^{-ikx},$$
$$\Psi_{x\to\infty} = be^{ikx}.$$

The coefficients a and b (the amplitudes of the transmitted and reflected waves) are connected by a relation arising from conservation of particle number:

$$1 - |a|^2 = |b|^2.$$

For small barrier penetrability ($|b| \to 0$), the solution to the left of the barrier is close to a standing wave. We have assumed above that the potential decreases sufficiently fast for $x \to \pm\infty$; if the potential actually tends to different values for $x \to \infty$ and $x \to -\infty$, the only difference is that the momentum of the transmitted wave is different from that of the incident wave, and the condition of conservation of current gives $1 - |a|^2 = |b^2| (k'/k)$, where k' is the momentum of the transmitted wave and k that of the incident one.

The coefficients a and b may be obtained by matching the quasiclassical solutions outside and inside the barrier, with the help of formulae (3.9) given on p. 157 .

---

* This formulation of the problem was first analysed by N. S. Krylov and V. A. Fok, Zh. Eksp. Teor. Fiz. 17, 93 (1947).

(It is very useful to start by finding the coefficients a and b for the case of a square barrier. Although this problem is solved in many books on quantum mechanics, we recommend the reader to carry out the calculation for himself.)

Consider the case of a barrier with low penetrability ($|b| \to 0$). As we see from formulae (3.9), the solution inside the barrier, after matching at the right-hand edge (cf. Fig. 29) has the form

$$\Psi_{II} = be^{i\frac{\pi}{4}} \sqrt{\frac{k_0}{|k|}}\, \frac{1}{i} \left\{ \exp \int_x^{x_2} |k|\, dx + \frac{i}{2} \exp\left(- \int_x^{x_2} |k|\, dx \right) \right\},$$

where k_0 is the momentum at infinity. Near $x = x_1$ the second term will be exponentially small compared to the first and consequently has only a small effect on the matching condition at the

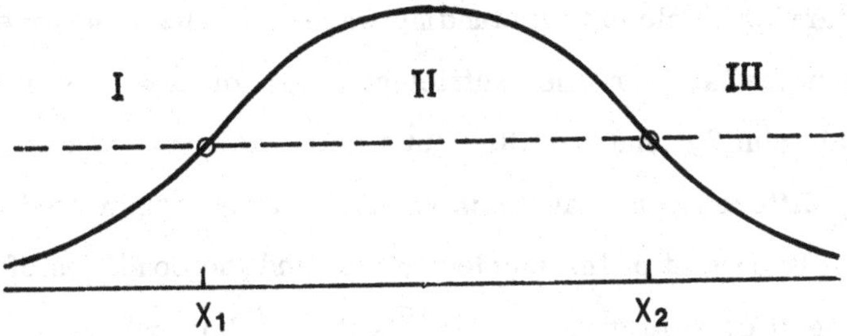

Fig. 29

boundary of regions I and II. Representing the solution inside the barrier in the form

$$\frac{B}{\sqrt{|k|}} \exp\left(- \int_{x_1}^{x} |k|\, dx \right),$$

where B is given by

$$B = be^{-\frac{i\pi}{4}}\sqrt{k_0}\exp\left(\int_{x_1}^{x_2}|k|dx\right),$$

we obtain after matching at the point $x = x_1$ the solution to the left of the barrier:

$$\Psi_I = \frac{2B}{\sqrt{k}}\cos\left(\int_{x_1}^{x} k\,dx - \frac{\pi}{4}\right).$$

The requirement that the coefficient of the incident wave be equal to 1 gives

$$|b|^2 = e^{-2\gamma}, \quad |a|^2 = 1 - e^{-2\gamma},$$

where $\gamma = \int_{x_1}^{x_2} |k|\,dx$. The quantity $e^{-2\gamma}$ is usually called the penetrability of the barrier.

The problem of field emission of electrons mentioned above reduces to the calculation of the penetrability of the barrier shown in Fig. 30. The depth of the potential well for electrons in the

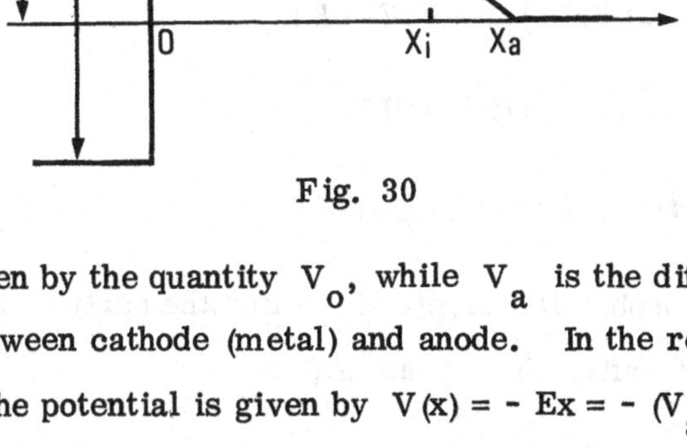

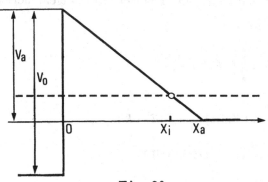

Fig. 30

metal is given by the quantity V_o, while V_a is the difference in potential between cathode (metal) and anode. In the region between 0 and x_a the potential is given by $V(x) = -Ex = -(V_a/x_a)x$. The problem is to find the penetrability of the barrier.

We now turn to the second possible formulation of the barrier-transmission problem. Suppose that at the initial time the particle is known to be in a state close to some stationary state which would be obtained for a completely impenetrable barrier (e.g. a barrier which far from the well tended to a height greater than the energy of the particle to be emitted.) Since the particle in fact has a positive energy E_0, its state is described by a superposition of eigenfunctions of the continuous spectrum, and the problem reduces to that of the spreading of such a wave packet. Since the state is only nonstationary by virtue of the finite penetrability of the barrier, it follows that for a weakly penetrable barrier the wave packet describing the initial state will spread only slowly. An energy level corresponding to such a state is called quasi-stationary (cf. p. 237).

We expand the initial state, which describes the particle inside the potential well, with respect to the eigenfunctions of the continuous spectrum:

$$\Psi_0(x) = \int_0^\infty C(E)\,\Psi_E(x)\,dE,$$

where

$$\int_{-\infty}^{\infty} \Psi_E(x)\,\Psi_{E'}(x)\,dx = \delta(E - E').$$

At time t the wave function will be

$$\Psi(x,t) = \int_0^\infty C(E)\,e^{-iEt}\,\Psi_E(x)\,dE.$$

We define the probability amplitude to find the particle in the initial state Ψ_0 after time t as $\mathcal{L}(t)$:

$$\mathcal{L}(t) = (\Psi(x,t);\ \Psi_0(x)) = \int_0^\infty |C(E)|^2 e^{-iEt} dE.$$

As we shall see below (p. 245) for energy E near E_0 the wave functions of the continuous spectrum have near the well the form

$$\Psi_E = \chi(E)\, \Psi_0(x),$$

that is, they can be decomposed into a product of a function of energy $\chi(E)$ and the function Ψ_0 which describes the stationary state in the case of an impenetrable barrier. The function $\chi(E)$ has a pole in the complex E-plane for $E = E_0 - i\Gamma_0$, where Γ_0, the width of the quasi-stationary level, is proportional to the penetrability of the barrier. It is easy to find the relation between $\chi(E)$ and the quantity C(E) which characterizes the energy distribution of the initial state. In fact we have

$$C(E) = \int \Psi_E^*(x)\, \Psi_0(x)\, dx = \chi^*(E).$$

Let us return now to the calculation of the quantity $\mathcal{L}(t)$ which determines the probability of finding the particle still in its initial state. Since the function C(E) has a pole near the real axis and obeys the normalization condition, we have for $|C(E)|^2 = |\chi(E)|^2$ the approximate expression

$$|C(E)|^2 \simeq \frac{\Gamma_0}{(E-E_0)^2 + \Gamma_0^2}\,\frac{1}{\pi}.$$

We shift the integration contour in the expression for $\mathcal{L}(t)$ into the lower half-plane (cf. Fig. 31). The integration over the regions labelled 2 gives a contribution which falls off exponentially as the

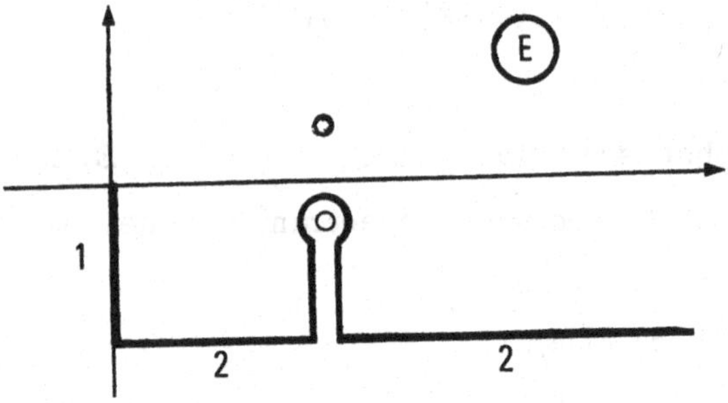

Fig. 31

contour is shifted out into the lower half-plane, and we are left with the residue at the pole and the integral over the negative real axis. As a result we have

$$\mathscr{L}(t) = e^{-iE_0 t} e^{-\Gamma_0 t} - i \int_0^\infty |C(-i\xi)|^2 e^{-\xi t} d\xi.$$

For large t the second term is determined by the behaviour of $|C(E)|^2$ at low energies. To estimate the quantity $|C(E)|^2$ in this case, we must be given the method of preparation of the initial state. For definiteness we shall discuss the case of α-decay, and consider some specific reaction leading to the formation of a parent nucleus which will subsequently decay into a daughter nucleus and an α-particle with energy E_0. In parallel with this process there also takes place, in the course of the same reaction, direct formation of daughter nuclei and α-particles. The presence, in the spectral decomposition of the initial state, of α-particles with energies E such that $(E - E_0)/\gamma \gg 1$ just corresponds to those daughter nuclei and α-particles which originated in the formation of the original state (i.e. in the course of the reaction). The number of particles with energy E far from E_0 is proportional to the barrier penetrability $e^{-2\gamma(E)}$, and in the three-dimensional case also to the density of final states, which is $\sim k^2\, dk/dE \sim \sqrt{E}$. Thus the number of α-particles of low energy occurring in the reaction process is given by

$$|C(E)|^2 = A\sqrt{E}e^{-2\gamma(E)}.$$

For $E \sim E_0$ this expression must go over into the formula given above. Introducing the quantity $\Gamma(E)$, we can write $|C(E)|^2$ in a form which will be applicable even at low energies:

$$|C(E)|^2 = \frac{\Gamma(E)}{(E-E_0)^2 + \Gamma^2(E)},$$
$$\Gamma(E) \simeq \sqrt{EE_0}\, e^{-2\gamma(E)}.$$

It is now easy to estimate the correction to $\mathcal{L}$. In the case of α-decay the barrier penetrability tends for $E \to 0$ to the value $e^{-2\gamma(E)} = \exp(-2\pi Z/v)$, and hence for $t \to \infty$ we can find the integral by the method of steepest descents. The reader is recommended to verify that the order of magnitude of the second term in $\mathcal{L}$ is given by

$$|\mathcal{L}_1| \sim \sqrt{E_0}\frac{Z^{2/3}}{t^{7/6}} \exp\left[-\frac{(2\pi)^{2/3}\, 3\sqrt{3}}{4} Z^{2/3} t^{1/3}\right].$$

For the case when the barrier penetrability tends to a finite limit for $E \to 0$, we have

$$|\mathcal{L}_1| \sim e^{-2\gamma(0)}\sqrt{E_0}/t^{3/2}.$$

As we see from the above expressions, the correction term describes the spreading of the wave packet of α-particles formed in the course of the initial reaction, and has no connection with the process of decay of the parent nucleus. Were we to determine the number of parent and daughter nuclei not by the number of α-particles but by (e. g.) the form of the atomic spectra observed, then the number of parent nuclei determined in this way would decrease simply as $e^{-2\Gamma_o t}$. For, if it is established at some time that the atomic spectrum corresponds to the parent nucleus, then the energy of the latter is ipso facto determined to within an atomic linewidth; hence the spectral decomposition of the α-particles contains no terms with low energy and at large t only the first term in $\mathcal{L}$ is left.*

Reflection Above a Barrier

Let us solve the problem of the reflection from a barrier of a particle moving with energy greater than the barrier height, so that the criterion of applicability of the quasiclassical approximation is everywhere fulfilled. It is easy to verify that no reflection of the wave occurs to any order in the "quasiclassicality" parameter $1/(k\ell)^2$. In fact the quasiclassical approximation for a wave travelling to the right has the form

$$A \exp\left[i \int_{x_1}^{x} \sqrt{k^2 + A''/A}\, dx \right].$$

Expansion of this solution in powers of A''/A gives again only waves travelling to the right. The reason for this result lies in the fact that the reflection coefficient decreases exponentially with increasing $(k\ell)^2$, while the quasiclassical solution is an asymptotic

---

* A more detailed analysis and references to earlier work may be found in A. Degasperis, L. Fonda and G. Ghirardi, Nuovo Cimento 21A, 471 (1974).

series and exponentially small terms are completely lost in it.

It is convenient to look for the solution of the Schrödinger equation

$$\Psi'' + k^2(x)\Psi = 0 \qquad (3.23)$$

in the form

$$\Psi = \sqrt{\frac{k_0}{k}} \exp\left(i \int_{-\infty}^{x} k\, dx\right) + \varphi,$$

where φ is a small correction term containing the reflected wave. Then we get

$$\varphi'' + k^2\varphi = f(x),$$

where

$$f(x) = -\left(\sqrt{\frac{k_0}{k}}\right)'' \exp\left[i \int_{-\infty}^{x} k\, dx\right].$$

We can express the solution of this inhomogeneous equation in terms of the two solutions Ψ_1 and Ψ_2 of the corresponding homogeneous equation (where Ψ_1 and Ψ_2 are respectively the waves travelling to right and to left):

$$\varphi = \frac{1}{\Delta}\left\{\Psi_1 \int^{x} \Psi_2 f\, dx - \Psi_2 \int^{x} \Psi_1 f\, dx\right\},$$

where

$$\Delta = \Psi_1'\Psi_2 - \Psi_1\Psi_2' = \text{const}$$

is the Wronskian. Thus the solution of the Schrödinger equation (3.23) has the form

$$\Psi(x) = \sqrt{\frac{k_0}{k}} \exp\left[i \int_{-\infty}^{x} k\,dx\right] +$$
$$+ \frac{1}{\Delta}\left\{\Psi_1 \int_{-\infty}^{x} \Psi_2 f\,dx - \Psi_2 \int_{\infty}^{x} \Psi_1 f\,dx\right\}.$$

The limits of integration are determined by the following considerations. In the limit $x \to \infty$ we must be left with only a wave travelling to the right; this is guaranteed by the function Ψ_1. For $x \to -\infty$ the function $\varphi(-\infty)$ should contain only a reflected wave. With the above choice of lower limit of integration we get

$$\Psi(x) \underset{x \to -\infty}{=} e^{ik_0x + i\alpha} + \frac{1}{\Delta} e^{-ik_0x} \int_{-\infty}^{\infty} \Psi_1 f\,dx.$$

Since Δ = constant, it can be calculated by taking $x \to \infty$. Then $\Psi_1 \to \exp(ik_0x)$, while the function Ψ_2, owing to the smallness of the reflection coefficient, tends approximately to $\exp(-ik_0x + i\alpha)$. Hence we get $|\Delta| = 2k_0$.

Thus the reflection coefficient is given by

$$R = \left|\frac{1}{\Delta} \int_{-\infty}^{\infty} \Psi_1 f\,dx\right|^2 = \left|\frac{1}{2k_0} \int_{-\infty}^{\infty} \Psi_1 f\,dx\right|^2. \tag{3.24}$$

Note that in deriving (3.24) we nowhere used the quasiclassical approximation.

In the case $V \ll E$, (3.24) gives just the usual perturbation formula. We can see this as follows. We have

$$\left(\sqrt{\frac{k_0}{k}}\right)' = -\frac{\sqrt{k_0}}{2\sqrt{k^3}} k'.$$

Since $k = 2\sqrt{E-V}$, it follows that $k' = -V'/k$, and so

$$f = -\left(\sqrt{\frac{k_0}{k}}\right)'' e^{iS} \approx \frac{V''}{2k_0^2} e^{ik_0x + i\alpha}$$

and

$$R = \left| \frac{1}{4k_0^3} \int_{-\infty}^{\infty} V'' e^{2ik_0x} dx \right|^2.$$

Integrating twice by parts, we get an expression for the reflection coefficient which is identical to that obtained in perturbation theory:

$$R = \left| \frac{1}{k_0} \int_{-\infty}^{\infty} V e^{2ik_0x} dx \right|^2.$$

In general the quantity R is determined by the behaviour of $\Psi_1 f$ in the neighbourhood of the singularity closest to the real axis. The argument of the exponential in R is determined by the distance of this singularity from the real axis; to find the pre-exponential factor it is necessary to use the exact solution for Ψ near the singularity,* similarly to the procedure employed above to match quasiclassical wave functions (see also the remark on p. 118).

2. THE THREE-DIMENSIONAL CASE

Spherically Symmetric Field

Let us consider quasiclassical motion in a spherically symmetric field. In this case the angular variables θ, φ separate

---

* V.L. Pokrovskii and I.M. Khalatnikov, Zh. Eksp. Teor. Fiz. 40, 1713 (1961). (English translation: Soviet Physics JETP, 13, 207 (1961).

from the variable r:

$$\Psi_{nlm} = R_{nl}(r) Y_{lm}(\theta, \varphi).$$

Consider first of all the radial part of the wave function. We introduce the function

$$u_{nl}(r) = r R_{nl}(r).$$

From the condition that $R_{n\ell}(r)$ should be finite at $r = 0$ it follows that $u_{n\ell}(0) = 0$. In the standard way we get for $u_{n\ell}$ the one-dimensional equation

$$u''_{nl} + k_{nl}^2 u_{nl} = 0,$$

where

$$k_{nl}^2 = 2\left[\varepsilon_{nl} - V(r) - \frac{l(l+1)}{2r^2}\right].$$

The condition $u_{n\ell}(0) = 0$ corresponds in the one-dimensional problem to an infinite potential wall at the origin.

As we increase ℓ for a given value of $V(r)$ the potential well in Fig. 32(b) becomes shallower and shallower and finally disappears altogether, so that for sufficiently large ℓ the system

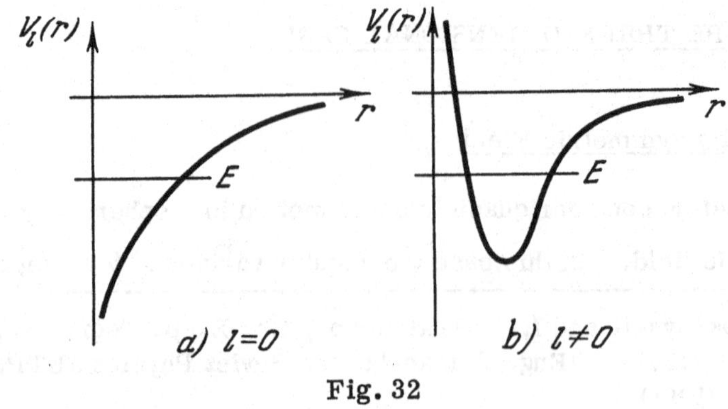

Fig. 32

has no discrete levels. Positive values of $E_{n\ell}$ correspond to the continuous spectrum.

Modification of the Centrifugal Potential

We will show that if the quasiclassical approximation is to be used the quantity $\ell(\ell+1)$ must be replaced by $(\ell+\frac{1}{2})^2$. This is known as the Langer correction.

Consider the region near the origin. In this region only the centrifugal potential is important and hence $k \sim \sqrt{\ell(\ell+1)}/r$, and so $d\lambda/dr \sim (\ell(\ell+1))^{-\frac{1}{2}}$. For not too large ℓ we have $d\lambda/dr \sim 1$, so that the quasiclassical condition is not fulfilled near the origin. Let us see how to obtain a solution which will be valid near the origin for arbitrary ℓ. To do this, we change both the independent and dependent variable in Schrödinger's equation, putting $r = e^{-x}$ and $u = we^{-x/2}$. Then the equation takes the form

$$\frac{d^2w}{dx^2} + 2\left[(\varepsilon_{nl} - V)\,e^{-2x} - \frac{(l+{}^1\!/_2)^2}{2}\right] w = 0.$$

The point $r = 0$ corresponds to $x = \infty$. For $x \to \infty$ the quasi-classical condition looks as follows

$$\frac{d\lambda}{dx} = \frac{d}{dx}\,\frac{1}{\sqrt{2\,(\varepsilon_{nl} - V)\,e^{-2x} - (l+{}^1\!/_2)^2}} \sim e^{-2x} \to 0.$$

Thus the condition $d\lambda/dx \ll 1$ is fulfilled independently of the value of ℓ. If now we formulate the Bohr quantization rule in terms of the variable x, and then translate it back into the variable r, it takes its usual form except that now the effective potential V_{eff} is given by $V + (\ell+\frac{1}{2})^2/2r^2$.

After this replacement the criterion of applicability of the quasiclassical method to find the energy levels (by the Bohr quantization rule) consists only in the condition $n_r \gg 1$ (i.e. the radial wave function must have many nodes); the condition $\ell \gg 1$ is not necessary. Obviously, the correction should not be added in the case $\ell = 0$.

Energy Levels in the Coulomb Potential

Let us find the energy levels in the Coulomb potential. The quantization condition has the form (for $\ell \neq 0$, corresponding to Fig. 32(b)

$$\int_{r_{min}}^{r_{max}} \sqrt{2\left(E_n + \frac{Z}{r} - \frac{(l+1/2)^2}{2r^2}\right)}\, dr = \left(n_r + \frac{1}{2}\right)\pi.$$

Here n_r is the radial quantum number and r_{min}, r_{max} the quasiclassical turning points. We have replaced the factor $\ell(\ell+1)$ in the centrifugal potential term by $(\ell + 1/2)^2$. The integral on the left-hand side of the above condition is equal to

$$\frac{Z\pi}{\sqrt{-2E_n}} - \left(l + \frac{1}{2}\right)\pi.$$

Thus,

$$E_n = -\frac{Z^2}{2n^2},$$

where $n = n_r + \ell + 1$ is the principal quantum number.

For $\ell = 0$ (Fig. 32(a)) the left-hand turning point is $r_{min} = 0$ and coincides with a singularity of the equation, so that it is not legitimate to replace the potential near the left-hand turning point by a linear one as we did above. Thus, the matching condition at the left-hand edge is altered. To find the energy levels for $\ell = 0$

we must find the exact solution near the singularity at $r = 0$ and match it with the quasiclassical solution far from the origin. For large quantum numbers (energies near zero) the $\ell = 0$ wave functions in the Coulomb field have the form*

$$R_{n0} = a_1 Z^{3/2} \frac{J_1(\sqrt{8Zr})}{\sqrt{Zr}},$$

where a_1 is a normalization coefficient. We therefore get for $r \gg 1/Z$

$$R_{n0} = a_0 \frac{\cos(\sqrt{8Zr} - 3/4\,\pi)}{r^{3/4}}.$$

On the other hand, the quasiclassical solution for R_{no} can be written

$$R_{n0} = a_2 \frac{\cos\left(\int_0^r p_r\,dr - C_1\pi\right)}{r\sqrt{p_r}} = a_2 \frac{\cos(\sqrt{8Zr} - C_1\pi)}{r\sqrt{p_r}}.$$

Thus, the phase appearing here is $C_1\pi = (3/4)\pi$ instead of $(1/4)\pi$ as in the usual case.

The function R_{no} can equally well be written in the form

$$R_{n0} = a_3 \frac{\cos\left(\int_r^{r_{\max}} p_r\,dr - \pi/4\right)}{r\sqrt{p_r}}.$$

where r_{max} is the right-hand turning point. As in Section 3.1, we then obtain from the condition that R_{no} should be single valued the quantization rule

---

* L.D. Landau and E.M. Lifshitz, Quantum Mechanics (Pergamon Press, Oxford, 1965), p. 122.

$$\int_0^{r_{max}} p_r\,dr = (n_r + 1)\,\pi = n\pi,$$

where n is the principal quantum number. Calculation of the integral gives

$$E_n = -\frac{Z^2}{2n^2}\,.$$

As we see, in this case (as in the case of the harmonic oscillator) the quasiclassical approximation gives the exact result for the energy levels.

Quasiclassical Representation of Spherical Functions

To calculate matrix elements of physical quantities for large angular momenta ℓ it is convenient to use the quasiclassical representation for the spherical functions

$$Y_{lm}(\theta, \varphi) = \frac{1}{\sqrt{2\pi}} P_{lm}(\cos\theta)\, e^{im\varphi}.$$

Here $P_{\ell m}(x)$ is the associated Legendre polynomial, which satisfies the equation

$$\frac{d}{dx}\left[(1-x^2)\frac{dP_{lm}}{dx}\right] + \left[l(l+1) - \frac{m^2}{1-x^2}\right]P_{lm} = 0.$$

We introduce the function $\theta_{\ell m}(x)$ by

$$P_{lm}(x) = \frac{1}{\sqrt{1-x^2}}\,\vartheta_{lm}(x).$$

Then the equation for $\theta_{\ell m}(x)$ is

$$(1-x^2)\,\vartheta''_{lm}(x) + \left[l(l+1) - \frac{m^2-1}{1-x^2}\right]\vartheta_{lm}(x) = 0.$$

With an accuracy of order $1/\ell^2$ the solution of this equation can be found in the quasiclassical form

$$\vartheta_{lm}(x) = \frac{a}{\sqrt{k}} \cos\left(\int_{x_1}^{x} k\,dx - \frac{\pi}{4}\right),$$

where

$$k^2(x) = \frac{1}{1-x^2}\left[\left(l+\frac{1}{2}\right)^2 - \frac{m^2}{1-x^2}\right].$$

The quantity $k(x)$ tends to zero at the points x_1, x_2, that is,

$$x_{1,2} = \mp\sqrt{1-\frac{m^2}{(l+1/2)^2}}.$$

Here we have replaced $\ell(\ell+1)$ by $(\ell+1/2)^2$, remembering that the quasiclassical approximation is valid to order $1/\ell^2$ (not $1/\ell$: cf. p. 162).

The coefficient a is determined by the normalization condition for the associated Legendre polynomials

$$\int_{-1}^{1} P_{lm}^2(x)\,dx = 1,$$

that is,

$$\frac{a^2}{2}\int_{x_1}^{x_2} \frac{1}{1-x^2}\frac{dx}{k} = 1 = \frac{a^2}{2}\frac{1}{l+\frac{1}{2}}\frac{\partial}{\partial l}\int_{x_1}^{x_2} k\,dx.$$

Calculating $\int_{x_1}^{x_2} k\,dx$, we find

$$\int_{x_1}^{x_2}\sqrt{\frac{1}{(1-x^2)}\left[\left(l+\frac{1}{2}\right)^2-\frac{m^2}{1-x^2}\right]}\,dx = \left(l-|m|+\frac{1}{2}\right)\pi. \tag{3.25}$$

This result was to be expected, since we know that the number of nodes of $P_{\ell m}(x)$ is $\ell - |m|$. From (3.25) we get

$$\frac{\partial}{\partial l}\int_{x_1}^{x_2} k\,dx = \pi.$$

and so

$$a^2 = \frac{2l+1}{\pi}.$$

Thus we finally obtain

$$P_{lm}(x) \underset{x_1<x<x_2}{=} \sqrt{\frac{2l+1}{\pi}}\frac{1}{\sqrt{1-x^2}}\frac{1}{\sqrt{k}}\cos\left(\int_{x_1}^{x} k\,dx - \frac{\pi}{4}\right),$$

where

$$k = \frac{1}{\sqrt{1-x^2}}\sqrt{\left(l+\frac{1}{2}\right)^2 - \frac{m^2}{1-x^2}}.$$

This quasiclassical representation of the spherical functions enables us to calculate approximately sums of the form

$$I_l(\theta) = \sum_{m=-l}^{l} |Y_{lm}(\theta, \varphi)|^2 \Phi(m).$$

Using the quantization condition (3.25) and replacing the summation by integration, we get

$$I_l(\theta) = \frac{l+{}^1/_2}{2\pi^2}\int_{-(l+{}^1/_2)|\sin\theta|}^{(l+{}^1/_2)|\sin\theta|}\frac{\Phi(m)\,dm}{(l+{}^1/_2)\sin\theta\sqrt{1-\frac{m^2}{(l+{}^1/_2)^2\sin^2\theta}}} =$$

$$= \frac{l+{}^1/_2}{2\pi^2}\int_{-1}^{1}\frac{dt}{\sqrt{1-t^2}}\Phi[t(l+{}^1/_2)\sin\theta].$$

In particular, for $\Phi(m) = 1,\ m^2,\ m^4$, we get

$$\sum_{m=-l}^{l} |Y_{lm}|^2 = \frac{l+1/2}{2\pi^2}\int_{-1}^{1}\frac{dt}{\sqrt{1-t^2}} = \frac{2l+1}{4\pi},$$

$$\sum_{m=-l}^{l} |Y_{lm}|^2 m^2 = \frac{l+1/2}{2\pi^2}\int_{-1}^{1}\frac{dt}{\sqrt{1-t^2}}\, t^2 (l+1/2)^2 \sin^2\theta = \frac{(l+1/2)^3}{4\pi}\sin^2\theta$$

$$\sum_{m=-l}^{l} |Y_{lm}|^2 m^4 = \frac{l+1/2}{2\pi^2}\int_{-1}^{1}\frac{dt}{\sqrt{1-t^2}}\, t^4 (l+1/2)^4 \sin^4\theta = \frac{3(l+1/2)^5}{16\pi}\sin^4\theta,$$

and also

$$\int_{-1}^{1} P_{lm}^2 x^2\, dx = \frac{1}{2}\left(1 - \frac{m^2}{(l+1/2)^2}\right).$$

We recommend the reader to use these formulae to calculate the levels of a deformed nucleus (cf. Chapter 2, Section 2).

The Thomas-Fermi Distribution in the Atom

The potential energy V of an electron can be described by the electrostatic equation

$$\nabla^2 V = -4\pi\rho \tag{3.26}$$

where ρ is the density of electrons at the point in question. This equation is actually an approximation, since in reality we have a many-particle problem, involving the nucleus and Z electrons, and the potential

$$V(\boldsymbol{r}, \boldsymbol{r}_i) = -\frac{Z}{r} + \sum_{i=1}^{Z}\frac{1}{|\boldsymbol{r}-\boldsymbol{r}_i|}$$

is an operator rather than a c number, that is V depends on the electron coordinates, while ρ clearly does not. Consequently the potential occurring in (3.26) is the true potential averaged over the motion of the electrons. Since the number of particles is large, the deviation of $V(\mathbf{r}, \mathbf{r}_i)$ from $V(r)$ will be small on the average (in spite of the fact that formally $V(\mathbf{r}, \mathbf{r}_i)$ may take arbitrary values).

Schrödinger's equation for the motion in this averaged potential reads

$$\nabla^2\Psi + 2(E - V)\Psi = 0 \tag{3.27}$$

where we can take the potential V to be spherically symmetric (the only factor which would tend to destroy the spherical symmetry is the presence of electrons in outer unfilled shells). Then we can write

$$\Psi_{nlm} = \frac{u_{nl}(r)}{r} Y_{lm}.$$

The electron density is given by the expression

$$\rho = 2\sum_{nlm} |\Psi_{nlm}|^2 = 2\sum_{nlm} \frac{u_{nl}^2}{r^2} |Y_{lm}|^2. \tag{3.28}$$

In view of the relation

$$\sum_{m=-l}^{l} |Y_{lm}|^2 = \frac{2l+1}{4\pi}$$

the summation over m leads to

$$\rho(r) = 2\sum \frac{u_{nl}^2(r)}{r^2} \frac{2l+1}{4\pi}. \tag{3.29}$$

In the quasiclassical approximation the radial wave functions have the form

$$u_{nl}(r) = \frac{a_{nl}}{\sqrt{k_{nl}}} \cos\left(\int_{r_1}^{r} k_{nl}\,dr - \frac{\pi}{4}\right), \tag{3.30}$$

where

$$k_{nl}^2(r) = 2\left(E_{nl} - V(r) - \frac{(l+1/2)^2}{2r^2}\right), \quad a_{nl}^2 \approx \frac{2}{\pi}\frac{\partial E_{nl}}{\partial n}.$$

The energy levels are found from the Bohr quantization rule

$$\int_{r_1}^{r_2} k_{nl}\,dr = \left(n + \frac{1}{2}\right)\pi. \tag{3.31}$$

In the case $\ell = 0$ the $(n + 1/2)$ in Eq. (3.31) should be replaced by n (cf. p. 193). Combining (3.29) and (3.30), we get

$$\rho(r) = \frac{1}{2\pi}\sum_{nl}\frac{a_{nl}^2}{k_{nl}r^2}(2l+1)\cos^2\left(\int_{r_1}^{r} k_{nl}\,dr - \frac{\pi}{4}\right). \tag{3.32}$$

In the quasiclassical approximation we can replace the mean value of the squared cosine by 1/2. Then we get from (3.32)

$$\rho(r) = \frac{1}{2\pi^2}\sum_{nl}\frac{\partial E_{nl}}{\partial n}\frac{2l+1}{r^2}\frac{1}{k_{nl}}. \tag{3.33}$$

We change the summation over n in (3.33) to integration, keeping ℓ fixed; since

$$\sum_n \frac{\partial E_{nl}}{\partial n}(\ldots) \to \int \frac{\partial E_{nl}}{\partial n}\,dn\,(\ldots) \to \int dE\,(\ldots),$$

we get

$$\sum_n \frac{\partial E_{nl}}{\partial n}\frac{1}{k_{nl}} = \int \frac{dE}{k_{nl}} = \int \frac{dE}{\sqrt{2(E - V_l)}} = \sqrt{2(E - V_l)}\,\Big|_{E_{min}}^{0},$$

where $E_{min} = V_\ell$ (for $E > V_\ell$ the wave function has the form $\cos\left(\int k\,dr - \pi/4\right)$ while for $E < V_\ell$ it decreases exponentially, and the contribution to ρ (r) from this region can be neglected). Thus we can write (3.33) in the form

$$\rho(r)=\frac{1}{2\pi^2}\sum_l \frac{2l+1}{r^2}\sqrt{-2V_l}. \tag{3.34}$$

Since we have $V_\ell = (1/2r^2)(\ell + 1/2)^2$, we get $2\, dV_\ell/d\ell = (2\ell+1)/r^2$, so that (3.34) becomes

$$\rho(r)=\frac{1}{\pi^2}\sum_l \frac{dV_l}{dl}\sqrt{-2V_l}=\frac{1}{\pi^2}\int\sqrt{-2V_l}\,dV_l= -\frac{2}{3\cdot 2\pi^2}(-2V_l)^{3/2}\Big|_{V_{l\min}}^{V_{l\max}}. \tag{3.35}$$

$V_{\ell_{min}}$ corresponds to $\ell = 0$, $V_{\ell_{min}} = V$, while the quantity $V_{\ell_{max}}$ (r) is zero. For states with ℓ larger than the maximum value so defined, the point r lies in the classically inaccessible region, so that such states give an exponentially small contribution to the density at the point r (see Fig. 33)

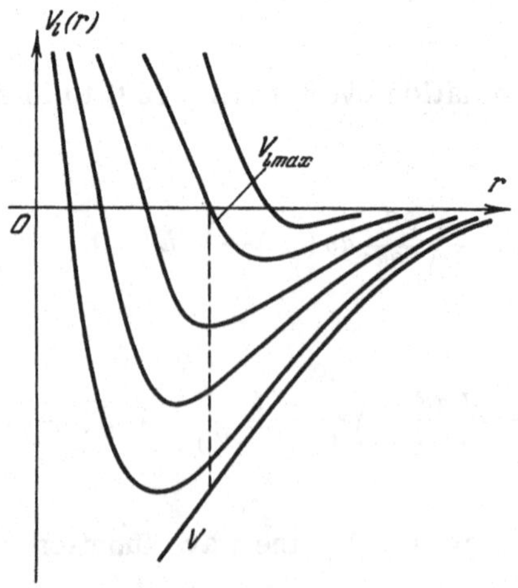

Fig. 33

Thus we finally have

$$\rho(r) = \frac{1}{3\pi^2}(-2V)^{3/2}. \qquad (3.36)$$

Setting $-V \equiv \varphi$, we can use (3.36) to write Poisson's equation (3.26) in the form

$$\nabla^2\varphi = 4\pi\rho = \frac{8\sqrt{2}}{3\pi}\varphi^{3/2} = C\varphi^{3/2} \qquad (3.37)$$

where $C \equiv 8\sqrt{2}/3\pi$. This equation is called the Thomas-Fermi equation.

When $r \to 0$, we have $\varphi \to Z/r$, the potential of the nucleus. It is convenient to substitute $\varphi(r) = Z\chi(r)/r$, with the boundary condition $\chi(0) = 1$. In spherical coordinates we then have

$$\nabla^2\varphi = \frac{1}{r}\frac{d^2}{dr^2}(r\varphi) = \frac{1}{r}\frac{d^2}{dr^2}(Z\chi) \qquad (3.38)$$

and so from (3.37)

$$\chi''(r) = C\frac{Z^{1/2}}{r^{1/2}}\chi^{3/2}. \qquad (3.39)$$

We introduce a new variable $x = \alpha Z^{1/3} r$ (where α will be chosen below) with a view to obtaining a universal equation for the function χ. Then we find from (3.39)

$$\alpha^2 Z^{2/3}\chi'' = CZ^{1/2}\frac{1}{x^{1/2}}\alpha^{1/2}Z^{1/6}\chi^{3/2},$$

so that if we choose $C = \alpha^{3/2}$, that is, $\alpha = (8\sqrt{2}/3\pi)^{2/3}$, we finally have

$$\left.\begin{aligned} &\frac{d^2\chi}{dx^2} = \frac{1}{\sqrt{x}}\chi^{3/2}, \\ &\chi(\infty) = 0,\ \chi(0) = 1. \end{aligned}\right\} \qquad (3.40)$$

Equation (3.40) can be solved numerically (an approximate

solution was given in Section 1.1). The solution determines the distribution of the electron density. The characteristic length describing this distribution is $x \sim 1$ in the dimensionless units used above, so that in ordinary units this "Thomas-Fermi radius" a_{TF} is $\sim a_o Z^{-1/3}$. The greater part of the atomic electrons are to be found within this radius. This result agrees with the approximate estimate given in Chapter 1 (p. 40).

Equation (3.40) is, of course, valid only for that region of the atom where the quasiclassical approximation is applicable.

> <u>Problem:</u> What are the minimum and maximum radii between which the Thomas-Fermi distribution is applicable ?
>
> <u>Solution</u>: $r_{min} \sim 1/Z, \; r_{max} \sim 1$

Estimates of Nuclear Matrix Elements

In nuclear theory we often have to calculate various sums of matrix elements. Here we consider which matrix elements give the largest contributions to such sums.

Let $U(r)$ be some quantity which changes appreciably over distances of the order of the nuclear radius. We will estimate the quasiclassical matrix element $U_{\lambda_1 \lambda_2} = (\varphi_{\lambda_1} \, U \varphi_{\lambda_2})$, where φ_λ is the wave function of a nucleon in the self-consistent field generated by all the other nucleons in the nucleus, and λ is the set of quantum numbers describing the state of the nucleon.

First consider a spherical nucleus. Then $\lambda \equiv n, \ell, j, m$, where n is the radial, ℓ the azimuthal and m the magnetic quantum number, and $j = \ell \pm \frac{1}{2}$. We will show that in a spherical nucleus neighbouring levels do not "combine", that is they give only a small value of the matrix element $U_{\lambda_1 \lambda_2}$. As a preliminary we estimate the spacing between neighbouring levels. In a deformed (nonspherical) nucleus levels with different components of angular momentum along the axis of symmetry are split in energy, so that if A is the number of nucleons in the nucleus and ϵ_F is the Fermi energy (that is, the difference in energy between the bottom of the self-consistent potential well and the levels occupied by the last nucleons), then the mean spacing between the single-particle levels is of order ϵ_F/A. The quantity ϵ_F does not depend on A to the same accuracy as the density n of nuclear matter is constant. To see this, note that in a system held together by forces acting between its particles, the range of the forces and the mean spacing between particles must be of the same order of magnitude. Hence all quantities characterizing the nucleus can be expressed in terms of one another by the use of dimensional estimates; in particular, $\epsilon_F \sim n^{2/3}$ (in a system of units where $M = \hbar = 1$).

In a spherical nucleus there exists degeneracy with respect to the projection of angular momentum, with a multiplicity of the order of the angular momentum $\ell \sim p_F R$, where R is the nuclear radius and p_F the momentum of a nucleon at the Fermi surface. Since $R = r_o A^{1/3}$, where r_o is a quantity of the order of the mean spacing between nucleons, we have $p_F R = p_F r_o A^{1/3} \sim A^{1/3}$. Thus the mean spacing in energy between neighbouring levels of a spherical nucleus is of order $\epsilon_F/A^{2/3}$.

Let us now first estimate the spacing between levels which differ only in their radial quantum numbers n, with $\delta n \sim 1$. For this purpose we differentiate the Bohr quantization condition $\int p_r\, dr \sim n$ (where p_r is the radial momentum) with respect to n, obtaining

$$1 \sim \int \frac{\partial p_r}{\partial n} dr = \int \frac{\partial p_r}{\partial \varepsilon_{nl}} \frac{\partial \varepsilon_{nl}}{\partial n} dr \sim \frac{\partial \varepsilon_{nl}}{\partial n} \int \frac{dr}{v} \sim \frac{\partial \varepsilon_{nl}}{\partial n} \frac{R}{v},$$

where $\epsilon_{n\ell}$ is the energy level, and $v = \partial \epsilon_{n\ell}/\partial p_r$ is the velocity of the nucleon. Hence we find

$$\frac{\partial \varepsilon_{nl}}{\partial n} \sim \frac{v}{R} = \frac{v}{r_0 A^{1/3}} \sim \frac{p_F v}{A^{1/3}} \sim \frac{\varepsilon_F}{A^{1/3}}.$$

Now let us estimate the difference in energy between levels which differ only in their azimuthal quantum numbers ℓ, with $\delta \ell \sim 1$. For this, we differentiate the relation $\int p_r\, dr \sim n$ with respect to ℓ, remembering that

$$p_r = \sqrt{2\left[\varepsilon_{nl} - V(r) - \frac{(l + 1/2)^2}{2r^2}\right]}.$$

where V(r) is the potential energy of the nucleon. In this way we get

$$\int \frac{dr}{p_r}\left(\frac{\partial \varepsilon_{nl}}{\partial l} - \frac{l}{r^2}\right) = 0,$$

i. e.

$$\frac{\partial \varepsilon_{nl}}{\partial l} \int_{r_{\min}}^{R} \frac{dr}{p_r} \sim l \int_{r_{\min}}^{R} \frac{dr}{r^2 p_r},$$

where $r_{\min}$ is found from the condition $p_r = 0$, i. e. $r_{\min} \sim \ell/p_F$.

In the integral $\int_{r_{min}}^{R} p_r^{-1}\,dr$ the whole range of integration is important, and we can therefore estimate it to be of order R/v. In the integral $\int_{r_{min}}^{R} r^{-2} p_r^{-1}\,dr$, on the other hand, the principal contribution comes from the region near r_{min}, and so it is of order of magnitude $1/r_{min}\,p_F \sim 1/\ell$. Therefore we have

$$\frac{\partial \varepsilon_{nl}}{\partial l}\frac{R}{v} \sim l\frac{1}{l} = 1,$$

i. e.

$$\frac{\partial \varepsilon_{nl}}{\partial l} \sim \varepsilon_F/A^{1/3}.$$

Thus, we have shown that

$$\frac{\partial \varepsilon_{nl}}{\partial n} \sim \frac{\partial \varepsilon_{nl}}{\partial l} \sim \varepsilon_F A^{-1/3}.$$

It follows that the level of a spherical nucleus closest to a given one (i. e. differing from it by an energy of order $\epsilon_F/A^{2/3}$) is obtained from it, as a rule, by large changes δn and $\delta \ell$ in n and ℓ such that

$$\delta\varepsilon_{nl} = \frac{\partial \varepsilon_{nl}}{\partial n}\,\delta n + \frac{\partial \varepsilon_{nl}}{\partial l}\,\delta l$$

should be a minimum. To ensure this we have to choose $\delta n,\ \delta\ell \sim A^{1/3}$. As a result, neighbouring levels of a spherical nucleus as a rule differ considerably in the numbers of nodes in the radial and angular parts of the wave function, and hence give only a small value of the matrix element $U_{\lambda_1\lambda_2}$. States with energies $|\epsilon_{\lambda_1} - \epsilon_{\lambda_2}| \gg \epsilon_F/A^{1/3}$ also give a small value of $U_{\lambda_1\lambda_2}$, since the functions φ_{λ_1} and φ_{λ_2} differ greatly in the number of nodes.

Thus, the matrix element $U_{\lambda_1\lambda_2}$ is appreciably different from zero only (a) when $\lambda_1 = \lambda_2$ (if there is no selection rule which makes it strictly zero in this case) or (b) when $|\epsilon_{\lambda_1} - \epsilon_{\lambda_2}| \sim \epsilon_F A^{-1/3}$, when the number of nodes of the functions φ_{λ_1} and φ_{λ_2} are only slightly different.

Everything we have just said about the "combination" of levels applies equally to a deformed nucleus, provided the operator U changes the quantum numbers n and ℓ. If U changes only the magnetic quantum number m (as is the case, for instance, for the angular momentum operator) then levels belonging to the same $n\ell$-multiplet, with neighbouring m-values, can combine. Let us estimate the spacing between them. We showed above (p. 101) that in the first order of perturbation theory the level splitting is given by

$$\varepsilon_{nlm} = \varepsilon_{nl} + \varepsilon_{nl}\beta\left(\frac{m^2}{l^2} - \frac{1}{3}\right),$$

where β is the magnitude of the deformation. Consequently we have

$$\frac{\partial \varepsilon_{nlm}}{\partial m} \sim \varepsilon_{nl}\beta\frac{m}{l^2} \sim \varepsilon_F \beta A^{-1/3},$$

that is, in this case neighbouring levels can combine (provided β is small).

Noncentral Potential

Let us consider the more general case when the variables are not separable and construct the quasiclassical solution for this case. We shall look for a solution of the equation

$$\nabla^2\Psi + k^2\Psi = 0$$

in the form $\Psi = Ae^{iS}$. Then we have

$$\nabla^2 A + 2i\nabla A\cdot\nabla S - A(\nabla S)^2 + iA\nabla^2 S + k^2 A = 0 \tag{3.41}$$

Separation of the real and imaginary parts of (3.41) gives the two equations

$$\nabla^2 A + k^2 A = A(\nabla S)^2 \tag{3.42}$$

$$2\nabla A\cdot\nabla S + A\nabla^2 S = 0 \tag{3.43}$$

Equation (3.43) is actually the law of conservation of particle number. In fact we have

$$j = \frac{1}{2i}(\Psi^*\nabla\Psi - \Psi\nabla\Psi^*) =$$
$$= \frac{1}{2i}[Ae^{-iS}\nabla(Ae^{iS}) - Ae^{iS}\nabla(Ae^{-iS})] = A^2\nabla S$$

and

$$\operatorname{div} j = \nabla(A^2\nabla S) = A^2\nabla^2 S + 2A\nabla A\cdot\nabla S = 0$$

Thus along the streamlines (lines of current) we have

$$j_1 d\sigma_1 = j_2 d\sigma_2, \tag{3.44}$$

where $d\sigma_1$ and $d\sigma_2$ are elements of surface normal to the streamlines (see Fig. 34).

If $\nabla^2 A/A \ll k^2$, then we get from (3.42) the Hamilton-Jacobi equation for the action S:

$$2[E - V(r)] = (\nabla S)^2. \tag{3.45}$$

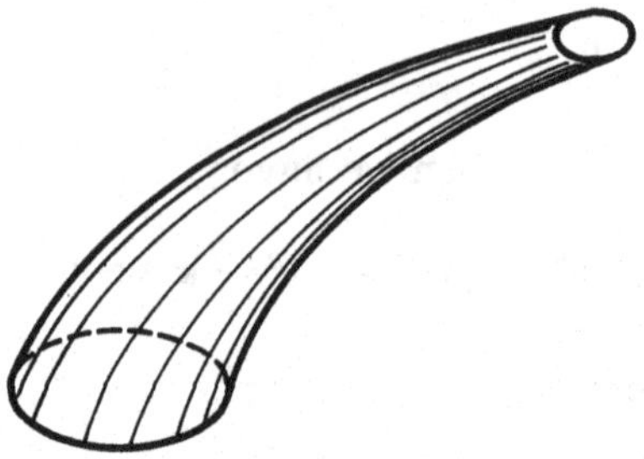

Fig. 34

The Quasiclassical Scattering Problem

Let us construct the quasiclassical wave function for the scattering problem. First of all we note that the solution of (3.45) is not unique; any point at $+\infty$, to which the particles tend after scattering, can be reached by them by at least two paths. This is illustrated in Fig. 35; here 1 and 2 are two points on the wavefront of the incoming plane wave. Therefore, rather than writing $\Psi = Ae^{iS}$, we must use the principle of superposition and write

$$\Psi = \sum_i A_i e^{iS_i},$$

where the sum is taken over all classical trajectories linking the points under consideration.

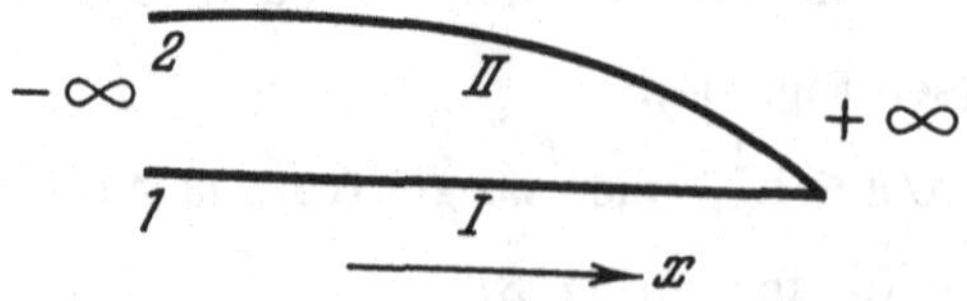

Fig. 35

We shall assume for simplicity that there are only two trajectories linking the wavefronts of the incoming and scattered waves at infinity. Along trajectory I, which passes far away from the scattering center, the lines of current are parallel to the x axis; A and ∇S are constant, with $|\nabla S| = p_0$. We may put the value of Ψ equal to 1 at every point of the wavefront of the incoming wave; then $A_1 = A_2 = 1$. Hence

$$\Psi_1 = \exp\left[i \int_{-L}^{x} p_0\, dx\right] = \exp(ip_0 x + ip_0 L),$$

where L is for the moment an arbitrary quantity. We now find the second solution Ψ_2. Along the second current line we have $A^2 \nabla S \cdot d\sigma = \text{const.}$ At $-\infty$ we have $A_2 = 1$, $|\nabla S| = p_0$, $d\sigma = \rho\, d\rho\, d\varphi$, where ρ is the impact parameter. Thus,

$$p_0 \rho\, d\rho\, d\varphi = A_2^2 p_0 r^2\, d\Omega. \tag{3.46}$$

Then we get

$$\Psi_2 \underset{r\to\infty}{\longrightarrow} A_2 e^{iS_2} = \sqrt{\frac{\rho\, d\rho\, d\varphi}{r^2\, d\Omega}} \exp\left[i \int_{-L}^{r} p\, dl\right],$$

and the complete solution to the problem has the form

$$\Psi \underset{r\to\infty}{\longrightarrow} \exp(ip_0 x + ip_0 L) + \sqrt{\frac{\rho\, d\rho\, d\varphi}{r^2\, d\Omega}} \exp\left[i \int_{-L}^{r} p\, dl\right]. \tag{3.47}$$

We multiply the right-hand side of this equation by the constant factor $\exp(-ip_0 L)$:

$$\Psi \underset{r\to\infty}{\longrightarrow} e^{ip_0 x} + \sqrt{\frac{\rho\, d\rho\, d\varphi}{r^2\, d\Omega}} \exp i\left[p_0 r + \int_{-L}^{r} p\, dl - \int_{-L}^{r} p_0\, dl_0\right],$$

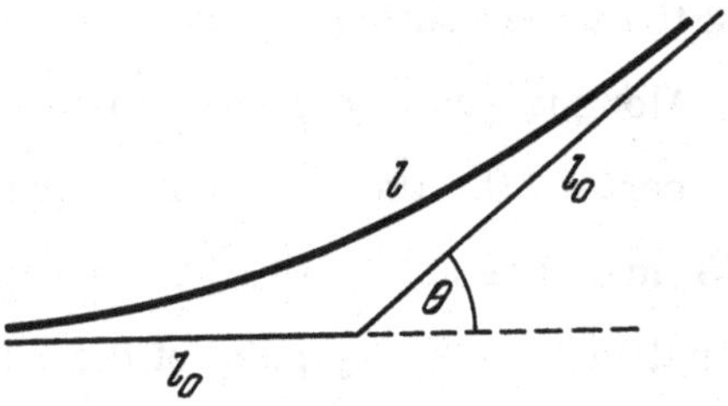

Fig. 36

or

$$\Psi \underset{r\to\infty}{\longrightarrow} e^{ip_0 x} + \frac{f}{r} e^{ip_0 r},$$

where

$$f = \sqrt{\left(\frac{d\sigma}{d\Omega}\right)_{c\ell}} e^{i\Phi}, \tag{3.48}$$

$$\Phi \equiv \int_{-\infty}^{\infty} p\, dl - \int_{-\infty}^{\infty} p_0\, dl_0. \tag{3.49}$$

where the trajectories ℓ and ℓ_0 are shown in Fig. 36.

The only regions important in expression (3.49) are those close to the scattering centre. Expression (3.49) needs correction in the case that the classical trajectory passes near points where the quasiclassical approximation breaks down. The set of all these points forms a surface, called the caustic surface, which separates the classically accessible from the classically inaccessible region. For instance, for a repulsive Coulomb potential the caustic surface is a paraboloid of rotation, while for an attractive Coulomb potential it is a straight line going from the scattering centre to $+\infty$.

Let us determine how the phase of the wave function changes near a point when the trajectory touches the caustic surface. We decompose the motion of the particle into its components along and perpendicular to the tangent to the caustic surface. The motion in the direction perpendicular to the surface is analogous to quasi-classical reflection from a turning point in the one-dimensional problem; as we saw, the phase difference of the incoming and reflected wave was equal to $\pi/4 - (-\pi/4) = \pi/2$. As for the motion along the tangent to the caustic surface, there is no change of phase. Consequently, we get for Φ instead of (3.49) the expression

$$\Phi = \int_{-\infty}^{\infty} p\,dl - \int_{-\infty}^{\infty} p_0\,dl_0 + i\nu\frac{\pi}{2},$$

where ν is the number of points at which the trajectory touches the caustic surface. For the Coulomb potential $\nu = 1$.

Cross Section for Scattering of a Proton on a Hydrogen Atom

In considering the problem of charge exchange (p. 139) we found that the scattering amplitude f_1 was given by $f_1 = (1/2)\,(f_s + f_a)$ and the charge exchange amplitude by $f_2 = (1/2)\,(f_s - f_a)$, where f_s and f_a are respectively the amplitudes for symmetric and antisymmetric scattering. Now the potential V_s is only slightly different from V_a, so we can take the classical cross section to be the same for V_s and V_a, while the phases Φ_s and Φ_a may be very different. (Φ_s and Φ_a are very large quantities, so that the difference $\Phi_s - \Phi_a$ may also be large.) Thus, according to (3.48),

$$f_1 = \frac{1}{2}\sqrt{\left(\frac{d\sigma}{d\Omega}\right)_{cl}}(e^{i\Phi_s}+e^{i\Phi_a}),$$

$$f_2 = \frac{1}{2}\sqrt{\left(\frac{d\sigma}{d\Omega}\right)_{cl}}(e^{i\Phi_s}-e^{i\Phi_a}).$$

The charge exchange cross section is given by

$$\left(\frac{d\sigma}{d\Omega}\right)_{c.e.} = |f_2|^2 = \frac{1}{4}\left(\frac{d\sigma}{d\Omega}\right)_{cl} |\exp(i\Phi_s) - \exp(i\Phi_a)|^2$$

$$= \left(\frac{d\sigma}{d\Omega}\right)_{cl} \sin^2\frac{1}{2}(\Phi_s - \Phi_a)$$

Since we have

$$\Phi \equiv \int_{-\infty}^{\infty} p\,dl - \int_{-\infty}^{\infty} p_0 dl_0,$$

we get

$$\Phi_s - \Phi_a = \int_{-\infty}^{\infty} (p_s - p_a)\,dl \approx$$

$$\approx \int_{-\infty}^{\infty} \frac{m_p(V_s - V_a)}{p}\,dl \approx \frac{1}{v_0}\int_{-\infty}^{\infty} (V_s - V_a)\,dl,$$

where m_p is the proton mass and v_o its velocity.

Let us consider small-angle scattering, and put $V_s - V_a \equiv V$. Then for the charge exchange cross section, we get the following expression:

$$\left(\frac{d\sigma}{d\Omega}\right)_{c.e.} = \left(\frac{d\sigma}{d\Omega}\right)_{cl} \sin^2\left[\frac{1}{2v_0}\int_{-\infty}^{\infty} V(\sqrt{\rho^2 + x^2})\,dx\right] \qquad (3.50)$$

where ρ is the impact parameter. This expression agrees with the cross section obtained in Section 2.4. For the scattering cross section (without charge exchange) we get

$$\left(\frac{d\sigma}{d\Omega}\right)_{sc} = \left(\frac{d\sigma}{d\Omega}\right)_{c\ell} \cos^2\left[\frac{1}{2v_0}\int_{-\infty}^{\infty} V(\sqrt{\rho^2+x^2})\, dx\right]$$

$$= \frac{1}{2}\left(\frac{d\sigma}{d\Omega}\right)_{c\ell} - \frac{1}{2}\left(\frac{d\sigma}{d\Omega}\right)_{c\ell} \cos\left[\frac{1}{v_0}\int_{-\infty}^{\infty} V(\sqrt{\rho^2+x^2})\, dx\right]$$

Thus we get not only the usual classical scattering but also the quantum mechanical oscillation of the scattering cross section as a function of deflection angle.

CHAPTER 4

THE ANALYTIC PROPERTIES OF PHYSICAL QUANTITIES

In recent years the development of the theoretical and experimental aspects of the physics of strongly interacting particles (hadrons) has required the creation of theoretical methods which do not assume that the interaction is weak and hence do not use perturbation theory. One of these methods is based on the analytic properties of physical quantities, for instance the analytic properties of the scattering amplitude as a function of energy and scattering angle.

In cases where the quantity in question has a singularity in the complex plane near the real axis, its behaviour on the real axis is determined by the character of this singularity. One of the first examples of this kind was the theory of reactions lending to the formation of slow particles (cf. below). In this case the pole in the amplitude for scattering, one against another, of the particles produced in the reaction determines the energy-dependence of the process. As we shall see, the existence in the system of interacting particles of levels with low energy simplifies the energy- and coordinate-dependence of the low-energy wave functions to such an extent that the problem of finding the energy spectrum is reduced to a simple algebraic equation.

The analytic properties allow us to establish a relation between the imaginary part of the scattering amplitude, which is determined by the absorption of the particles in question, and the real part (a so-called dispersion relation). From such relations arise a large number of experimentally observable consequences in different areas of physics.

For real values of the parameters no physical quantity can tend to infinity. That is to say, on the physically attainable regions of the real axis all physical functions are finite, since all quantities which are actually obtained as a result of observation are smooth functions of the parameters. Singularities of physical quantities on the real axis arise only as a result of judicious idealization of the system in question. Let us clarify this point by an example. We know that the scattering cross-section of a particle in a screened Coulomb potential $er^{-1}e^{-ar}$ is in the Born approximation proportional to $(q^2 + a^2)^{-2}$, where q is the momentum transfer. Thus, the scattering amplitude has a singularity in the complex plane of q. If we consider a more idealized system, namely one without screening, then $a \to 0$ and the singularity goes over on to the real q-axis. Now suppose that we are interested in the scattering by an unscreened charge. Then in the idealized expression, the scattering amplitude has a singularity at $q = 0$. In this case the infinity in the cross section is eliminated by the finite width d of the beam of incident particles; the fact that the beam is restricted leads to an indeterminacy $\Delta p_{\perp} \sim d^{-1}$ in the transverse momentum of the particles and thereby to the unresolvability of the small scattering angles which correspond to $q \to 0$. However, by improving the experiment (i.e. by increasing d) we may approach arbitrarily close to the singularity at $q = 0$, which indicates that the idealization is indeed a reasonable one.

Another example is the threshold singularities which arise from the density of final states (cf. below, p.259). For instance, the cross section for the knock-out of a nucleon from a nucleus is proportional to $\sqrt{E - E_0}$, where E is the energy of the incident particle and E_0 the threshold energy. This energy-dependence comes from the density of final states:

$$f_0(E - E_0) = \\ = \int \delta(E_p - E) \frac{d^3p}{(2\pi)^3} \sim \int_{E_0}^{\infty} \delta(E_p - E) \sqrt{E_p - E_0}\, dE_p = \\ = \sqrt{E - E_0}.$$

In a realistic experimental situation, however, the energy of the incident particles is not fixed but fluctuates around the energy E, with say a Gaussian distribution. Hence in the production cross section it is not f_0 which occurs, but rather the quantity

$$f(E - E_0) = \sqrt{\pi\alpha} \int_{E_0}^{\infty} e^{-\alpha(E-E_1)^2} \sqrt{E_1 - E_0}\, dE_1 = \\ = \sqrt{\pi\alpha} \int_{0}^{\infty} e^{-\alpha(E-E_0-x)^2} \sqrt{x}\, dx.$$

The production cross section is proportional to this quantity. On calculating the integral we find

$$f(\varepsilon) = \frac{\pi}{2\sqrt{\alpha}} (2\alpha)^{-1/4} e^{-\alpha\varepsilon^2/2} \mathcal{D}_{-3/2}(-\varepsilon\sqrt{2\alpha}).$$

where $\mathcal{D}$ is the parabolic cylinder function (Weber function) and $\epsilon = E - E_0$. The function $f(\epsilon)$ has no singularity at the threshold point $\epsilon = 0$. Using the asymptotic forms of the $\mathcal{D}$-function, we find that for $\epsilon \to -\infty$, $f(\epsilon) \sim e^{-\alpha\epsilon^2}$, while for $\epsilon \to +\infty$, $f(\epsilon) \sim \epsilon^{1/2}$. The graph of $f(\epsilon)$ is shown in Fig. 37.

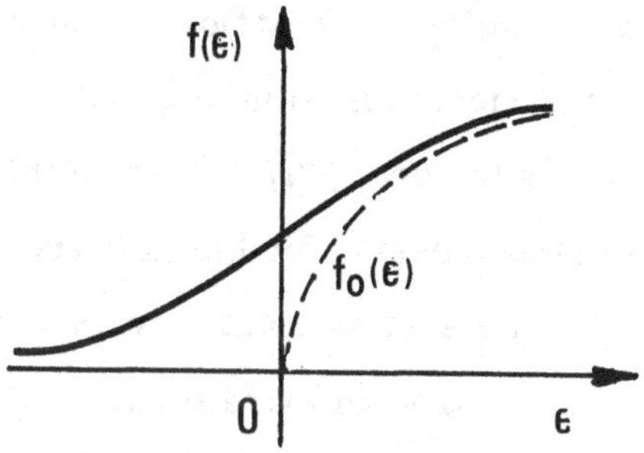

Fig. 37

When the energy of the incident particles is fixed more and more precisely ($\alpha \to \infty$) the curve of $f(\epsilon)$ tends to the limiting curve shown by the dashed line in Fig. 37, which has a singularity for $\epsilon = 0$. Thus the square-root singularity at $\epsilon = 0$ appears as a result of idealization of the real-life scattering problem.

In the complex plane of the parameters physical functions may have various singularities; in each case there must be physical reasons for their appearance. From the analyticity properties we can draw important conclusions about the relations between different physical quantities. In the following we shall first consider some simple physical examples, and then find the solutions of a series of problems which cannot be investigated without using the analytic properties.

Dependence of the Moment of Inertia of a Nucleus on Deformation

The moment of inertia of a spherical quantum-mechanical system must be set equal to zero; for, to observe rotation, we need a marker on the surface of the system. Such a marker would correspond to an excited state, in which the system would no longer be spherical. As the deformation is decreased, the energy of the rotational terms increases to infinity, which means the moment of inertia tends to zero. To elucidate this point we obtain a quantum mechanical expression for the moment of inertia. Let us go over to a coordinate system rotating with angular velocity Ω. Then the Hamiltonian acquires an extra term

$$H' = -M\Omega,$$

where M is the angular momentum operator of the system. If we treat H' as a small perturbation, it is easy to obtain an expression for the mean value of the angular momentum:

$$\langle M\rangle = 2\sum \frac{|M_{0S}|^2}{E_S - E_0}\Omega \equiv J\Omega.$$

where we have defined the moment of inertia as the ratio $\langle M\rangle/\Omega$. If at the angular velocities in question we can neglect excitation of the internal degrees of freedom of the system, then the rotational energy will be given by

$$E = \frac{M^2}{2J} = \frac{j(j+1)}{2J}$$

with the above moment of inertia:

$$J = 2\sum \frac{|M_{0S}|^2}{E_S - E_0}\,.$$

From this expression it follows that for a spherical system, where M has only diagonal matrix elements, $J = 0$.

Consider now the moment of inertia J induced by a small deformation δ. What is the first term in the expansion of J of powers of δ? We can easily verify that the hypothesis $J = A\delta$ is incorrect. Suppose $J = A$ for positive δ (i.e., a "rugby-football" type deformation). Since J must be everywhere positive, it follows that for negative δ ("pancake" type deformation) $J = A|\delta| = A\sqrt{\delta^2}$. Consequently the moment of inertia, regarded as a function of δ^2, would have a branch point at the origin. However, it is easy to see that there is no reason for the occurrence of a branch point. Let us go over from the weakly deformed nucleus to the spherical one by making the appropriate coordinate transformation. The result is to add to the Hamiltonian a small correction H' proportional to the deformation δ. For a finite system perturbation theory gives for the moment of inertia a convergent series in δ, and we can therefore conclude that there can be no singularity at $\delta = 0$. Thus we can expand J in powers of δ, and the expansion begins with the quadratic term:

$$J = B\delta^2 + \cdots$$

Dependence of the Frequency of Sound on the Wave Vector

It is well known that there are systems in which for small wave numbers k (i.e. long wavelengths $\lambda = 1/k$) sound-wave excitations can be propagated with frequency $\omega = ck$, where c is the speed of sound. Since ω is a scalar, while $\underset{\sim}{k}$ is a vector, this relation must be interpreted as $\omega = c\sqrt{k^2}$, so that the frequency ω regarded as a function of k^2 has a branch point at $k^2 = 0$. We shall explain the reason for the appearance of a branch point in this case.

The equations which describe the state of the system are invariant under the transformation $t \to -t$, provided there is no dissipation. Thus the equation to determine the frequency always contains even powers of ω. For instance, in classical dynamics Newton's equation for a system of particles has the form (for small vibrations)

$$m\ddot{u}_n = -\sum_m F_{mn}u_m.$$

where m is the mass of the n-th particle, u_n its displacement, and $F_{mn}u_m$ the force exerted on the n-th particle by the m-th. Putting $u_n = u_{on}e^{i\omega t}$, we get the dispersion relation for the square of the frequency ω^2 in the form

$$m\omega^2 u_{on} = \sum_m F_{mn}u_{0m}.$$

Consequently it is the square of the frequency ω^2, rather than ω itself, which must be an analytic function of the parameters of the problem. Thus

$$\omega^2 = a + c^2k^2 + bk^4 + \dots$$

The deformation of the system corresponding to a sound wave in the limit $k \to 0$ must go over into a displacement of the system as a whole, and the energy corresponding to such a displacement is zero. Hence for sound waves $a = 0$ ("Goldstone's theorem") and consequently for small k the frequency is given by $\omega = c\,|k|$.

1. ANALYTIC PROPERTIES OF THE DIELECTRIC CONSTANT

In some cases, as we shall see, the analyticity of physical quantities is a consequence of causality. Consider, for instance, the analytic properties of the dielectric constant. The electric displacement $\mathcal{D}(t)$ induced in the system by an electric field $\mathcal{E}$ is determined by the amplitude of the field at all preceding moments of time $t - \tau$, $\tau > 0$. Thus,

$$\mathcal{D}(t) = \int_0^\infty \mathcal{K}(\tau)\, \mathcal{E}(t - \tau)\, d\tau; \tag{4.1}$$

where $\mathcal{K}(\tau)$ is a real and decreasing quantity, since it relates two physical quantities displaced in time by an amount τ. The relation (4.1) is a consequence only of the requirement that cause should precede effect. Let us take the Fourier transforms of $\mathcal{E}$ and $\mathcal{D}$:

$$\mathcal{E} = \int \mathcal{E}_\omega e^{-i\omega t} d\omega, \qquad \mathcal{D} = \int \mathcal{D}_\omega e^{-i\omega t} d\omega.$$

Then we find from (4.1)

$$\mathcal{D}_\omega = \mathcal{E}_\omega \int_0^\infty \mathcal{K}(\tau)\, e^{i\omega\tau}\, d\tau,$$

whence the dielectric constant is given by

$$\varepsilon_\omega = \frac{\mathcal{D}_\omega}{\mathcal{E}_\omega} = \int_0^\infty \mathcal{K}(\tau)\, e^{i\omega\tau}\, d\tau. \tag{4.2}$$

The quantity ϵ_ω is analytic in the upper half-plane of the complex variable ω. For if we write $\omega = \omega_0 + i\omega_1$, where $\omega_1 > 0$, the integrand of (4.2) contains the decreasing exponential $e^{-\omega_1\tau}$:

consequently the integral (4.2) converges and defines an analytic function.

It is easy to see that the analytic properties of $1/\epsilon$ are the same as those of ϵ itself. Consider two cylinders of the material in question, one long and one short, and apply to each of them an external field $\mathcal{E}_0$ (Fig. 38). In the case of the long

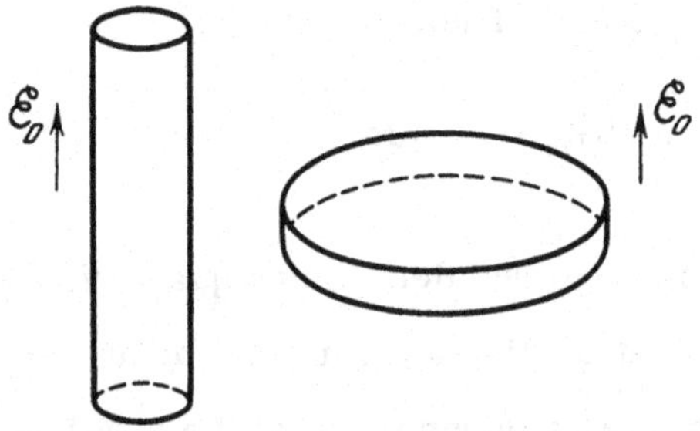

Fig. 38

cylinder the field $\mathcal{E}$ inside the cylinder is equal to the external field $\mathcal{E}_0$ and so the displacement $\mathcal{D}$ is given by $\mathcal{D} = \epsilon\,\mathcal{E}_0$. In the case of the short cylinder it is the displacement which is equal to the external field $\mathcal{E}_0$, so that the field inside the cylinder is given by $\mathcal{E} = \mathcal{E}_0/\epsilon$. (This is a result of the well-known boundary conditions on the normal and tangential components of $\mathcal{D}$ and $\mathcal{E}$). In the first case, therefore, we have

$$\mathcal{D} = \int_0^\infty \mathcal{K}_1(\tau)\,\mathcal{E}_0(t-\tau)\,d\tau, \qquad \varepsilon = \int_0^\infty \mathcal{K}_1(\tau)\,e^{i\omega\tau}\,d\tau.$$

In the second case, the internal field $\mathcal{E}$ is determined by the value of $\mathcal{E}_0$, which is the same as that of $\mathcal{D}$, at all previous times, so that we get

$$\mathcal{E} = \int_0^\infty \mathcal{K}_2(\tau)\,\mathcal{D}(t-\tau)\,d\tau,$$

and so

$$\frac{1}{\varepsilon} = \int_0^\infty \mathcal{K}_2(\tau)\,e^{i\omega\tau}\,d\tau.$$

Thus the analytic properties of $1/\epsilon$ are the same as those of ϵ, and it follows that the dielectric constant can have neither poles nor zeros in the upper half-plane of ω.

We may use the analyticity of the dielectric constant to express the real part in terms of the imaginary part. Consider the complex plane of the variable ω^2. With respect to this variable the region of analyticity of $\epsilon(\omega^2)$ is the first Riemann sheet; all the singularities lie on the second sheet. The cut corresponding to the transition from one sheet to the other lies along the real positive ω^2 axis from the origin to infinity. (See Fig. 39). To relate the real and imaginary parts of the dielectric constant, we use Cauchy's theorem:

$$f(z) = \frac{1}{2\pi i}\oint_C \frac{f(z')\,dz'}{z'-z}.$$

where C is any closed contour enclosing z but not including any singularities of f(z). We can apply this theorem to the function $\epsilon(\omega^2)$ by taking C as the contour shown in Fig. 39. It is convenient to choose the function f so that the integral over the infinitely large circle C_0 tends to zero. Since, as we shall see below, $\epsilon \to 1$ for $\omega \to \infty$, we must take $f = \epsilon - 1$. Thus we get

$$\varepsilon(\omega^2) - 1 = \frac{1}{2\pi i} \int\limits_{C_1+C_2} \frac{\varepsilon(\omega_1^2) - 1}{\omega_1^2 - \omega^2} d\omega_1^2. \tag{4.3}$$

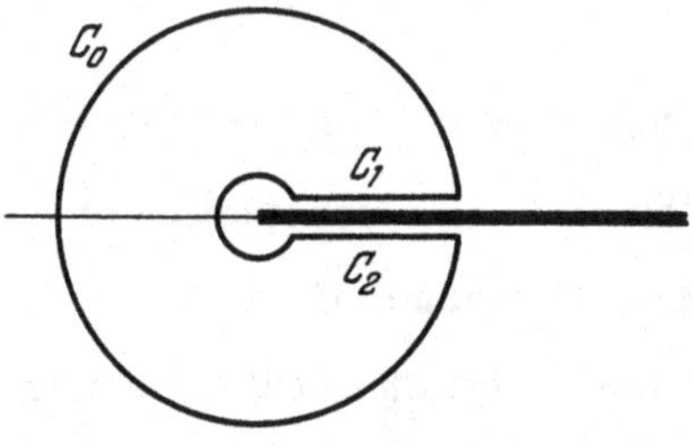

Fig. 39

Here the integral reduces to the passage around the cut. Let us denote by $\epsilon_1(\omega^2)$ the value of ϵ on the upper side of the cut, and by $\epsilon_2(\omega^2)$ its value on the lower side; we now relate these two quantities. On the upper side of the cut put $\omega = \omega_0$, where ω_0 is real. Then on the lower side we have $\omega = \omega_0 e^{\pi i} = -\omega_0$. Hence

$$\varepsilon_1(\omega_0^2) = \int\limits_0^\infty \mathcal{K}(\tau) e^{i\omega_0\tau} d\tau$$

and

$$\varepsilon_2(\omega_0^2) = \int\limits_0^\infty \mathcal{K}(\tau) e^{-i\omega_0\tau} d\tau.$$

Since $\mathcal{K}(\tau)$ and ω_0 are real, we therefore find that

$$\varepsilon_2(\omega_0^2) = [\varepsilon_1(\omega_0^2)]^*. \tag{4.4}$$

Substituting (4.4) into (4.3) and noting that on the upper side of the cut we have $\omega^2 \to \omega^2 + i\delta$ (where $\delta \to +0$) we get

$$\varepsilon_\omega = \varepsilon_1(\omega^2) = 1 + \frac{1}{\pi}\int_0^\infty \frac{\operatorname{Im}\varepsilon(\omega_1^2)\,d\omega_1^2}{\omega_1^2 - \omega^2 - i\delta}. \tag{4.5}$$

It is easy to verify that with the above choice of δ the calculation of the imaginary part of (4.5) leads to an identity. Thus, we can reconstruct the whole dielectric constant ϵ from a knowledge of its imaginary part.

The imaginary part of the dielectric constant, which governs the absorption of electromagnetic radiation by the system, is non-zero only when the frequency of the wave is equal to an eigen-frequency ω_n of the system, to within an error of the order of the widths of the levels in question. This will become obvious from the example considered below. In other words, neglecting the widths, we can write $\operatorname{Im}\epsilon$ in the form

$$\operatorname{Im}\varepsilon(\omega^2) = \sum_n \pi f_n \delta(\omega^2 - \omega_n^2). \tag{4.6}$$

Substituting (4.6) in (4.5), we get

$$\varepsilon(\omega^2) = 1 + \sum_n \frac{f_n}{\omega_n^2 - \omega^2 - i\delta}. \tag{4.7}$$

If $\omega^2 \gg \omega_n^2$, that is, if the wavelength of the light is small compared to atomic dimensions, then the structure of the atom is unimportant and the quantity ϵ goes over into the dielectric constant of an ideal electron gas:

$$\varepsilon \underset{\omega\to\infty}{=} 1 - 4\pi ne^2/m\omega^2. \tag{4.8}$$

where n is the number of electrons per unit volume and m the electron mass. Formula (4.8) will be illustrated by considering the example below. Comparing this expression with (4.7), we find

$$\sum_n f_n = 4\pi ne^2/m.$$

the so-called f-sum rule.

Analytic Properties of the Dielectric Constant in a Simple Model

Consider a medium consisting of oscillators with frequency ω_0 (it is easy to discuss also the more general case of a set of oscillators with different frequencies). Such oscillators are a crude model for the atomic electrons. The equation of motion for an oscillator subjected to an electric field $\mathcal{E}(t)$ has the form

$$\ddot{r} + h\dot{r} + \omega_0^2 r = \frac{e}{m}\mathcal{E}.$$

where h is the damping coefficient. Expanding $\underset{\sim}{r}$ and $\underset{\sim}{\mathcal{E}}$ in Fourier integrals by the formulae $\underset{\sim}{r} = \int \underset{\sim}{r}_\omega e^{-i\omega t} d\omega$ and $\underset{\sim}{\mathcal{E}} = \underset{\sim}{\mathcal{E}}_\omega e^{-i\omega t} d\omega$, we get

$$-\omega^2 r_\omega - ih\omega r_\omega + \omega_0^2 r_\omega = \frac{e}{m}\mathcal{E}_\omega,$$

and so

$$r_\omega = \frac{e}{m}\mathcal{E}_\omega \frac{1}{\omega_0^2 - \omega^2 - ih\omega}.$$

We can calculate the dipole polarization of the medium $\mathcal{P}_\omega$:

$$\mathcal{P}_\omega \equiv ner_\omega = \frac{ne^2}{m}\mathcal{E}_\omega \frac{1}{\omega_0^2 - \omega^2 - ih\omega}.$$

where n is the number of oscillators per unit volume. The die-electric permittivity ϵ_ω (dielectric constant) is found by using the well-known formulae of electromagnetic theory:

$$\varepsilon_\omega = \frac{\mathcal{D}_\omega}{\mathcal{E}_\omega} = \frac{\mathcal{E}_\omega + 4\pi\mathcal{P}_\omega}{\mathcal{E}_\omega} = 1 + \frac{4\pi ne^2}{m}\frac{1}{\omega_0^2 - \omega^2 - ih\omega}.$$

Hence for $\omega^2 \gg \omega_0^2$ we obtain the formula (4.8) found above.

If the system is composed of oscillators of different types, then we must sum over the various possible frequencies ω_n in the above formula:

$$\varepsilon_\omega = 1 + \frac{4\pi e^2}{m}\sum_n \frac{n_n}{\omega_n^2 - \omega^2 - ih_n\omega}. \tag{4.9}$$

where n_n is the number of oscillators per unit volume with energy ω_n. In this case, if we let ω tend to ∞, we get

$$\varepsilon_\omega \to 1 - \frac{4\pi e^2}{m\omega^2}\sum_n n_n,$$

Comparing this expression with (4.8), we obtain the f-sum rule in the form $\sum_n n_n = n$, where n is the total number of electrons per unit volume. We see that in the limit $\omega \to \infty$ the properties of the model chosen do not affect the results, as we should expect.

The damping coefficient h_n is determined by the intensity of the transitions from the n-th state into other states. We saw on p. 64 that $h_n \ll \omega_n$. As can be seen from (4.9) the damping coefficient determines the imaginary part of the dielectric constant.

Let us check that ϵ_ω has poles and zeros only in the lower half-olane of ω. It follows from (4.9) that the poles of ϵ_ω occur at the points $\omega = \pm (\omega_n^2 - ih_n\omega)^{1/2} \approx \pm|\omega_n| - ih_n/2$. Since $h_n > 0$, it follows that $\mathrm{Im}(\omega) < 0$. To find the zeros of ϵ_ω we first consider the case of a single oscillator with frequency ω_0

and damping coefficient h_0. Equating the expression (4.9) to zero, we obtain

$$\omega = \pm\sqrt{\omega_0^2 + \frac{4\pi e^2 n}{m} - ih_0\omega} \approx$$
$$\approx \pm\sqrt{\omega_0^2 + \frac{4\pi e^2 n}{m}} - i\frac{h_0}{2}.$$

whence $\mathrm{Im}(\omega) < 0$. In the case of several oscillators, we can see from (4.9) that the real parts of the zeros will lie between the real parts of the poles. Since all the h_n have the same sign, they will shift the zeros of ϵ_ω in the same direction as in the case of a single oscillator. Thus, we have verified all the analytic properties of the dielectric constant in our simple model.

Consider the structure of the function $\mathcal{K}(t)$ in this model. We have:

$$\varepsilon_\omega = \int_0^\infty \mathcal{K}(\tau)\, e^{i\omega\tau} d\tau = 1 + \frac{4\pi e^2}{m} \sum_n \frac{n_n}{\omega_n^2 - \omega^2 - ih_n\omega}.$$

Hence

$$\mathcal{K}(\tau) = \delta(\tau) + \frac{4\pi e^2}{m} \sum_n \frac{\sin(\sqrt{\omega_n^2 - h_n^2/4}\,\tau)}{\sqrt{\omega_n^2 - h_n^2/4}} e^{-h_n\tau/2}.$$

We see that the quantity h_n determines the damping of $\mathcal{K}(t)$, and also the shift of the eigenfrequencies $(\omega_n \to (\omega_n^2 - h_n^2/4)^{1/2})$. We notice that since $\mathcal{K}(t)$ decreases exponentially, the displacement $\mathcal{D}(t)$ is determined by the values of the field $\mathcal{E}(t - \tau)$ at a distance in time $\tau \sim 1/h$. We also notice that the imaginary part of the dielectric constant is an odd function of time and the real part an even one, as is required by time-reversibility.

2. ANALYTIC PROPERTIES OF THE SCATTERING AMPLITUDE

Unitarity as a Consequence of the Superposition Principle and the Conservation of Probability

Let us investigate the analytic properties of the scattering amplitude. By definition, the wave function Ψ_{out} of the system after scattering is related to the wave function Ψ_{in} before scattering by the so-called S-matrix: $\Psi_{out} = S\,\Psi_{in}$. We shall show that the conservation of probability and the superposition principle together imply that the S-matrix has the property of unitarity.

The matrix elements of the S-matrix S_{ac} are the amplitudes for transition from the state a to the state c. Consequently, from the conservation of probability $\sum_c |S_{ac}|^2 = 1$; this equality may be written symbolically in the form $(S^+S)_{aa} = 1$. The quantum-mechanical principle of superposition implies that the wave function $|\tilde{a}>$ of an arbitrary state may be written in the form $|\tilde{a}> = \alpha|a> + \beta|b> + \ldots$, where $|a>$, $|b>$ ... form a set of basis states. The condition $(S^+S)_{aa} = 1$ may be written in the form

$$|\alpha|^2 (S^+S)_{aa} + |\beta|^2 (S^+S)_{bb} + \ldots$$
$$\ldots + \alpha\beta^* (S^+S)_{ab} + \alpha^*\beta (S^+S)_{ba} + \ldots = 1.$$

Since $(S^+S)_{aa} = (S^+S)_{bb} = \ldots = 1$ and $|\alpha|^2 + |\beta|^2 + \ldots = 1$, it follows that

$$\alpha\beta^* (S^+S)_{ab} + \alpha^*\beta (S^+S)_{ba} + \ldots = 0.$$

and since the coefficients $\alpha, \beta \ldots$ are arbitrary, we finally find $(S^+S)_{ab} = 0$. We may combine this equation with the condition of conservation of probability and write $(S^+S)_{ab} = \delta_{ab}$, or in operator form $S^+S = 1$. Thus, the S-matrix possesses the property of unitarity.

It is convenient to separate out from the S-matrix the unit matrix, which describes the process in which no actual scattering takes place: $S = 1 + iT$. Then the unitarity condition on the S-matrix takes the form $(1 - iT^+)(1 + iT) = 1$, or

$$T^+T = i(T^+ - T).$$

Written out in terms of the matrix elements this becomes

$$2\mathrm{Im}\, T_{ab} = \sum_c T^*_{ca} T_{cb}. \tag{4.10}$$

We define the scattering amplitude f by the relation $T = (4\pi^2/M) f \delta(E - E')$, where M is the mass of the scattered particle (assumed nonrelativistic) and the delta-function expresses the law of conservation of energy in scattering. Then the diagonal elements $(a = b)$ of equation (4.10) can be written in the form

$$2\mathrm{Im}\, f_{aa} = \frac{4\pi^2}{M} \sum_c |f_{ac}|^2 \frac{\delta(E - E_c)\,\delta(E_c - E')}{\delta(E - E')}. \tag{4.11}$$

The left-hand side contains the zero-angle (forward) scattering amplitude. We can replace the factor $\delta(E_c - E')$ in the numerator of the right-hand side by $\delta(E - E')$, after which it cancels with the δ-function in the denominator. Then the right-hand side of eqn. (4.11) contains the sum $\sum_c |f_{ac}|^2 \delta(E-E_c)$. The intermediate states c are characterised by some momentum $\underset{\sim}{p}'$. Replacing the summation over $\underset{\sim}{p}'$ by integration $(\sum_{p'} \to \int \frac{d\underset{\sim}{p}'}{(2\pi)^3})$, we get

$$\int |f_{pp'}|^2 \delta(E_p - E_{p'}) \frac{dp'}{(2\pi)^3} = \frac{1}{8\pi^3} \int |f(\theta)|^2 d\Omega \frac{p^2}{\frac{dE}{dp}} .$$

where we used the notation $f_{ab} = f_{pp'} = f(\theta)$. Now the quantity $dE/dp = v$ is the velocity of the incident particle, and the integral $\int |f(\theta)|^2 d\Omega = \sigma$ is the total scattering cross section. Hence we get from (4.11)

$$2\,\mathrm{Im}\, f(0) = \frac{4\pi^2}{8\pi^3 M} \sigma \frac{p^2}{v} ,$$

or

$$\mathrm{Im}\, f(0) = \frac{p}{4\pi} \sigma. \qquad (4.12)$$

This relation is called the optical theorem; it relates the total cross section to the imaginary part of the forward scattering amplitude. We shall give below an alternative proof of the optical theorem, based on the conservation of particle number.

The Dispersion Relation

We can relate the real and imaginary parts of the scattering amplitude, as we did for the dielectric constant. Consider, for instance, the scattering of light by a charged system. We write down the operator relation B = SA, where A describes the amplitude of the incident wave and B that of the scattered wave. In the special case in which the amplitude A is taken at time $t = -\infty$

and B at time $t = +\infty$, the operator S is equivalent to the S-matrix introduced above. Note that the dielectric constant is just a special case of the operator S, that is, the case when the quantities A and B are position-independent.

Let us make the assumption that up to time $t = 0$ the amplitude of the incident wave on some reference plane placed in front of the scattering system is zero. Then we can assert of B that $B(t) = 0$ for $t < t_1$, where $t_1 > 0$. Going over to Fourier components, we have

$$A_\omega = \int_0^\infty A(t)\, e^{i\omega t}\, dt, \quad B_\omega = \int_{t_1}^\infty B(t)\, e^{i\omega t}\, dt.$$

Since $A(t) = 0$ for $t < 0$, A_ω has no singularities in the upper half-plane of ω. By the same argument, B_ω cannot have singularities in this half-plane either. Then it follows from the relation $B_\omega = SA_\omega$ that the quantity S can have singularities only at the zeros of A_ω. But S must be independent of the particular form of A_ω, and in particular of the position of its poles; hence, the zeros of A_ω must coincide with the zeros of B_ω, and we conclude that S cannot have any singularities in the upper half-plane. As we saw above, the operator S is linearly related to the scattering amplitude of light $f(\omega^2)$. The quantity $f(\infty)$ is nonzero (it is just the sum of the scattering amplitudes of light on the free particles of the medium). Hence the dispersion relation must be applied not to $f(\omega^2)$ itself, but to the difference $f(\omega^2) - f(\infty)$. It is analogous to the dispersion relation for the dielectric constant ϵ, and reads:

$$f(\omega^2) = f(\infty) + \frac{1}{\pi} \int_0^\infty \frac{\operatorname{Im}[f(\omega_1^2) - f(\infty)]}{\omega_1^2 - \omega^2 - i\delta}\, d\omega_1^2. \quad (4.13)$$

Resonance Scattering at Low Energies

For large values of r the wave function in the scattering problem has the asymptotic form

$$\Psi \underset{r\to\infty}{\longrightarrow} e^{ipz} + \frac{f}{r} e^{ipr}.$$

Consider first spherically symmetric scattering. The spherically symmetric part of Ψ, which describes the $\ell = 0$ scattering, is given for large r by the expression

$$\int \Psi \frac{d\Omega}{4\pi} \underset{r\to\infty}{=} \frac{\sin pr}{pr} + \frac{f_0}{r} e^{ipr} = \frac{u}{r}.$$

where f_0 is the spherically symmetric partial-wave amplitude. The function u can be written in the form

$$u = -\frac{1}{2ip} e^{-ipr} + \frac{1+2ipf_0}{2ip} e^{ipr}. \tag{4.14}$$

The particle current arising from the expression (4.14) is the difference of the currents formed from the two terms in u separately (in the expression for the current the cross terms cancel). If the scattering centre neither absorbs nor emits particles, these two currents must be equal; the condition for their equality is $|1 + 2ipf_0|^2 = 1$, or in other words

$$f_0 - f_0 = 2ipf_0f_0^*, \quad \text{Im}\, f_0 = p\,|f_0|^2, \tag{4.15}$$

whence we get

$$\text{Im}\frac{1}{f_0} = -p, \quad f_0 = \frac{1}{g_0(p^2) - ip}, \tag{4.16}$$

where $g_0(p^2)$ is some real function of p^2. At low energies it can be expanded in a series; as we shall see, in the case of scattering from a potential well the expansion parameter is the ratio of the energy to the depth of the well. Formula (4.15) is a special case of the optical theorem, eqn. (4.12), which relates the imaginary part of the forward scattering amplitude to the total cross section.

Now let us investigate the behaviour of an arbitrary partial-wave amplitude f_ℓ at low energies. We have by definition

$$f(\theta) = \sum_{l=0}^{\infty} f_l (2l+1) P_l(\cos\theta).$$

By introducing in eqn. (4.10) the scattering amplitude in place of T_{ab} and replacing the sum of c by an integration over the intermediate momentum $\underset{\sim}{p}'$ (but not setting a = b as in (4.14)) we find

$$\operatorname{Im} f(\vartheta) = \\ = \frac{p}{4\pi} \int f^*(\gamma) f(\gamma')\, d\gamma'.$$

Here $\gamma = (\theta, \varphi)$ and $\gamma' = (\theta', \varphi')$ are the angles shown in Fig. 40. Expanding the amplitudes $f(\vartheta)$, $f(\gamma)$ and $f(\gamma')$ in partial-

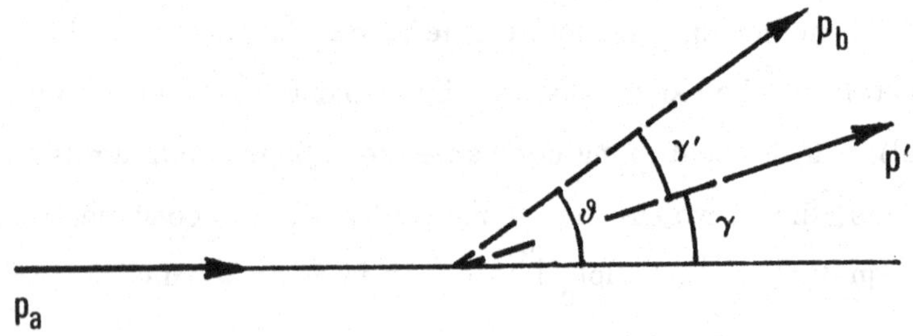

Fig. 40

wave amplitudes, we get

$$\sum_l (2l+1) \operatorname{Im} f_l P_l(\cos\vartheta) = \\ = \frac{p}{4\pi} \sum_{ll'} (2l+1)(2l'+1) f_l^* f_{l'} \int P_l(\cos\theta) P_{l'}(\cos\theta')\, d\gamma'.$$

According to the addition theorem for Legendre polynomials we have

$$P_l(\cos\theta) = P_l(\cos\vartheta)\, P_l(\cos\theta') + \ldots$$

The omitted terms contain expressions of the form $\cos m(\varphi - \varphi')$, which give zero after integration over φ'. Using the property of orthonormality of the Legendre polynomials, we finally get $\mathrm{Im}\, f_\ell = p\, |f_\ell|^2$, whence

$$f_l = \frac{1}{g_l(p^2) - ip}.$$

Let us return to the case of spherically symmetric (s-wave) scattering. We will see below (p. 240) that for small p^2 it is in fact s-wave scattering which dominates. Expanding $g_0(p^2)$ in a Taylor series and introducing the notation $-g_0(0) = \varkappa$, we find

$$f_0 \approx \frac{1}{-\varkappa - ip + \alpha p^2}. \tag{4.17}$$

The quantity f_0 has a pole at the point $p \approx i\varkappa$ (the second root of the quadratic equation $-\varkappa - ip + ap^2 = 0$ corresponds to values of the momentum p so large that the Taylor expansion of g_0 is no longer applicable). If we consider f as a function of the energy $p^2/2$ (where we put the mass of the particle equal to unity), then there is a square-root branch point at $p^2 = 0$. To make $f_0(p^2)$ single-valued, we make a cut in the p^2 plane from the origin along the real positive axis (see Fig. 41). This cut separates the p^2 plane into two sheets. If $\varkappa > 0$, the pole of f_0 lies on the negative real axis of the first sheet, while if $\varkappa < 0$ it is on the same axis but on the second sheet.

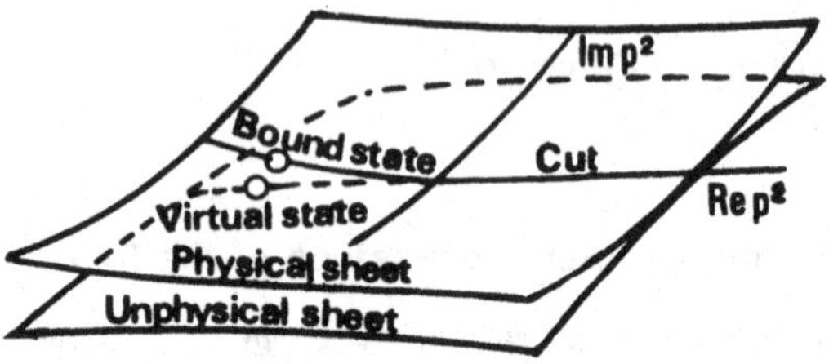

Fig. 41

The quantity $\varkappa$ may be expressed in terms of the energy of a weakly-bound state in the scattering potential (if such a state exists). Substituting (4.17) in (4.14), we get

$$u \underset{r\to\infty}{\simeq} -\frac{1}{2ip}\left(e^{-ipr} - \frac{-\varkappa + ip}{-\varkappa - ip}\, e^{ipr}\right).$$

The function u may be analytically continued into the region of negative energies, which corresponds to imaginary values of p. Suppose the energy of the bound state is $-E_0$, and define $\varkappa_0 = (2E_0)^{1/2}$. Then for $p = \pm i\varkappa_0$ the wave function must just turn into the bound-state wave function. Hence, in the expression for u for $p = \pm i\varkappa_0$ the term involving an increasing exponential must vanish. For $p = i\varkappa_0$ we get

$$u = \frac{1}{2\varkappa_0}\left(e^{\varkappa_0 r} - \frac{-\varkappa - \varkappa_0}{-\varkappa + \varkappa_0}\, e^{-\varkappa_0 r}\right).$$

To be able to neglect the increasing exponential we must put $\varkappa = \varkappa_0 > 0$. Thus, in the case of a bound state the scattering amplitude has a pole at $p = i\varkappa_0$:

$$f \approx -\frac{1}{\varkappa_0 + ip}. \tag{4.18}$$

The position of the pole is fixed by the energy of the bound state. This formula is meaningful so long as the energy of the bound state is small compared to the depth of the potential well. In this case, when the scattered particles have low energy, there occurs a resonance enhancement of the effective scattering cross section, and we can neglect the effect of potential scattering and of the other levels.

Since $\varkappa_0 > 0$, the pole corresponding to the bound state lies on the first sheet of the complex plane of p^2. It can be shown that all poles on the first sheet represent bound states; for this reason the first sheet is called the "physical" sheet (cf. Fig. 41).

In the case $\varkappa < 0$ the pole of the scattering amplitude lies on the second ("unphysical") sheet and corresponds to a so-called virtual state. For instance, the neutron-proton singlet scattering amplitude has a pole at an energy of about -70 keV which corresponds to a virtual state. Such poles show up as resonances in the low-energy scattering, but do not correspond to bound states.

There may be other singularities on the unphysical sheet; in contrast to the case of the physical sheet, they may occur at any point in the complex plane. Particular interest attaches to poles situated near the positive real axis of p^2 at the points $p^2 = p_0^2 - i\gamma$ ($\gamma > 0$). Each of these poles will strongly affect the scattering amplitude at energies close to $p_0^2/2$ and lead to resonance scattering with a width γ. These poles define the so-called quasi-stationary states; the width γ is equal to the inverse lifetime of the state (see p. 182).

Problem: By a procedure similar to the one in the text, use the law of conservation of particle number to obtain the optical theorem in the case of absorption represented by a cross section σ_c.

Solution:

$$\mathrm{Im}\, f_0 = \frac{p}{4\pi}\left(4\pi\,|f_0|^2 + \sigma_c\right).$$

Nonresonant Scattering at Low Energies

Let us investigate the analytic properties of the scattering amplitude $f(p, \theta)$ at low energies. First we establish the conditions for the function $f(p, \theta)$ to be analytic at $p = 0$. For this purpose we write down the equation which relates $f(p, \theta)$ to the wave function for the scattering problem, $\Psi_p = e^{i\mathbf{p}\cdot\mathbf{r}}\, u_p(r)$ where u_p is a function which modulates the plane wave. As we shall see, the singularities of $f(p, \theta)$ for $p = 0$ are determined by the behaviour of Ψ_p and V at large distances r, where we have

$$\Psi_p \to e^{i\mathbf{pr}} + \frac{f}{r} e^{ipr}$$

and

$$u_p \to 1 + \frac{f}{r} e^{i(pr - \mathbf{pr})},$$

i. e. $u_p \approx 1$ for $r \to \infty$. The question of analyticity for $p \to 0$ may be discussed in the Born approximation. In fact we have

$$f(p, \theta) = -\frac{1}{2\pi}\int e^{-i\mathbf{qr}} V(r)\, u_p(r)\, d\mathbf{r} \sim -\frac{1}{2\pi}\int e^{-i\mathbf{qr}} V(r)\, d\mathbf{r},$$

where

$$q^2 = 2p^2 - 2\mathbf{p}\mathbf{p}'.$$

If $V(r)$ decreases exponentially (or faster) as $r \to \infty$, then all the derivatives of f with respect to p are finite (since the corresponding integrals converge), which means that f is analytic at $p = 0$. Suppose on the other hand that $V(\mathbf{r})$ decreases for $r \to \infty$ according to a power law; then differentiate $f(p, \theta)$ with respect to p a sufficient number of times. At each differentiation of the exponential $e^{-i\mathbf{q}\cdot\mathbf{r}}$ the integrand acquires an extra factor proportional to r, which softens the decrease of this expression for $r \to \infty$. Thus, sufficiently high derivatives will diverge for $p \to 0$. Thus, for a potential which decreases only as a power law for $r \to \infty$, the scattering amplitude has a singularity at $p = 0$. However, if the power according to which the potential decreases is large, this has no substantial effect on the low-energy scattering, since the singularity shows up only in very high derivatives of the scattering amplitude.

Let us now find the p-dependence of the scattering amplitude for small p in the case of a potential which decreases sufficiently fast (exponentially or faster) at infinity. For this purpose it is convenient to go over from the variables p, θ to the variables $\mathbf{p}\cdot\mathbf{p}'$ and p^2. Then by definition of the partial-wave scattering amplitudes we have

$$f = \sum_{l=0}^{\infty} f_l P_l\left(\frac{\mathbf{p}\mathbf{p}'}{p^2}\right).$$

where, as we know, the function P_ℓ is a polynomial of the ℓ-th degree; in fact,

$$P_l\left(\frac{pp'}{p^2}\right) = C_1\left(\frac{pp'}{p^2}\right)^l + C_2\left(\frac{pp'}{p^2}\right)^{l-2} + \ldots$$

Let us consider the behaviour of the amplitude for fixed $\underset{\sim}{p}.\underset{\sim}{p}'$ when $p^2 \to 0$; that is, we consider the analytic continuation of the amplitude into the unphysical region of angles $\cos\theta \to \infty$ and require analyticity in this region. This requirement is equivalent to that of analyticity in p^2 for fixed q^2 (q = momentum transfer); the latter can easily be proved for sufficiently fast decreasing potentials, and is particularly obvious in the Born approximation, in which f is completely independent of p^2 at fixed q^2.

For a fixed value of $\underset{\sim}{p}.\underset{\sim}{p}'$ and $p^2 \to 0$, the function f has no singularity with respect to p only if

$$f_l = d_1 p^{2l} + d_2 p^{2l+1} + \ldots$$

Thus, if there is no physical reason for d_1 to be zero, then at small p the amplitude f_ℓ is proportional to $p^{2\ell}$. Here $d_1 \sim R^{2\ell+1}$, where R is the length characterizing the potential. The phase shift δ_ℓ is related to f_ℓ by the standard relation*

$$f_l = \frac{(e^{2i\delta_l} - 1)}{2ip}.$$

whence we get for small p that $\delta_\ell \sim p^{2\ell+1}$.

Scattering by a Potential Well

As an example to illustrate the analytic properties of the

---

* L.D. Landau and E.M. Lifshitz, Quantum Mechanics, Pergamon, Oxford 1965, p. 472.

scattering amplitude, we shall consider the scattering of a low-energy particle by a potential well. We assume that the well has a sufficiently sharp edge and denote its effective radius by R. We shall consider energies E of the incident particle low enough that the condition $pR \ll 1$ is fulfilled, where $p = \sqrt{2E}$. As we have just seen, when this condition is fulfilled only s-wave scattering is important.

Outside the well the function $u(r) = r\,\Psi(r)$ has a form identical to its asymptotic expression:

$$u(r) = \frac{\sin pr}{p} + f_0 e^{ipr}.$$

For $r \sim R$ this gives

$$u(r) \approx r + f_0(1 + ipr).$$

Let us denote the logarithmic derivative of the wave function inside the well for $r \sim R$ by $g_0(E)$. Matching the logarithmic derivatives of the wave function inside and outside the well, we find

$$g_0(E) = \frac{1 + ipf_0}{f_0},$$

whence

$$f_0 = \frac{1}{g_0(E) - ip}.$$

This result constitutes an alternative derivation of formula (4.16).

The general nature of $g_0(E)$ may be seen from the example of a square well. Denote the depth of the well by U_0; then inside the well we have $u(r) = A \sin kr$, where $k = [2(U_0 + E)]^{1/2}$. Consequently we have

$$g_0(E) = k \cot kR.$$

The expression on the right-hand side of this equation may be expanded in a series in integral powers of the small quantity E/U_0. Thus, $g_0(E)$ is indeed an analytic function of the energy E.

If inside the well there is a bound state with energy $-E_0$, then the wave function outside the well has the form $u(r) = B\exp(-\sqrt{2E_0}\, r)$. Matching the logarithmic derivatives as above, we find $g_0(-E_0) = -\sqrt{2E_0} \equiv -\varkappa_0$. If the energies E and E_0 are small compared to U_0, then we can assert that $g_0(E) \approx g_0(-E_0) = -\varkappa_0$, and we get back the result (4.18). Small values of $g_0(E) = k \cot kR$ are attained when $kR \approx \pi/2$; hence we get the standard condition for the appearance of a bound state in a square well, $U_0 > \pi^2/8R^2$.

An analogous technique may be used for the case of a well with a barrier.* The result has the form

$$f_0 = \frac{1}{-ip - \varkappa + \alpha p^2}.$$

Here $\alpha \sim -R_1 e^{-\xi}$, where $\xi = 2\int_{R_1}^{R_2} \sqrt{2(V-E)}\, dr$ is the exponent in the expression for the penetrability of the barrier, and R_1, R_2 are the classical turning points.

Analytic Properties of the Wave Function

According to Poincarè's theorem, singularities of the solution to a linear differential equation can occur only at the singularities of the coefficients of this equation (apart from possible

---

* A. B. Migdal, A. M. Perelomov and V. S. Popov, Ya. Fiz. 14, 829 (1971) (English translation: Soviet Journal of Nuclear Physics 14, 488 (1972).)

singularities at infinity). For instance, the harmonic-oscillator potential $V = \alpha x^2$ for Schrödinger's equation in one dimension has no singularities in any finite region of space, and hence the solution of Schrödinger's equation can have a singularity only at infinity. The same is true for the three-dimensional case ($V = \alpha r^2$); for example, the groundstate wave function Ψ is a constant times $\exp(-\sqrt{2\alpha}\, r^2)$.

Consider the analytic properties of the wave function $\Psi(r)$ with respect to r for a particle moving in an arbitrary spherically symmetric potential with angular momentum zero. Since no direction is singled out in this system, the wave function Ψ must be analytic in the variable r^2; it can have singularities only at points where the potential is singular as a function of r^2.

However, the singular points of the coefficients are not necessarily singular points of the solution. For example, in the equation

$$u_l'' + 2\,(E - V_l)\,u_l = 0,$$

where $V_\ell = V + \ell(\ell + 1)/2r^2$, the point $r = 0$ is a singular point of a coefficient in the equation, namely V_ℓ. Suppose that for small r one may neglect the quantity V in comparison with $\ell(\ell + 1)/r^2$. Then the solution of Schrödinger's equation takes the form

$$-r^2 u_l'' + l\,(l + 1)\,u_l = 0,$$

whence $u_\ell \sim r^{\ell+1}$ or $u_\ell \sim r^{-\ell}$. Thus there exists a solution (namely $u_\ell \sim r^{\ell+1}$) which is regular at the point $r = 0$.

The Coulomb potential $V = Z/r = Z/(x^2 + y^2 + z^2)^{1/2}$ has, for fixed y and z, a square-root branch point with respect to the variable x. Consequently, the wave function may also have a singularity at this point. For instance, the groundstate of the particle is given, as is well known, by

$$\Psi \sim \exp[-\sqrt{x^2+y^2+z^2}].$$

In particular, the origin is a singular point. We saw above (p. 62) that this singularity of the Coulomb wave functions determines the nature of the energy-dependence of the photoeffect cross section.

In the following we shall return once more to the analyticity properties of the wave function, but as a function of energy rather than of r.

Single-Particle Wave Functions of the Continuous Spectrum at Low Energy

We will obtain an expression for the wave functions of the continuous spectrum at low energies ϵ_p ; this will be useful in later calculations. In the calculation of matrix elements there usually enter functions evaluated at distances of the order of the radius R of the potential well. We shall now show that if $pR \ll 1$, then the wave function $\varphi_{\underset{\sim}{p}}(r)$ can be written as the product of a factor depending only on p with the function $\varphi_0(r)$ which satisfies the Schrödinger equation for $\underset{\sim}{p} = 0$. The argument is that $\nabla^2 \varphi_0$ is of order φ_0/R^2, so that in the equation for $\varphi_{\underset{\sim}{p}}(\underset{\sim}{r})$

$$\nabla^2\varphi_p(r) + 2(\varepsilon_p - V)\varphi_p(r) = 0$$

the quantity $\epsilon_p = p^2/2$ may be neglected if $pR \ll 1$.

Thus, we shall represent $\varphi_{\underset{\sim}{p}}$ in the form*

$$\varphi_p(r) = \chi(p)\,\varphi_0(r). \tag{4.19}$$

For simplicity we restrict ourselves to spherically symmetric states. For φ_0 we choose the solution of Schrödinger's equation which is finite at the origin and at infinity satisfies the condition

$$\varphi_0(r) \underset{r\to\infty}{=} \frac{1}{r}.$$

As the binding energy ϵ_0 tends to zero the function φ_0 differs from the bound-state wave function only by a normalization factor. For two particles of identical mass (for instance, for the case of the deuteron) we easily find

$$\varphi_d = a\varphi_0, \qquad a = \frac{\sqrt[4]{M\varepsilon_0}}{\sqrt{2\pi}}.$$

We will assume that the function $\varphi_{\underset{\sim}{p}}(\underset{\sim}{r})$ is normalized in the interval $d\underset{\sim}{p}$; then outside the well the s-wave part of $\varphi_{\underset{\sim}{p}}(\underset{\sim}{r})$ has the form

$$\frac{u_p(r)}{r} = \frac{\sin(pr + \delta_0(p))}{pr}.$$

As we saw above, for small p the s-wave partial-wave scattering amplitude f_0 has a pole at the point $p = i\varkappa$ (where $\varkappa > 0$ for a real level and $\varkappa < 0$ for a virtual one). In the neighbourhood of this pole we also get a resonance increase in the

---

* A similar property of the continuous-spectrum wave functions for energies near the energy of a quasistationary level has been used in the paper by V. M. Galitskii and V. F. Chel'tsov (Nucl. Phys. 56, 86 (1964)); see also the reference on p. 242.

phase shift $\delta_0(p)$, which is related to f_0 by the equation

$$|f_0| = \frac{1}{p}|\sin\delta(p)|.$$

Consequently there is a wide region $R \lesssim r \ll 1/p$ where $\delta_0(p) \gg pr.$; thus, we have

$$u_p \approx \frac{1}{p}\sin\delta_0(p),$$

while $r\varphi_0$ in this region is equal to 1 (because of the normalization condition).

Thus, using (4.17), we find

$$|\chi(p)|^2 = \frac{1}{p^2}\sin^2\delta_0(p) = |f_0|^2 = \frac{1}{(\alpha p^2 - \varkappa)^2 + p^2}. \tag{4.20}$$

Far from the pole, i.e. for $p^2 \sim U_0$ (where U_0 is the depth of the well), we have $|\chi(p)|^2 \sim 1/U_0$.; near the pole (e.g. for $\alpha p^2 - \varkappa = 0$) we have $|\chi(p)|^2 \sim 1/E$. (Below we shall consider some examples of the ways in which we can use this property of the wave functions). Thus, for small p and $\varkappa$ the continuous-spectrum wave functions are enhanced at short distances by a factor $\sqrt{U_0/E}$.

The fact that the wave function $\varphi_{\underset{\sim}{p}}(\underset{\sim}{r})$ can be decomposed into a product of two terms, one depending only on $\underset{\sim}{r}$ and the other only on $\underset{\sim}{p}$, permits us to calculate matrix elements extremely simply; the integration over $\underset{\sim}{r}$ simply reduces to taking the average with respect to the function φ_0.

3. THE USE OF ANALYTICITY PROPERTIES IN PHYSICAL PROBLEMS

Theory of Nuclear Reactions with the Formation of Slow Particles

We shall give one more example of the way in which we can solve a complex problem by using the singularities of the wave function. In this case the pole whose existence must be taken into account is not in the scattering amplitude from a potential well, as it was in the problem of the previous section, but in the scattering amplitude of two nucleons produced as a result of a reaction (Migdal, 1950: Watson, 1952).

Consider a nuclear reaction which results in the production, along with other particles, of two nucleons with small relative energy. Such a case occurs, for instance, in the production of a π-meson in the collision of two nucleons with energy near the threshold energy for pion production, or in the disintegration of a deuteron by a neutron when the energy of the emerging proton is near its maximum value and the relative energy of the neutrons is sufficiently small. Apart from unimportant factors, the cross section for such a process will be proportional to the square of a matrix element containing the wave function of relative motion of the pair of nucleons in question:

$$\sigma \sim |(\Phi, \varphi_p\ (\boldsymbol{r}_1 - \boldsymbol{r}_2))|^2. \tag{4.21}$$

The quantity Φ contains integrals over the coordinates of the other particles participating in the reaction and at small relative momentum p is insensitive to the relative energy of the two nucleons in question. Since the important distances in the integral (4.21) are of order $|\underset{\sim}{r}_1 - \underset{\sim}{r}_2| \sim r_0$, where r_0 is the range of the forces (so that $p r_0 \ll 1$), we may use for φ_p the relation (4.19). Hence, with the normalization of φ_p chosen in (4.20), the cross section is of the form

$$d\sigma = A_1 |f|^2 \, dp = A \frac{dp}{E + \varepsilon_0}, \tag{4.22}$$

where $E = p^2/M$ is the energy of relative motion (the reduced mass of the nucleons is $M/2$). Here we used the expression (4.17) for the scattering amplitude S of two nucleons. The quantity ϵ_0 depends on the type and total spin of the nucleons: for a neutron-proton system with spin 1 ϵ_0 is 2.2 MeV (the deuteron binding energy), while for spin 0 we have $\epsilon_0 \approx 70$ keV (the energy of the virtual state). For two neutrons there is a pole in the amplitude only for spin 0 ($\epsilon_0 \simeq 70$ keV, cf. below); two neutrons with spin 1 cannot be in an S-state because of the Pauli principle, and hence have only a small nonresonance scattering amplitude (as we saw on p.240 , the $\ell \neq 0$ amplitude decreases with decreasing p). In the case of two protons formula (4.22) becomes more complicated because of their Coulomb repulsion.

Since we have $d\underset{\sim}{p} \sim E^{1/2}\, dE$, the distribution with respect to relative energy has the form

$$dW_E = \text{const} \frac{\sqrt{E}\, dE}{E + \varepsilon_0}. \tag{4.23}$$

The graph of the function dW_E/dE is shown in Fig. 42; the reaction cross section has a maximum for $E = \epsilon_0$.

Let us now find the dependence of the reaction probability on the angle between the emerging nucleons. Denote by $p_{\parallel}$ the

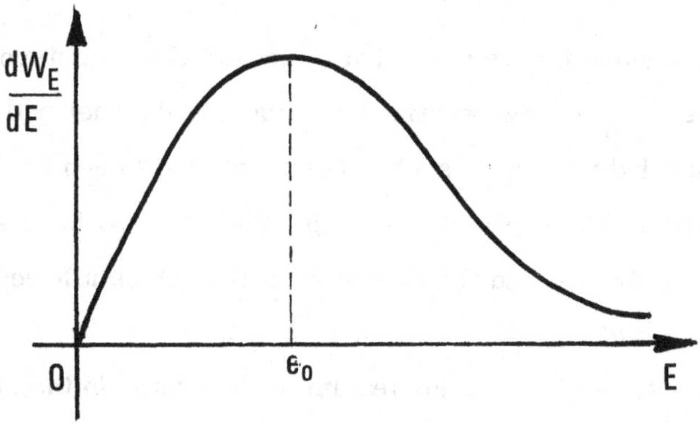

Fig. 42

component of $\underset{\sim}{p}$ along the direction of the total momentum $\underset{\sim}{P}$ of the two nucleons, and by $p_\perp$ the component perpendicular to $\underset{\sim}{P}$. The angle between the momenta $(\underset{\sim}{p} + \underset{\sim}{P})/2$ and $(\underset{\sim}{p} - \underset{\sim}{P})/2$ of the first and second nucleon respectively is given by the relation

$$\sin\theta = \frac{|(p+P)\times(-p+P)|}{|p+P|\cdot|-p+P|} \approx \frac{2p_\perp}{P} \ll 1.$$

where we assumed that $p \ll P$. In terms of the variables $p_\parallel$, $p_\perp$ the phase volume (density of states) $d\underset{\sim}{p}$ is proportional to $p_\perp \, dp_\perp \, dp_\parallel$. Substituting in (4.22), we find

$$dW_p = \text{const}\,\frac{\theta\,d\theta\,dp_\parallel}{M\varepsilon_0 + p_\parallel^2 + \frac{P^2\theta^2}{4}}.$$

and integration over $dp_\parallel$ gives

$$dW_\theta = \text{const}\,\frac{\theta\,d\theta}{\sqrt{\theta^2 + \varepsilon_0/E_0}}, \tag{4.24}$$

where $E_0 = P^2/4M$ is the energy of centre-of-mass motion of the two nucleons. Thus, the characteristic angles are of order $(\epsilon_0/E_0)^{1/2}$.

The above theory predicted the possibility of determining the constant ϵ_o for two neutrons by studying the energy distribution of the third particle; such experiments have been successfully carried out on the reaction $d + n \rightarrow p + 2n$ (V.K. Voitovetskii et al., 1965). Let us find the proton spectrum which allowed the determination of ϵ_o.

Denote by $\underset{\sim}{P}_p$ the momentum of the proton in the centre-of-mass frame of the three nucleons; then the momentum of the two-neutron system is $-\underset{\sim}{P}_p$. The total energy of the system, E_t, is made up of the proton energy, the energy of centre-of-mass motion of the two neutrons and their relative energy E:

$$E_t = \frac{1}{2M} P_p^2 + \frac{1}{4M} P_p^2 + E = \frac{3}{2} E_p + E.$$

The maximum possible proton energy E_p^m, corresponding to $E=0$, will be $E_p^m = 2E_t/3$. The probability distribution as a function of E_p is given by the expression obtained from (4.23) by going over to the variable E_p:

$$dW_{E_p} = \text{const} \frac{\sqrt{E_p^m - E_p}\, dE_p}{\varepsilon_0 + \frac{3}{2}(E_p^m - E_p)}. \tag{4.25}$$

A comparison of this distribution with the experimentally observed one showed that $\epsilon_o \simeq 70$ keV. It was moreover shown that this must represent a virtual level, since otherwise the proton distribution would contain a monochromatic line at energy $E_p = E_p^m + \epsilon_o$ corresponding to a bound state of the two neutrons, and this was not observed experimentally.

Thus the energy of the virtual level of two neutrons with spin 0 is the same as the energy of the virtual level of a neutron and a proton with the same spin, in agreement with the hypothesis of isotopic invariance of the nuclear forces. The theory enables us to calculate the relative probability of emission of a free neutron and proton and emission of a deuteron; according to (4.21) the ratio of these probabilities is given by

$$dW = \frac{|(\Phi, \varphi_p(r_1 - r_2))|^2}{|(\Phi, \varphi_d(r_1 - r_2))|^2} \frac{d^3p}{(2\pi)^3}. \tag{4.26}$$

Using the relations $\varphi_p = \chi(p)\,\varphi_o$, $\varphi_d \simeq a\,\varphi_o$ and the normalization of the bound-state wave function φ_d given in the last section (p. 245) we find

$$d\sigma_{n,p}^{\uparrow\uparrow} = \sigma_d \frac{1}{8\pi^2 \sqrt{\varepsilon_0}} \frac{\sqrt{E}\,dE}{E + \varepsilon_0}, \tag{4.27}$$

where $d\sigma_{n,p}^{\uparrow\uparrow}$ corresponds to a free neutron and proton with parallel spins. For a crude estimate of the ratio of the corresponding total cross sections, we assume that (4.27) is valid over the whole region of values of E up to E_m. In this way we find

$$\frac{\sigma_0}{\int d\sigma} \simeq 4\sqrt{\frac{\varepsilon_0}{E_m}}. \tag{4.28}$$

<u>Interacting Particles in a Potential Well</u>

The simple form of the wave functions at low energy found in the last section allows us to solve the problem of the motion of two interacting particles in a potential well*. The

---

* A. B. Migdal, Ya. Fiz. XVI, 8 (1972) (English translation: Soviet Journal of Nuclear Physics <u>16</u>, 5 (1973).

Hamiltonian for this problem has the form

$$H = H_1(r_1) + H_2(r_2) + H'(r_1, r_2). \tag{4.29}$$

We assume that we know the eigenfunctions $\varphi_\lambda^{(1)}(\mathbf{r}_1)$, $\varphi_\lambda^{(2)}(\mathbf{r}_2)$ of the one-particle problem, which satisfy the equation

$$H_{1,2}(r)\varphi_\lambda^{(1,2)}(r) = \varepsilon_\lambda^{(1,2)}\varphi_\lambda^{(1,2)}(r), \tag{4.30}$$

and assume that there exists a bound state with angular momentum zero and energy ϵ_0 near zero. Then the functions $\varphi_p^{(\ell)}(\mathbf{r})$ with angular momentum ℓ and $pr \ll 1$ can be written in a form similar to the formulae of the last section:

$$\varphi_p^{(l)}(r) = \varphi_0^{(l)}(r)\chi^{(l)}(p),$$

where $\chi^0(p) \equiv \chi(p)$ is given by expression (4.20). For small $\mathbf{r}$ the bound-state wave function φ_{ϵ_0} differs from $\varphi_0^{(0)}$ only by the normalization factor a (cf. below).

We expand the eigenfunction $\Psi(\mathbf{r}_1, \mathbf{r}_2)$ which is the solution of the equation

$$H\Psi = E\Psi, \tag{4.31}$$

with respect to the eigenfunctions of the problem without the interaction H':

$$\begin{aligned}\Psi = &\sum_{l, l' \neq 0} \int dp\, dp' C_{ll'}(p, p')\varphi_p^{(l)}(r_1)\varphi_{p'}^{(l')}(r_2) + \\ &+ \sum_{l \neq 0} \int dp \{C_{l0}(p)\varphi_p^{(l)}(r_1)\varphi_{\varepsilon_0}(r_2) + C_{0l}(p)\varphi_{\varepsilon_0}(r_1)\varphi_p^{(l)}(r_2)\} + \\ &+ C_{00}\varphi_{\varepsilon_0}(r_1)\varphi_{\varepsilon_0}(r_2).\end{aligned} \tag{4.32}$$

where we have assumed for simplicity that the two particles have an identical set of wave functions φ_λ. Eqn. (4.32) may be written symbolically

$$\Psi = \Sigma C_\alpha \Psi_\alpha^0. \tag{4.33}$$

Then Schrödinger's equation (4.31) takes the form

$$(E - E_\alpha^0) C_\alpha = (\Psi_\alpha^0 H' \Psi) = \sum_\beta (\Psi_\alpha^0 H' \Psi_\beta^0) C_\beta. \tag{4.34}$$

In this equation we cannot use the simple form of the functions $\Psi_\alpha^0 (r_1, r_2)$ which is valid for small $r_{1,2}$, since in the integrals $C_\alpha = (\Psi_\alpha^0, \Psi)$ both large and small distances are important. It is therefore more convenient to write down an equation for the quantity $A_\alpha = (\Psi_\alpha^0\ H'\ \Psi)$, in which, as we see, only small distances $r_{1,2} \sim R$ are important. We find from (4.34)

$$A_\alpha = \sum_\beta C_\beta (\Psi_\alpha^0 H' \Psi_\beta^0) = \sum_\beta \frac{(\Psi_\alpha^0 H' \Psi_\beta^0)}{E - E_\beta^0} A_\beta. \tag{4.35}$$

Since at small distances states with angular momentum $\ell = 0$ are enhanced owing to the presence of the bound state (cf. (4.20)), we may keep in the expansion (4.32) only states with $\ell = 0$. (It can be shown that taking the $\ell \neq 0$ terms into account leads to appreciable corrections only in the case of a resonance interaction between the particles (Dyugaev, 1974)). Then Ψ takes the simple form

$$\Psi = C \varphi_0 (r_1) \varphi_0 (r_2), \tag{4.36}$$

and the system of functions Ψ_α^0 is given by the expression:

$$\Psi_\alpha^0 = \begin{cases} \chi (p_1) \chi (p_2) \varphi_0 (r_1) \varphi_0 (r_2) & \text{(two particles in the continuous spectrum)} \\ \chi (p) a \varphi_0 (r_1) \varphi_0 (r_2) & \text{(one free particle)} \\ a^2 \varphi_0 (r_1) \varphi_0 (r_2) & \text{(both particles bound)} \end{cases} \tag{4.37}$$

Here $\varphi_p(r)$ is normalized in the interval dp (i. e. $\varphi_p(r) \to (2/\pi)^{1/2} r^{-1} \sin(pr + \delta))$, and so the normalization of φ_o differs from that on p. 245 by the factor $(2/\pi)^{1/2}$ p.

All the matrix elements of the form $(\Psi^o_\alpha H' \Psi^o_\beta)$ are determined by the behaviour of Ψ^o_α at small distances from the potential well. Indeed, if we assume that the interaction H' decreases sharply when $|\underset{\sim}{r}_1 - \underset{\sim}{r}_2| \gg r_o$, we get

$$(\Psi^0_\alpha H' \Psi^0_\beta) \sim$$
$$\sim \int d\boldsymbol{r}_1 d\boldsymbol{r}_2 \varphi_{\lambda_1}(r_1) \varphi_{\lambda_2}(r_2) H'(\boldsymbol{r}_1 - \boldsymbol{r}_2) \varphi_{\lambda_3}(r_1) \varphi_{\lambda_4}(r_2) \sim$$
$$\sim \int dr\, r^2 (\varphi_0(r))^4 \int d(\boldsymbol{r}_1 - \boldsymbol{r}_2) H'(\boldsymbol{r}_1 - \boldsymbol{r}_2).$$

(Because φ_o decreases as r^{-1}, the first integral in the above expression is determined by small distances from the well ($r \sim R$), where the resonance character of the functions φ_λ dominates).

Writing $\varphi_{\epsilon_o}(r)$ in the form

$$\varphi_{\varepsilon_0}(r) \underset{r \gg R}{=} \frac{a e^{-\varkappa r}}{r} \sqrt{\frac{2}{\pi}} p \quad (\varkappa = \sqrt{2m\varepsilon_0})$$

and assuming that $\varkappa R \ll 1$, we find

$$a^2 \simeq \frac{\pi \varkappa}{p^2}. \tag{4. 38}$$

When we substitute (4. 36), (4. 37) and (4. 38) in eqn. (4. 35), the constant C cancels and we get

$$1 = H_0' \left\{ \frac{a^4}{E - 2\varepsilon_0} + 2a^2 \int \frac{\chi(p)\, dp}{E - \varepsilon_0 - \varepsilon_p} + \int \frac{\chi(p_1) \chi(p_2)}{E - \varepsilon_{p_1} - \varepsilon_{p_2}} dp_1 dp_2 \right\}, \tag{4. 39}$$

where $H_o' = (\varphi_o(r_1) \varphi_o(r_2)\ H' \varphi_o(r_1) \varphi_o(r_2))$

Thus, by using the resonance character of the functions $\varphi_p^{(o)}$ we have been able to reduce the complex integral equation

(4.35) to a simple algebraic equation for the energy E (eqn. (4.39)). Consideration of this equation in the problem of the motion of two nucleons in the potential well of the nucleus leads to the conclusion that whenever there is a one-particle level with energy near zero, there must exist at the surface of the nucleus a bound state of two neutrons (or two protons).

Theory of Direct Reactions

When a classical particle executes a finite motion, it remains all the time within the range of the forces confining it. In quantum mechanics this is not so: a system of bound particles may, as it were, decay into free particles for a short time. Such temporary decays are called virtual transitions. The possibility of virtual transitions produces phenomena which are at first sight paradoxical; for instance, a proton incident on a nucleus will in some fraction of cases knock a particle out of it in just the same way as if elastic scattering of the proton had taken place on a free rather than a bound particle. In particular, in the reaction (p, 2p) the angle between the directions of the emerging protons is near 90°, as is the case in scattering of free particles of identical mass when one was initially at rest. Even more remarkable is the deuteron knock-out reaction (p, pd), in which one can also observe kinematic correlations characteristic of the elastic pd-scattering on free deuterons. Since the binding energy of the deuteron is considerably less than the energy of interaction of nucleons with the nucleus, the deuteron certainly cannot exist as a stable structure within the nucleus. However, this does not exclude formation of a free deuteron for a short time as a result of a virtual transition. A proton incident on the nucleus, on colliding with such a deuteron, will transfer energy and momentum to it according to the laws of elastic scattering. Thus the cross sections for the reactions (p, 2p), (p, pd) etc. can be approximately expressed in terms of the amplitudes for virtual decay and for the elastic scattering of the free particles. Below we shall see what are the conditions for this approximation to hold. These reactions are a special case of a wide class of processes which go under the general name of direct reactions; their defining characteristic is that almost all the energy (and momentum) brought into the nucleus is transferred to some one single particle, while the rest of the nucleus takes no part in the process.

From a theoretical point of view direct reactions are characterized by the fact that the amplitudes for these reactions, regarded as analytic functions of the kinematic variables, have singularities in the momentum transfer which are close to the physical region.*

Let us take as an example the above-mentioned reaction (p, 2p). Suppose that a proton of sufficiently high energy E(p) is incident on the nucleus X(N, Z) and there occurs the reaction

$$X\,(N,\,Z) + \mathrm{p} \rightarrow Y\,(N,\,Z-1) + 2\mathrm{p}. \tag{4.40}$$

---

* The analytic properties of direct-reaction amplitudes were first investigated by I. S. Shapiro, Zh. Eksp. Teor. Fiz. 41, 1616 (1961) (English translation: Soviet Physics JETP 14, 1148 (1962); cf. also I. S. Shapiro, Uspekhi Fiz. Nauk 92, 549 (1967) (Soviet Physics - Uspekhi 10, 515 (1968).)

Then the direct-reaction amplitude will be determined by the formula for the amplitude for transition via the intermediate state:

$$A(p, q; p', q') = \frac{\Phi(q) F(p, q; p', q')}{E_X - E_Y - E(q)}, \qquad (4.41)$$

where $\Phi(q)$ is the amplitude for a virtual transition of the nucleus X to the nucleus Y plus a free proton, with momentum q, $F(p, q: p', q')$ is the scattering amplitude of the incident proton on the virtual proton and E_X, E_Y are the internal energies of nucleus X and nucleus Y respectively. It should be emphasized that formula (4.41) is a quite general quantum-mechanical formula for the amplitude of transition via a given intermediate state, and in no way assumes that the interaction between the particles is small. Taking account of the recoil energy of nucleus Y leads to a change of the effective mass and energy of the virtual proton: $E(q) = q^2/2M_{eff}$, where $M_{eff} = MM_Y/(M + M_Y)$. The quantity $E_Y - E_X = E_0 > 0$ is the binding energy of the proton in the nucleus Y. The function $\Phi(q)$ is defined by the matrix element

$$\Phi(q) = (\Psi_X^{N, Z}(r_j, r), \Psi_Y^{N, Z-1}(r_j) e^{iqr}),$$

where r_j indicates the coordinates of all the nucleons except the ones considered specially (we have omitted spin suffices).

For a crude estimate of the dependence of $\Phi(q)$ on q we may assume that the wave functions of all the remaining nucleons in nucleus Y differ only slightly from the wave functions of these nucleons in nucleus X. We assume for simplicity that nucleus X differs from nucleus Y only in the presence of a proton in some state λ near the Fermi surface. Then we have

$$\Phi(q) = (\varphi_\lambda(r)\, e^{iqr}). \tag{4.42}$$

To illustrate the form of $\Phi(q)$ we consider the case when the orbital angular momentum of the state is equal to zero. Assuming that the nuclear potential in which the proton moves has the form of a square well of radius $R \gg 1/p_F$ (where p_F is the momentum at the Fermi surface) we have $\varphi_\lambda = (2\pi R)^{-1/2} r^{-1} \sin p_F r$ and

$$\Phi(q) = \sqrt{\frac{2\pi}{R}} \int_0^R 2 \sin p_F r \cdot \sin qr \cdot dr = \\ = \sqrt{2\pi R} \left\{ \frac{\sin(q - p_F) R}{(q - p_F) R} - \frac{\sin(q + p_F) R}{(q + p_F) R} \right\}.$$

This function has a maximum for $q = p_F$. Moreover, the (negative) denominator of expression (4.41) decreases fast for $E(q) > E_0$ (= 5-10 MeV), and hence in (4.41) the important values of q are small ones, for which the scattering amplitude F is only slightly different from the scattering amplitude on a proton at rest. Thus, formula (4.41) allows us to express the cross section for the process in question in terms of the proton-proton scattering cross section. It is obvious that to calculate $\Phi(q)$ we need to make some assumption about the character of the wave functions of the nuclei X and Y; however, for sufficiently small E_0 the q-dependence of the reaction cross section at small q is determined by the resonance denominator in (4.41) independently of the form of $\Phi(q)$. The only approximation we used to write eqn. (4.41) was the assumption that other reaction mechanisms, proceeding other than through a single virtual state, give a smaller contribution. This is guaranteed by the smallness of the denominator in (4.41).

More complex reactions can be discussed in a similar way.

Threshold Singularities of the Scattering Amplitude

Consider the singularities of the scattering amplitude near the threshold for production of a given particle. The production cross section σ is proportional to the density of final states of the particle $d\underset{\sim}{p}'/(2\pi)^3$. Using the law of conservation of energy, we find

$$\sigma \sim \int \delta(E - I - E') \frac{dp'}{(2\pi)^3} \sim \sqrt{E - I}. \tag{4.43}$$

where I is the reaction threshold and E the energy of the incident particle.

Let us now find the elastic scattering cross section σ_s near the reaction threshold. To do this we find the form of the S-matrix in this region. Because of the low energy of the incident particle, the inelastic channel shows up only in the part of the S-matrix corresponding to $\ell = 0$ scattering (Wigner 1948: Baz' 1957). We call this the S_0-matrix.

In terms of this S_0-matrix eqn. (4.14) may be written

$$u = \frac{S_0 e^{ipr} - e^{-ipr}}{2ip}.$$

By calculating the particle current through a sphere of radius R and dividing by the incident current density, we find the cross section for production (or absorption) of particles:

$$\sigma = \frac{\pi}{p^2}(1 - |S_0|^2). \tag{4.44}$$

Comparing (4.43) and (4.44), we get

$$|S_0| \underset{E>I}{=} 1 - C_1\sqrt{E-I},$$

where $C_1 > 0$. For $E < I$ there are no inelastic processes and so $|S_0|_{E<I} = 1$. The last two relations may be written near threshold in the form

$$S_0 = (1 - C_1\sqrt{E-I})\, e^{2i\delta_0}, \tag{4.45}$$

where δ_0 is a real phase. For $E<I$ this expression gives $|S_0| = 1$ up to terms of order $(|I-E|)^{1/2}$.

The elastic scattering cross section σ_s is the sum of two terms, one of which is connected with the $\ell \neq 0$ partial-wave amplitudes and may be replaced by a constant in the region of the threshold. The second term is equal to $\pi p^{-2}|1 - S_0|^2$. Substituting in this the expression for S_0, we find

$$\sigma_s = \begin{cases} \text{const} + \dfrac{2\pi\sin^2\delta_0}{p^2} C_1\sqrt{E-I}, & E>I, \\ \text{const} + \dfrac{\pi\sin 2\delta_0}{p^2} C_1\sqrt{I-E}, & E<I. \end{cases}$$

Depending on the magnitude of the phase S_0, we can get two different types of behaviour of σ_s; they are shown in Fig. 43.

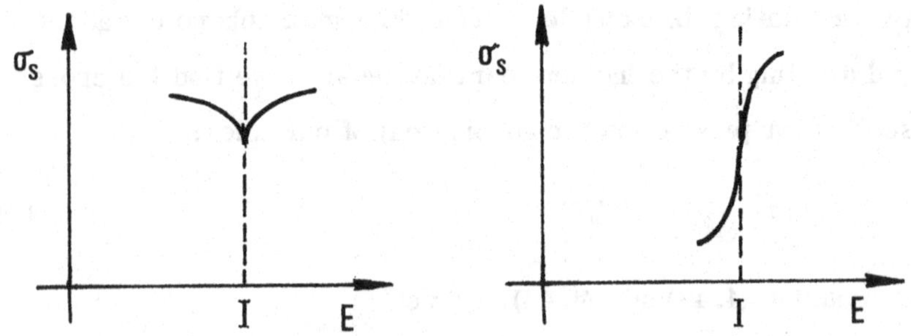

Fig. 43

CHAPTER 5

METHODS IN THE MANY-BODY PROBLEM

The solution of Schrödinger's equation for systems composed of a large number of strongly interacting particles is an unsolved problem. Indeed, even the classical three-body problem cannot be solved in general form. Fortunately, the problem of obtaining the Ψ -function of such a system is not only insoluble but unnecessary; such a detailed description of the many-body problem is not required by any experiment that is possible in real life. Any real experimental set-up contains a comparatively small number of particle indicators, and hence, in cases where a large number of particles take part in a given process, can give only their averaged characteristics. Any attempt to determine the coordinates of every particle of such a system would lead (because of the uncertainty relation) to a complicated excited state and would indeed change the properties of the system. Thus, the description of a macroscopic system by means of a wave function is an inadequate approach*: we need

---

* N. S. Krylov in his book "Raboty po Osnovaniyu Statistiki" uses this idea as the foundation of statistical physics.

instead methods of incomplete description, which determine only the relations between averaged quantities. An example of such a description is hydrodynamics, the equations of which determine only the average velocity of the particles at each point (i. e. the velocity field). A second example, closer to our problem, is the kinetic equation, which allows us to find the distribution of particles with respect to velocity and position. A knowledge of the one-particle distribution function $f(\underset{\sim}{r}, \underset{\sim}{p}, t)$, which depends on the coordinate $\underset{\sim}{r}$ and velocity $\underset{\sim}{p}$ of the particles, allows us to calculate the mean values of additive quantities such as the density $n(\underset{\sim}{r}) = \sum_i \delta(\underset{\sim}{r} - \underset{\sim}{r}_i)$ or the momentum per unit volume $\underset{\sim}{j}(\underset{\sim}{r}) = \sum_i \underset{\sim}{p}_i \, \delta(\underset{\sim}{r} - \underset{\sim}{r}_i)$. The two-particle distribution function $f(\underset{\sim}{r}_1, \underset{\sim}{r}_2; \underset{\sim}{p}_1, \underset{\sim}{p}_2; t)$ allows us to find the correlations between the coordinates and the velocities of two particles and to determine the averages of quantities which depend on the coordinates of two particles, for instance the mean value of the pairwise interaction energy $V = \sum_{ik} V(\underset{\sim}{r}_i - \underset{\sim}{r}_k)$. The periodic or weakly damped solutions of the equation for the distribution function give the eigenfrequencies of the system.

As we shall see, the Green's function method to be developed below includes this method of classical description of a many-particle system and allows us to translate it into quantum-mechanical language. However, even such an incomplete description of the system demands approximation methods. For even if we are interested in the behaviour (say) of only two particles, we inevitably have to deal with intermediate states in which, as a result of the interaction, a number of particles play a role; each of these sets yet more particles into motion, and the problem is insoluble without recourse to approximate methods.

The simplest case is where the interaction between particles can be taken as small compared to their mean kinetic energy and perturbation theory applied. We already met an example of this kind in the use of the Thomas-Fermi method to find the field in a heavy atom. Here, the Schrödinger equation is solved for an electron moving in the self-consistent field of the other particles, which are taken to be in their ground state. The parameter ζ which indicates the applicability of this approximation is the ratio of the interaction energy of two electrons ($\sim Z^{1/3}$) to the kinetic energy of an electron ($\sim Z^{4/3}$, see p.40), that is $\zeta \sim 1/Z$. The next step is to take into account the effect of quantum fluctuations of the density on the motion of the particle in question.

A second method of approximation in the many-body problem is possible in the case where the interaction between two particles is not small, but the particles are on average so far apart that interactions involving three particles may be neglected. This is the so-called "gas approximation", which is realized in the case of a gas of strongly interacting particles. The parameter ζ of this approximation is the quantity $fn^{1/3}$, where f is the two-particle scattering amplitude and n the density; that is $\zeta \sim f/r_o$, where r_o is the mean interparticle spacing.

In the most interesting physical systems (metals, other solids, liquid helium, the atomic nucleus) the conditions of applicability are fulfilled neither for perturbation theory nor for the gas approximation, and in such cases other methods must be used. First of all we must investigate the character of the lowest excited states of the system with definite values of the constants of the motion, e.g., in a homogeneous sytem, states with a definite value of momentum; then the more complicated excitations may be considered as a gas of such "elementary excitations".

Let us illustrate this idea with the example of the excited states of a solid; we assume it is an insulator. Then the electrons do not participate in the low-energy excitations and all weakly excited states reduce to sound waves. The application of quantum mechanics to sound waves (i.e. to the harmonic-oscillator problem) leads to the result that the energy of a wave with given wave vector p comes in finite "packets" $\epsilon_n(p) = (n + \frac{1}{2})\omega$, $\omega = cp$, where c is the speed of sound. The "elementary excitation" here is the lowest $(n = 1)$ excited state of the system, which has energy $\omega = \epsilon_1 - \epsilon_0$ and momentum p, and an arbitrary weakly excited state of the system may be regarded as a gas of these elementary excitations (the phonons). The nonlinear terms in the equations of elasticity correspond to interactions between phonons. Such elementary excitations may be called "quasiparticles"; the corresponding quasiparticles of the electromagnetic field is the photon.

Thus, the proper method of investigating a system of strongly interacting particles consists in taking as the object of investigation not the real particles composing the system, but the quasiparticles; since the number of the latter is small in weakly excited states, the gas approximation may be used for them.

We shall see below that the lowest-lying excited states of a Fermi system have a very simple nature even for strong interparticle interactions. First of all, there exist so-called one-particle excitations, which are analogous to the excitations in an ideal Fermi gas. In the latter case the excited states correspond to a transition of a particle from a state with energy less than the Fermi energy to an unoccupied state above the Fermi surface, or in other words to the creation of a particle and a hole on the background of the Fermi sea. The excitations in a real Fermi system also correspond to formation of particles and holes, but with properties different from those of free particles and holes. In particular, these quasiparticles have a mass different from the free-particle mass. In other words, the one-particle excitations in a real Fermi system are equivalent to the excitations of an ideal gas composed of quasiparticles with a Fermi distribution with respect to energy.

From a physical point of view these results are very natural. A particle moving in a medium will set in motion the nearby particles. For weak excitation, when the energy of the particle is near the Fermi energy, the general nature of the distribution of the particles which are set in motion depends only weakly on the state of the initial particle. Thus, in the case of weak excitation, the original particle and its environment emerges as a single stable entity, which we can call a quasiparticle. Since spin is conserved, the spin of the whole conglomerate forming the quasiparticle is the same as the spin of the original particle. Consequently, when the quasiparticles emerge as single entities, they must, like any other particles of spin $\frac{1}{2}$, obey the Pauli principle. Thus, in all cases where only a small number of quasiparticles and quasiholes take part, they behave exactly like the excitations in an ideal Fermi gas.

In an infinite system determination of the one-particle

excitation spectrum requires the introduction of just one uncalculated constant, the quasiparticle effective mass. In a finite system, however, characterization of the one-particle excitations requires us to introduce not only the effective mass of the quasiparticles but also the parameters of the effective potential well in which they move. For a system with short-range forces of range r_0 the necessary parameters are the depth, dimensions and form of the well, and also the width $\delta\,(\sim r_0)$ of the layer over which the density drops off from its value inside the system to zero.

As well as the one-particle excitations, there exist in a system of interacting particles also so-called collective excitations, which may be interpreted as bound states of a quasiparticle and a quasihole. An example of such an excitation is sound waves in an infinite system. To determine the spectrum of the collective excitations one must introduce the interaction between quasiparticles, which, as we shall see, is very different from the interaction between two real particles.

For most physical applications (transition intensities, magnetic and quadrupole moments, etc.) it is necessary to know the changes induced in the system by the action of an external field. As the theory will show, the problem of finding the reaction of the system to an external field reduces to the problem of the behaviour in the external field of a gas of quasiparticles in a potential well. It turns out to be adequate to take into account only binary collisions between quasiparticles; the multiple collisions between particles are treated exactly by the theory but lead only to a change in the interaction between the quasiparticles and in the "charge" for the interaction of the quasiparticles with the external field. In most cases, this "charge" can be found from general considerations (from the laws of conservation of charge, energy, momentum, etc.).

These results also have a very simple intuitive explanation. Suppose the system is acted on by a not too strong external field, so that the change in the energy of a given particle in the field is small compared to its kinetic energy. Then the state of the system will correspond to the appearance of a few quasiparticles and quasi-holes against the background of the Fermi distribution; their total number will only be a small fraction of the total number of particles in the system. If the mean distance between the particles is of the order of the range of the forces, then the mean distances between the quasiparticles will be considerably larger than the force range and so the quasiparticles will form a gas, so that we can neglect cases where three or more quasiparticles collide simultaneously.

What of the "charge" of the quasiparticles in relation to the external field ? This "charge" describes the interaction with the field of the complex of particles which forms the quasiparticle. Suppose, for instance, an electric field is applied to a nucleus. This will act only on the protons. Now since charge is conserved in the interaction of a proton with the other particles of the nucleus, the whole complex which forms a "proton quasiparticle" will have the same charge as the original proton. In this case, the charge of the quasiparticle is equal to the charge of the original particle; in the case of other external fields, for instance, for a magnetic field, the effective interaction of a quasiparticle with the field is different from that of a particle. We may see this as follows: whereas a neutron moving in a free space can interact with a magnetic field only through its intrinsic magnetic moment, the motion of a neutron quasiparticle also sets protons into motion and hence leads to an electric current, so that the interaction with the external field is changed. Thus, a neutron quasiparticle can have an orbital magnetic moment (a magnetic moment associated with its orbital motion). If there were no interparticle interaction, an orbital magnetic moment could be associated only with the protons.

For an infinite homogeneous Fermi system the theory of interacting quasiparticles described above was formulated by Landau in 1958.

The quasiparticle method as applied to nuclear theory proceeds as follows. We first prove that for weak excitation the nucleus may be considered as a gas of quasiparticles in a potential well with an interaction between them which can be described by a few universal constants. This interaction is not small and must be handled exactly. The only approximation made is that for weak excitation, when the number of quasiparticles is small, only binary collisions between them are taken into account.

Then for most observable nuclear phenomena we can obtain formulae which, with the aid of machine solution of the relevant equations, can be expressed in terms of the universal constants of the theory. The constants determining the interaction between quasiparticles, and the parameters of the potential well, cannot be calculated without assuming the interaction between particles to be weak. In the case of the nucleus the interaction between particles certainly cannot be taken to be weak, and so these constants must be found from a comparison of theory and experiment.

The quasiparticle method described above is most effectively carried out by the technique of Green's functions and graphical representation of processes. Below, we shall first explain this method by using some simple examples and then use it to solve various problems.

As we shall expound it, the basic principle of the graphical method consists in describing processes by diagrams representing the space-time development of the process, and then establishing by simple examples the correspondence between elements of the graphs and analytic expressions; this then allows us to read off the meaning of any arbitrary graph composed of these elements. Such an approach is not only simple, but allows us to emphasize the qualitative aspects of the calculations.*

1. THE QUASIPARTICLE METHOD AND GREEN'S FUNCTIONS

The Transition Amplitude

To develop the quasiparticle method quantitatively it is sufficient to obtain an equation for the small number of particles which actually participate in the phenomenon in question; Schrödinger's equation, by contrast, describes the behaviour of the whole system and leads to insoluble difficulties. To get such an incomplete description of the system it is convenient to go over

* A more formal account of these questions may be found in the following books: A. A. Abrikosov, L. P. Gor'kov, I. E. Dzyaloshinskii, Quantum Field Theoretic Methods in Statistical Physics, Pergamon, Oxford, 1965: A. B. Migdal, Theory of Finite Fermi Systems and the Atomic Nucleus, Interscience, New York 1967: A. B. Migdal, Nuclear Theory: the Quasiparticle Method, W. A. Benjamin, New York 1968.

from the wave function Ψ to the transition amplitude (Green's function). In contrast to the Ψ-function, which depends on the coordinates of all the particles, the transition amplitude is a function only of the coordinates of the particles in the initial and final state. Consider first of all the example of a single particle. Instead of Schrödinger's equation

$$i\frac{\partial\Psi(r, t)}{\partial t} - H\Psi(r, t) = 0$$

we can use the equation for the Green's function $G(\underset{\sim}{r}, t:\underset{\sim}{r}', t')$

$$i\frac{\partial G}{\partial t} - HG = i\delta(r - r')\delta(t - t'). \tag{5.1}$$

The Green's function represents the amplitude for transition of the particle from point $\underset{\sim}{r}'$ at time t' to point $\underset{\sim}{r}$ at time t; the squared modulus of the amplitude gives the transition probability. We can easily verify this by using the Green's function to express the wave function Ψ at time $t + \tau$ in terms of Ψ at time t:

$$\Psi(r, t+\tau) = \int G(r, t+\tau;\ r', t)\,\Psi(r', t)\,dr'. \tag{5.2}$$

Then it can be seen from (5.2) that $\Psi(\underset{\sim}{r}, t+\tau)$ obeys the Schrödinger equation, and moreover, if $G(\underset{\sim}{r}, t+0:\underset{\sim}{r}', t) = \delta(\underset{\sim}{r}-\underset{\sim}{r}')$, goes over to $\Psi(\underset{\sim}{r}, t)$ for $\tau \to 0$.

Formula (5.2) contains G only for $\tau > 0$. For $\tau < 0$ we put $G = 0$. Then we get from (5.1)

$$G(r, t+0;\ r', t) = \delta(r - r'),$$

which is just what is required for (5.2). In the absence of an external field it follows from symmetry considerations (i. e. from the uniformity and isotropy of space and the uniformity of time) that

$$G(r,\ t;\ r',\ t') = G(|r - r'|,\ t' - t).$$

Suppose a system of eigenfunctions is given by the relations

$$H\varphi_\lambda(\boldsymbol{r}) = \varepsilon_\lambda\varphi_\lambda(\boldsymbol{r}), \quad H = \frac{p^2}{2m} + V(r).$$

A typical form of the potential well V for a nucleon moving in a nucleus is shown in Fig. 44, where R is the nuclear radius and r_0 the width of the "diffuse region", that is, the region over which V goes over from its constant value inside the nucleus to the value outside.

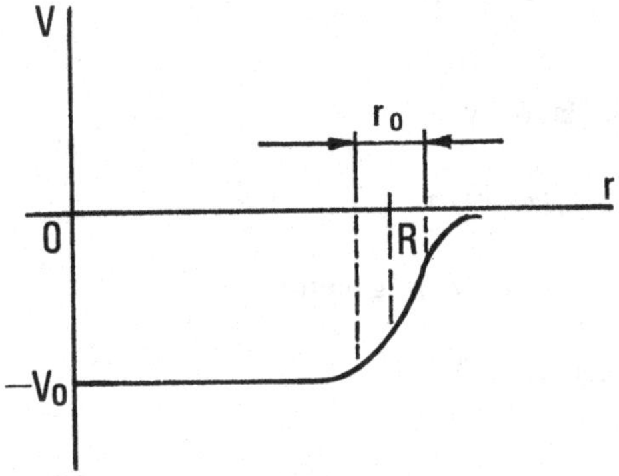

Fig. 44

We write the wave function of the particle in the form $\Psi(\underset{\sim}{r}, t) = \sum_\lambda C_\lambda(t)\,\varphi_\lambda(r)$; then eqn. (5.2) takes the form

$$C_\lambda(t+\tau) = \sum_{\lambda'} G_{\lambda\lambda'}(\tau)\,C_{\lambda'}(t),$$

$$G_{\lambda\lambda'} = \int d^3r\, d^3r' G(\boldsymbol{r}, \boldsymbol{r}', \tau)\,\varphi^*_\lambda(\boldsymbol{r})\,\varphi_{\lambda'}(\boldsymbol{r}').$$

Since φ_λ is an eigenfunction, transitions to other states do not take place and we have $C_\lambda(t+\tau) = e^{-i\epsilon_\lambda\tau}\,C_\lambda(t)$, that is

$$G_{\lambda\lambda'}(\tau) = G_\lambda(\tau)\,\delta_{\lambda\lambda'} = e^{-i\varepsilon_\lambda\tau}\delta_{\lambda\lambda'}\theta(\tau), \tag{5.3}$$

where

$$\theta(\tau) = \begin{cases} 1, & \tau > 0, \\ 0, & \tau < 0. \end{cases}$$

This result can also be easily obtained directly from the equation for G.

Going over to a Fourier transform with respect to τ, we get

$$G_\lambda(\varepsilon) = \frac{1}{\varepsilon - \varepsilon_\lambda + i\delta}, \quad \delta = +0. \tag{5.4}$$

where $G_\lambda(\epsilon)$ is defined by

$$G_\lambda(\varepsilon) = \frac{1}{i}\int G_\lambda(\tau)\,e^{i\varepsilon\tau}\,d\tau.$$

Hence the inverse transform is given by

$$G_\lambda(\tau) = \int e^{-i\varepsilon\tau} G_\lambda(\varepsilon)\,\frac{i\,d\varepsilon}{2\pi}.$$

The sign of δ is chosen so as to make $G_\lambda(\tau)$ zero for $\tau < 0$. It is easy to verify that the choice of sign for δ is correct by going back to the τ-representation:

$$G_\lambda(\tau) = \int \frac{e^{-i\varepsilon\tau}}{\varepsilon - \varepsilon_\lambda + i\delta}\,\frac{i\,d\varepsilon}{2\pi}$$

and shifting the integration contour into the upper half-plane of ϵ, as we did on p.92.

In the "mixed" $(\mathbf{r}, \epsilon)$ representation we have

$$G(\boldsymbol{r}, \boldsymbol{r}', \varepsilon) = \sum_{\lambda\lambda'} G_{\lambda\lambda'}(\varepsilon)\,\varphi_\lambda(\boldsymbol{r})\,\varphi^*_{\lambda'}(\boldsymbol{r}') = \sum_\lambda \frac{\varphi_\lambda(\boldsymbol{r})\,\varphi^*_\lambda(\boldsymbol{r}')}{\varepsilon - \varepsilon_\lambda + i\delta}. \tag{5.5}$$

The sum over λ contains a summation over bound states and an integration over the continuous spectrum. The function $G(r, r', \epsilon)$ has poles at values of ϵ equal to the bound-state energies ϵ_λ.

The Green's function is simply related to the S-matrix of the scattering problem for the potential $V(r)$, as introduced on p. 229. If at time $t = -\infty$ the wave function in the momentum representation had the form $C_{\underset{\sim}{p}} = e^{-iE_{\underset{\sim}{p}}t}$, and at time $t' \to +\infty$ takes the form $\sum_{\underset{\sim}{p}'} C_{\underset{\sim}{p}'}^{(\underset{\sim}{p})} e^{-iE_{\underset{\sim}{p}'}t'}$, then the matrix element $S_{\underset{\sim}{p}\underset{\sim}{p}'}$ of the S-matrix is $C_{\underset{\sim}{p}'}^{(\underset{\sim}{p})}$. On the other hand, (5.2) gives

$$C_{p'}^{(p)} = \underset{t\to-\infty,\, t'\to\infty}{G(p', p, t', t)}\, e^{iE_{p'}t' - iE_p t} = S_{pp'}.$$

It then follows from the expressions connecting the S-matrix with the scattering amplitude (p. 230) that

$$e^{iE_{p'}t' - iE_p t} \underset{t'\to-\infty,\, t'\to\infty}{G(p', p, t', t)} = -2\pi i \delta(E_p - E_{p'}) A(p, p'),$$

where $A(\underset{\sim}{p}, \underset{\sim}{p}')$ is the scattering amplitude with the energy normalisation for the wave functions; this is related to the conventional amplitude $f(\underset{\sim}{p}, \underset{\sim}{p}')$ by

$$f(p', p) = -\frac{m}{2\pi} A(p', p). \tag{5.6}$$

One-Particle Green's Functions in a System of Noninteracting Particles (Quasiparticle Green's Functions)

Let us find the Green's function of a particle $G_{\lambda\lambda'}(\tau)$, that is, the transition amplitude from a state with one particle λ to a state with one particle λ' in a system of noninteracting particles. For this purpose it is only necessary to take account of the Pauli principle in (5.3), that is, to exclude transitions into occupied states. (For simplicity we consider only the case of fermions here). Hence we must include in the Green's function a factor $1 - n_\lambda$, where n_λ, given by

$$n_\lambda = \begin{cases} 1, & \varepsilon_\lambda < \varepsilon_F \\ 0, & \varepsilon_\lambda > \varepsilon_F \end{cases}$$

is the number of particles in the state λ. Thus we get

$$G^+_{\lambda\lambda'}(\tau) = (1 - n_\lambda)\,\delta_{\lambda\lambda'} \begin{cases} e^{-i\varepsilon_\lambda \tau}, & \tau > 0, \\ 0, & \tau < 0. \end{cases} \tag{5.7}$$

Next let us find the hole transition amplitude. Since the number of places available for holes in the level λ is proportional to n_λ, we obtain similarly to the case of a particle

$$G^-_{\lambda\lambda'}(\tau) = n_\lambda \delta_{\lambda\lambda'} \begin{cases} e^{-i\bar{\varepsilon}_\lambda \tau}, & \tau > 0, \\ 0, & \tau < 0. \end{cases} \tag{5.8}$$

where $\bar{\epsilon}_\lambda$ is the energy of a hole, or more precisely the difference in the energy of the system after and before the appearance of the hole.

In many cases it is convenient to introduce a particle Green's function $G_\lambda(\tau)$ which is defined both for $\tau > 0$ and for $\tau < 0$ and combines eqns. (5.7) and (5.8):

$$G_\lambda(\tau) = \begin{cases} G^+_\lambda(\tau), & \tau > 0, \\ -G^-_\lambda(-\tau), & \tau < 0. \end{cases} \tag{5.9}$$

In Fourier-transformed form eqns. (5.7-9) take the form

$$G_\lambda^+(\varepsilon) = \frac{1-n_\lambda}{\varepsilon - \varepsilon_\lambda + i\delta},$$

$$G_\lambda^-(\varepsilon) = -\frac{n_\lambda}{\varepsilon - \varepsilon_\lambda^- + i\delta}, \tag{5.10}$$

$$G_\lambda(\varepsilon) = G_\lambda^+(\varepsilon) - G_\lambda^-(-\varepsilon) = \left[\frac{1-n_\lambda}{\varepsilon - \varepsilon_\lambda + i\delta} + \frac{n_\lambda}{\varepsilon + \varepsilon_\lambda^- - i\delta}\right].$$

Eqns. (5.10) display an important property of the Green's functions G^+ and G^-: each of them has a pole at a value of ϵ corresponding to the energy of a particle and a hole respectively.

It follows from what we said in the introduction to this chapter that the Green's function of a quasiparticle (quasihole) in a system of interacting particles has the same form as above; we need only replace the energy of a particle (hole) by the energy of a quasiparticle (quasihole). Below (p. 296) we shall explain our initial formulae (5.7) and (5.8) for a Fermi system and obtain analogous expressions for Bose particles.

For the ground state, when

$$n_\lambda = \begin{cases} 1 \text{ for } \epsilon_\lambda < \epsilon_F \\ 0 \text{ for } \epsilon_\lambda > \epsilon_F \end{cases}$$

the last of formulae (5.10) can be written in the form

$$G_\lambda(\varepsilon) = \frac{1}{\varepsilon - \varepsilon_\lambda + i\delta \operatorname{sign}(\varepsilon - \varepsilon_F)}. \tag{5.10'}$$

The Green's Function in a System of Interacting Particles

We have found the Green's function of a free particle and the one-particle Green's function in a system of noninteracting fermions in their groundstate. It would be easy to find also the Green's function which describes the behaviour of two or more particles or holes in a system of noninteracting particles. However, our problem is to handle the interactions between the particles. The basic principle of the Green's function method lies in the fact that to discuss a many-particle system there is no need to introduce Green's functions referring to very large numbers of particles. The relation (5.1) may, indeed, be easily generalized to the case of many particles (we need only take $\underset{\sim}{r}$ to denote the whole collection of coordinates of all the particles); however, actually to find the resulting Green's function $G(\underset{\sim}{r}_1, \ldots, \underset{\sim}{r}_N; t; \underset{\sim}{r}'_1, \ldots \underset{\sim}{r}'_N; t')$ is just as impossible in a many-body system as to find the wave function. In cases where the phenomenon in question effectively involves only a small number of particles, there is no need to consider all the particles of the system; as we shall see, almost all processes in a many-body system which can be investigated experimentally can be described by the one- and two-particle Green's functions.

We define the one-particle Green's function in a system of interacting particles by the expression

$$G^+(r, t; r', t') \underset{t>t'}{=} (\Phi_0 \Psi(r', t') \Psi^+(r, t) \Phi_0), \tag{5.11}$$

where Φ_0 is the exact groundstate eigenfunction, and $\Psi(\underset{\sim}{r}, t)$ is the second-quantized operator in the Heisenberg representation,

i. e.

$$\Psi(r, t) = e^{iHt}\Psi(r)\, e^{-iHt},$$

where H is the Hamiltonian operator of the system including the interaction terms, and $\Psi(\underset{\sim}{r})$ can be expressed in terms of the annihilation operators for a particle in the various states $\varphi_\lambda(\underset{\sim}{r})$:

$$\Psi(r) = \sum_\lambda a_\lambda \varphi_\lambda(r).$$

Below we shall verify that the expression (5.11) has a simple meaning, in fact it gives the amplitude for transition of a particle from the state $(\underset{\sim}{r}', t')$ to the state $(\underset{\sim}{r}, t)$. The squared modulus of this quantity gives the transition probability. An analogous expression can be written down for the hole Green's function

$$G^-(r, t; r', t') \underset{t>t'}{=} (\Phi_0 \Psi^+(r, t)\, \Psi(r', t')\, \Phi_0). \tag{5.12}$$

(annihilation of a particle is equivalent to creation of a hole). Both the above expressions are defined only for $t > t'$; we may formally combine them into a single Green's function which describes a particle for $\tau > 0$ and a hole for $\tau < 0$, just as we did above (p. 274) for a system of free particles:

$$G(r, t; r', t') = \begin{cases} G^+(r, t; r', t'), & t > t', \\ \pm G^-(r', t'; r, t), & t < t'. \end{cases} \tag{5.13}$$

Here the plus sign corresponds to bosons and the minus sign to fermions. It is easy to see that in the case of noninteracting particles expressions (5.11) and (5.12) reduce to the corresponding formulae of the last section; we recommend the reader to carry out this calculation for himself.

Relations (5.11) and (5.12) may be written in the form

$$G(x, x') = \langle T\Psi(x)\Psi^{+}(x')\rangle,$$

where the symbol $\langle \cdots\cdot \rangle$ denotes averaging over the ground-state, $x = (\underset{\sim}{r}, t)$, and the operator T (the time-ordering operator) denotes that quantities to the right of it are to be ordered so that the times in their arguments form a decreasing sequence. For Fermi systems at times $t' > t$ (when Ψ and Ψ^{+} change places) we add an overall minus sign.

The two-particle Green's functions may be defined similarly: in place of $\Psi(1)\ \Psi^{+}(2)$ there enter the products $\Psi(1)\ \Psi(2)\ \Psi^{+}(3)\ \Psi^{+}(4)$. Below we shall see how to calculate the particle Green's functions and how they are connected with the quasiparticle Green's functions.

Analytic Properties of the One-Particle Green's Function

We confine ourselves for simplicity to the case of a homogeneous infinite system. Then, in view of the homogeneity and isotropy of the system in space and its homogeneity in time, we have

$$G(\boldsymbol{r}, t; \boldsymbol{r}', t') = G(|\boldsymbol{r} - \boldsymbol{r}'|, t - t'). \tag{5.14}$$

We go over to the Fourier transform with respect to $\underset{\sim}{r}_1 = \underset{\sim}{r} - \underset{\sim}{r}'$. From (5.11) and (5.12) we obtain for the quantity

$$G(\boldsymbol{p}, \tau) = \int d^3r_1 G(\boldsymbol{r}_1, \tau)\, e^{-i\boldsymbol{p}\boldsymbol{r}_1}$$

the expression

$$G(p, \tau) = \begin{cases} \langle a_p e^{-iH\tau} a_p^+ \rangle e^{iE_0\tau} & \tau > 0, \\ \pm \langle a_p^+ e^{iH\tau} a_p \rangle e^{-iE_0\tau} & \tau < 0. \end{cases} \tag{5.15}$$

Writing the operators occurring in $G(p, \tau)$ in the energy representation, we have:

$$G(p, \tau) = \begin{cases} \sum_s |(a_p^+)_{s0}|^2 \exp\{-i(E_s - E_0)\tau\} & \tau > 0, \\ \pm \sum_s |(a_p)_{s0}|^2 \exp\{i(E_s - E_0)\tau\} & \tau < 0. \end{cases} \tag{5.16}$$

Since the operator $a_{\underset{\sim}{p}}^+$ increases the momentum of the system by an amount $\underset{\sim}{p}$, and the number of particles in the system by 1, the summation for $\tau > 0$ is taken over all states with momentum $\underset{\sim}{p}$ and particle number $N + 1$ (assuming that the ground state had particle number N and momentum zero). In the same way, the summation for $\tau < 0$ is taken over states with particle number $N - 1$ and momentum $-\underset{\sim}{p}$.

Let us write

$$E_s(N+1) - E_0(N) = \\ = \varepsilon_s(N+1) + E_0(N+1) - E_0(N) = \varepsilon_s + \mu,$$

where $\mu = E_0(N+1) - E_0(N)$ is the chemical potential. The excitation energy $\epsilon_s = E_s(N+1) - E_0(N+1)$ is positive by definition. In the same way,

$$E_s(N-1) - E_0(N) = \\ = \varepsilon_s(N-1) - E_0(N) + E_0(N-1) = \varepsilon_s' - \mu'.$$

The quantities ϵ_s' and μ' are the same as ϵ_s and μ to an accuracy of order 1/N.

We introduce the functions

$$\begin{aligned} A(p, E)\,dE &= \sum_s |(a_p^+)_{s0}|^2, \quad E \leqslant \varepsilon_s \leqslant E + dE, \\ B(p, E)\,dE &= \sum_s |(a_p)_{s0}|^2, \quad E \leqslant \varepsilon_s \leqslant E + dE \end{aligned} \tag{5.17}$$

and go over in expression (5.16) to the Fourier transform with respect to τ. Then we have

$$G(p, \varepsilon) = -\int_0^\infty dE \left\{ \frac{A(p, E)}{E - \varepsilon + \mu - i\delta} \pm \frac{B(p, E)}{E + \varepsilon - \mu - i\delta} \right\}. \tag{5.18}$$

Formula (5.18) is the spectral decomposition for the one-particle Green's function of a system compared of a finite number of fermions*. This formula allows us to obtain a relation between the real and imaginary parts of the function $G(\underset{\sim}{p}, \epsilon)$: from the identity

$$\frac{1}{E - \varepsilon + \mu - i\delta} = P \frac{1}{E - \varepsilon + \mu} + i\pi\delta(E - \varepsilon + \mu)$$

it turns out that

$$\operatorname{Im} G(p, \varepsilon) = \pi \begin{cases} -A(p, \varepsilon - \mu), & \varepsilon > \mu, \\ \mp B(p, \mu - \varepsilon), & \varepsilon < \mu. \end{cases} \tag{5.19}$$

Thus, for fermions the imaginary part of $G(p, \epsilon)$ changes sign at the point $\epsilon = \mu$, while for bosons it is negative for all p and ϵ. Using (5.18) and (5.19), we easily find

---

* The analogous decomposition in quantum field theory was obtained by Lehmann in 1954.

$$G(p, \varepsilon) = \frac{1}{\pi} \int_{-\infty}^{\infty} \frac{\operatorname{Im} G(p, \varepsilon')\, d\varepsilon'}{\varepsilon' - \varepsilon - i\delta}. \tag{5.20}$$

which is analogous to the relation (4. 5) on p. 225.

We will now establish the relation between the one-particle Green's function and the excitation spectrum. The function $G(\underset{\sim}{p}, \tau)$ has a simple physical meaning: suppose at the initial time the system is in the state $\Phi(0) = a_p^+ \Phi_o$, where Φ_o is the groundstate of the N-particle system (the physical "vacuum"). At time $\tau > 0$ the wave function of the system is

$$\Phi(\tau) = e^{-iH\tau} a_p^+ \Phi_0.$$

The function $G(\underset{\sim}{p}, \tau)$ is then the probability amplitude to find the system in the state $\Phi(0)$ at time τ. To see this, write

$$(\Phi(0), \Phi(\tau)) = (\Phi_0 a_p e^{-iH\tau} a_p^+ \Phi_0) = G(p, \tau). \tag{5.21}$$

An analogous relation holds for $\tau < 0$. According to (5.16) and (5.18), for $\tau > 0$ we have

$$(\Phi(0), \Phi(\tau)) = e^{-i\mu\tau} \int_0^{\infty} A(p, E)\, e^{-iE\tau}\, dE. \tag{5.22}$$

In the absence of interactions we have for p greater than p_F (we have $\mu = \epsilon_F$)

$$A(p, E) = \delta(E + \varepsilon_F - \varepsilon^0(p))$$

and so

$$(\Phi(0)\, \Phi(\tau)) = e^{-i\varepsilon^0(p)\tau}.$$

When the interactions between the particles are included the δ-function in $A(\underset{\sim}{p}, E)$ is replaced by a function having a sharp maximum near $E = \epsilon(p) - \mu$, where $\epsilon(p)$ is the quasiparticle energy.

Now consider the behaviour of the Green's function for large positive times. Suppose that the nearest singularity to the real axis of the analytic continuation of $A(\underset{\sim}{p}, E)$ into the lower half-plane is a simple pole at $E = \epsilon(p) - \mu - i\gamma$. Then, shifting the integration contour in (5.22) into the lower half-plane, we get

$$G(\mathbf{p}, \tau) = e^{-i\mu\tau} \int_C A e^{-iE\tau} \, dE. \tag{5.22'}$$

The integration contour C is shown in Fig. 45. The non-exponential term in the function $G(\underset{\sim}{p}, \tau)$, which comes from the integration along the imaginary axis near $E = 0$, is of order of magnitude

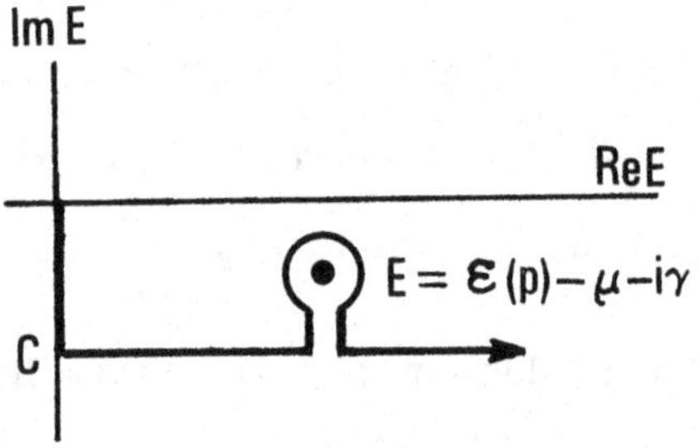

Fig. 45

$(\gamma/\epsilon(p))^2$ for $\tau \geq 1/\gamma$. Thus we get

$$G(\mathbf{p}, \tau) = Z e^{-i\varepsilon(\mathbf{p})\tau - \gamma\tau} + O\,[(\gamma/\varepsilon(\mathbf{p}))^2]. \tag{5.23}$$

This result may be interpreted as follows: in the state $\Phi(0)$ there is present with amplitude Z a wave packet representing a quasiparticle with energy $\epsilon(p)$ and damping γ. The values of $\epsilon(p)$ and γ are determined by the position of the pole in $A(\underset{\sim}{p}, E)$.

Had we considered the case $\tau < 0$, we would have obtained a similar relation for a quasihole. Thus, $G(p, \epsilon)$ may be written in the form

$$G(p, \varepsilon) = Z\left[\frac{1-n_p}{\varepsilon-\varepsilon(p)+i\delta} + \frac{n_p}{\varepsilon-\varepsilon(p)-i\delta}\right] + G_{\mathrm{Reg}} \equiv \equiv ZG_Q + G_{\mathrm{Reg}}, \tag{5.24}$$

where G_Q is the Green's function of the quasiparticle. This relation establishes the connection between the particle and quasiparticle Green's functions.

Calculation of Observable Quantities

The function G enables us to calculate the averages over the groundstate of operators which have the form of a simple sum over all particles, that is, the form

$$A = \sum_i A_i(\xi_i, p_i), \tag{5.25}$$

(where ξ_i represents all the space and spin variables). Such an operator is, for instance, the particle density at the point $\underset{\sim}{r}$, which is given by

$$n(r) = \sum_i \delta(r - r_i),$$

or the total orbital angular momentum

$$L = \sum_i [r_i \times p_i].$$

The operator A, when written in the language of second quantization, has the form

$$A = \int \Psi^+(\xi) A(\xi, p) \Psi(\xi) d\xi, \tag{5.25'}$$

and hence its mean value in the groundstate of the system can be expressed in terms of G for $t = t' - 0$:

$$G(\xi, \xi', \tau) \underset{\tau \to -0}{=} \mp(\Phi_0 \Psi^+(\xi') \Psi(\xi) \Phi_0) \tag{5.26}$$

(where Φ_0, as above, denotes the exact groundstate).

The expectation value of the operator A will be equal to

$$\langle A \rangle = \mp \int \{A(\xi, p) G(\xi, \xi', (\tau = -0))\}_{\xi'=\xi} d\xi \equiv \\ \equiv \mp \operatorname{Sp} AG_{\tau=-0}. \tag{5.27}$$

Thus, $G_{\tau \to -0}$ is identical with the density matrix up to a factor ± 1. In fact for fermions

$$\rho(\xi', \xi) = (\Phi_0 \Psi^+(\xi') \Psi(\xi) \Phi_0) = -G_{\tau=-0}; \tag{5.28}$$

while for bosons

$$\rho(\xi', \xi) = G_{\tau \to -0}. \tag{5.28'}$$

To determine the expectation values of operators of the form

$$B = \sum_{i, k} B_{ik}(\xi_i, p_i; \xi_k, p_k), \tag{5.29}$$

such as the interaction energy of the particles, a knowledge of the two-particle Green's function is necessary. This quantity is defined similarly to G:

$$G_2(1, 2; 3, 4) = (\Phi_0 \, T\Psi(1)\Psi(2)\Psi^+(3)\Psi^+(4)\Phi_0). \tag{5.30}$$

where the operator T means that all quantities standing to the right of T are ordered so that the times in the arguments of Ψ and Ψ^+ form a decreasing sequence; in front of the whole expression there is a plus or minus sign (for fermions) depending on whether the time-ordered expression is obtained from the one explicitly written in (5.30) by an even or an odd permutation. The function G_2 gives the transition amplitude for the case in which the initial and final states correspond (depending on the relations between the times t_1, t_2, t_3, t_4) either to two particles, or to two holes or to a particle and a hole. G_2 also includes the case in which at the initial time there is a single particle and in the final state two particles and one hole. For brevity we shall speak explicitly of two particles (or holes); it is understood that the remaining N-2 particles are in the groundstate at both initial and final times. The functions G and G_2 also contain information about the scattering amplitude in the field of any scattering centre situated in the medium, and about the scattering amplitude of two interacting particles. The scattering of two particles in the medium is determined by the function $G_2(\underset{\sim}{p}_1 t_1, \underset{\sim}{p}_2 t_2; \underset{\sim}{p}_3 t_3, \underset{\sim}{p}_4 t_4)$ for $t_1, t_2 \rightarrow -\infty$ and $t_3, t_4 \rightarrow +\infty$ (cf. p.273). When the distance in time between the initial times t_1, t_2 and the time of the interaction becomes large, the wave packets describing particles with momenta $\underset{\sim}{p}_1$ and $\underset{\sim}{p}_2$ are damped in such a way that by the time the collision takes place only those terms are left which correspond to quasiparticles with the same momenta (p. 282). After the period of interaction, when $t_3, t_4 \rightarrow \infty$, again only quasi-particles with momenta $\underset{\sim}{p}_3$ and $\underset{\sim}{p}_4$ are left. Hence it is more convenient to discuss the scattering problem in terms of quasi-particles right from the start. The same applies to the problem of scattering in the field of a scattering centre.

While the one- and two-particle Green's functions contain the most important information on the system, there do sometimes occur questions which require a knowledge of the three- or four-particle Green's functions; for example, they would be required to calculate the binding energy in a system with non-pairwise interactions.

The Fermion Momentum Distribution

It follows from formula (5.26) that the distribution of particles with respect to momentum can be expressed in terms of the Green's function:

$$n(p) \underset{\tau\to -0}{=} -\int G(p, \varepsilon)\, e^{-i\varepsilon\tau} \frac{i\, d\varepsilon}{2\pi}. \qquad (5.31)$$

It is impossible to take the limit $\tau = 0$ in this expression, because as is obvious from (5.24), $G \sim 1/\epsilon$ for $\epsilon \to \infty$ and the integral $\int G(p, \epsilon) d\epsilon$ taken along the real axis is divergent. However, for any finite negative value of τ it is possible to replace the integral along the real axis by a closed contour C which is composed of the real axis and an infinite semicircle in the upper half-plane, and thereafter put $\tau = 0$. Thus we get

$$n(p) = -\int_C G(p, \varepsilon) \frac{i\, d\varepsilon}{2\pi}.$$

As we have seen, the Green's function has a pole at the point $\epsilon = \epsilon(p) - i\gamma$:

$$G(p, \varepsilon) = \frac{Z}{\varepsilon - \varepsilon(p) + i\gamma(p)} + G_{\mathrm{Reg}}(p, \varepsilon),$$

where $G_{\mathrm{Reg}}(\underset{\sim}{p}, \epsilon)$ is a function which is regular near the pole. The damping $\gamma(\underset{\sim}{p})$ changes sign for $p = p_F$: $\gamma > 0$ for $p > p_F$, while $\gamma < 0$ for $p < p_F$. Hence for $p < p_F$ there is a pole inside the contour C, while for $p > p_F$ it goes over into the lower half-plane, so that its contribution is excluded from the integral over C. Hence we get

$$n(p_F - 0) - n(p_F + 0) = Z.$$

Since $0 \leq n(p) \leq 1$, the Green's function renormalization factor (as the factor Z is called) satisfies $0 < Z \leq 1$. The momentum distribution of the particles is shown in Fig. 46 (Migdal, 1957).

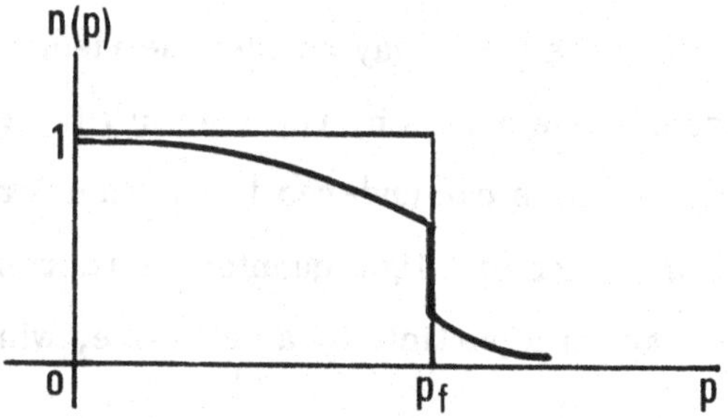

Fig. 46

Thus the investigation of the analytic properties of the Green's function has allowed us to obtain an important physical result: in spite of the interaction between particles, which gives a scatter in the momenta of the particles, there remains a "memory" of the free-particle Fermi distribution in the shape of a discontinuity in the function n(p). Fig. 46 shows also the quasi-particle distribution (the thin line); naturally, this distribution is meaningful only for p close to p_F, since only there is the quasi-particle concept applicable.

2. THE GRAPH METHOD

Graphical Representation of Processes

A much-used method of obtaining various relations in the many-body problem and in field theory is the method of Feynman graphs, which consists in representing all the processes of interest by diagrams which stand for complicated analytical expressions in much the same way as Chinese ideograms stand for whole phrases. One begins by representing the various physical processes which particles can undergo in diagram form. For instance, the propagation of a light quantum is represented by a dashed line, and that of a particle by a solid line, while a graph such as

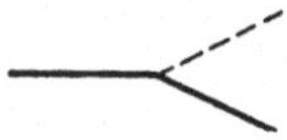

indicates that a charged particle, say an electron, has emitted a light quantum. The kink in the continuous line indicates that after the emission of the light quantum (photon) the electron has acquired a new momentum.

Suppose we have two noninteracting particles

If they interact, we draw a diagram of the following type:

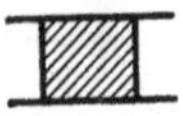

If the interaction takes place via a photon (i. e. if it is of Coulomb form) then we join the two solid lines by a dashed line:

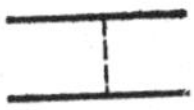

If the two particles involved are nucleons and the interaction takes place through the exchange of a π-meson, then we draw a wavy line between the particle lines:

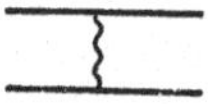 .

This graph means that the two nucleons interact once with one another. If they interact twice, then we draw the graph as follows:

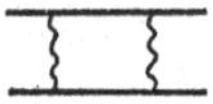

A graph like

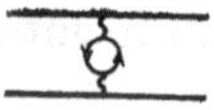

represents a more complicated process - one nucleon emits a π-meson which then decays into a nucleon and an antinucleon. These two particles are subsequently reconverted into a π-meson, which is absorbed by the second nucleon. We can go on in the same way to represent more complex processes which the particles undergo.

To associate with these graphs a quantitative, as well as merely an illustrative meaning, we shall interpret a given graph as representing the transition amplitude from one state at the initial time to another state at the final time. The square of the transition amplitude gives the probability of occurrence of the final state at the final time. Thus, for instance, the photon-emission graph drawn above represents the amplitude for the transition of a charged particle of momentum $\underset{\sim}{p}$ to a state containing a photon of momentum $\underset{\sim}{q}$ and a particle of momentum $\underset{\sim}{p}-\underset{\sim}{q}$.

According to the principle of superposition, the total transition amplitude is just the sum of all possible physically different transition amplitudes. The exact meaning of this assertion will be explained below with simple examples. First let us try to use the graph method to obtain in an intuitive manner the relation which expresses the two-particle scattering amplitude in terms of the interaction potential. According to the superposition principle, the scattering amplitude is represented graphically by a sum of graphs:

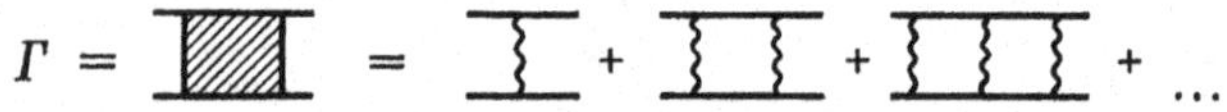

The first graph on the right-hand side represents a single interparticle interaction, the second corresponds to a double interaction, and so on. Between the interaction events we have the transition amplitude appropriate to two noninteracting particles. We shall associate with the first graph the interaction potential between the particles:

 ,

and with each straight line a Green's function, that is, the transition amplitude G appropriate to a free particle. Then the second graph is conventionally written

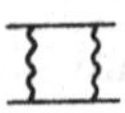 $= UGGU$,

since the transition amplitude for two free particles is just the product of the Green's functions of the two particles. Thus, we get for the scattering amplitude the series

$$\Gamma = U + UGGU + UGGUGGU + \ldots$$

The expression formed by the factors standing to the right of UGG in the second and subsequent terms is again just the series which gives Γ. So we get for Γ the equation

$$\Gamma = U + UGG\Gamma.$$

The function G occurring in this equation is the free-particle transition amplitude with which we are already familiar.

Of course, the operation we have just carried out is not a derivation of the equation, but rather an argument to orient us. To establish the meaning which must be attached to the symbolic multiplication in these expressions we must compare the equation we have obtained with the corresponding expression found in the usual way from the solution of Schrödinger's equation. It is obvious that the above expression for Γ is nothing but a symbolic way of writing the standard quantum-mechanical equation for the scattering amplitude in the energy normalization. In the centre-of-mass frame we have

$$\Gamma(p_1, p_2) = U(p_1, p_2) + \int U(p_1, p') \frac{\Gamma(p', p_2)}{\varepsilon_{p_1} - \varepsilon_{p'} + i\gamma} \frac{d^3p'}{(2\pi)^3}. \qquad (5.32)$$

In a similar way we can relate $\tilde{G}$, the Green's function of a particle in an external field, to G, the free-particle Green's function. The Green's function in the field, $\tilde{G}$, is represented by the sum of the partial transition amplitudes:

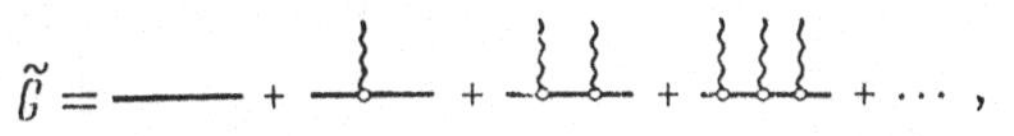

where a point with a wavy line attached represents an instance of action of the external field V. Collecting up all the graphs in $\tilde{G}$ which stand to the right of V, we get back $\tilde{G}$ itself; thus,

$$\tilde{G} = G + GVG + GVGVG + \ldots = G + GV\tilde{G}. \qquad (5.33)$$

In the simple cases we have just considered we could of course have managed without graphs. Expressions (5.33) and (5.32) can be easily obtained directly from eqn. (5.1) for G and the analogous expression for the two-particle Green's function. Let us write equation (5.33) in operator form by introducing the operator $G^{-1} = \partial/\partial t + iH_0$, where H_0 is the free-particle Hamiltonian. From (5.1) we get

$$G^{-1}\tilde{G} + iV\tilde{G} = I, \qquad (5.33')$$

and hence

$$\tilde{G} = G + G(-iV)\tilde{G}. \qquad (5.33'')$$

Since we have taken for the operator I the expression $\delta(\mathbf{r} - \mathbf{r}')\,\delta(t - t')$, the multiplication of operators must be interpreted as follows:

$$A(x_1, x_2) = BC = \int B(x_1, x')\,C(x', x_2)\,d^4x',$$

where $x = (\mathbf{r}, t)$.

Thus, the analytic form of equation (5.33) is

$$\tilde{G}(x_1, x_2) = G(x_1, x_2) + \int G(x_1, x')\,(-iV(x'))\,\tilde{G}(x', x_2)d^4x'$$

so that the graph indicating interaction with the field must be read as follows:

 $= -iV$.

If in the expressions (5.33') and (5.33") we understand by $\tilde{G}$ the two-particle Green's function corresponding to Schrödinger's equation for two particles, and by V the interaction potential $U(x_1 - x_2)$, we obtain in first order

$$G_2^{(1)}(x_1, x_2; x_1', x_2') =$$
$$= \int d^4x_3\, d^4x_4 G(x_1, x_3)\, G(x_2, x_4)(-i)\, U(x_3 - x_4) \times$$
$$\times\, G(x_3, x_1')\, G(x_4, x_2') = \quad (5.34)$$

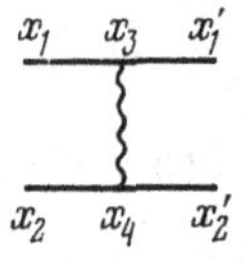

which indicates that:

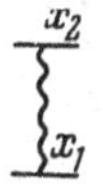

$$= -iU(x_1 - x_2)\,. \quad (5.35)$$

For a non-retarded interaction we have $U(x_1 - x_2) = U(\mathbf{r}_1 - \mathbf{r}_2)\, \delta(t_1 - t_2)$.

It is easy to obtain an expression for G_2 in the (λ, ϵ) representation. We have

$$G_2^{(1)}(\lambda_1\varepsilon_1, \lambda_2\varepsilon_2, \lambda_3\varepsilon_3, \lambda_4\varepsilon_4) =$$
$$= -i \sum_{\lambda_3\lambda_4} \int G_{\lambda_1}(\varepsilon_1)\, G_{\lambda_2}(\varepsilon_2)\, (\lambda_1\lambda_2 | \bar{U}(\omega) | \lambda_3\lambda_4)\, G_{\lambda_3}(\varepsilon_1 + \omega) \times$$
$$\times\, G_{\lambda_4}(\varepsilon_2 - \omega)\, \delta(\varepsilon_1 + \varepsilon_2 - \varepsilon_3 - \varepsilon_4) \frac{d\omega}{2\pi}\,,$$

where

$$\bar{U}(\omega) = \int d\tau e^{-i\omega\tau} U(\tau).$$

In the case of the interaction of free particles $G_2^{(1)}$ takes a very simple form indeed. In the $(\mathbf{p}, \epsilon)$ representation we have (where $p = (\mathbf{p}, \epsilon)$ and $q = (\mathbf{k}, \omega)$)

$$G_2^{(1)} = -i \int G(p_1)\, G(p_2)\, U(q) \times$$
$$\times\, G(p_1 + q)\, G(p_2 - q) \frac{d^4 q}{(2\pi)^4}\, \delta(p_1 + p_2 - p_3 - p_4)$$
$$(q = p_3 - p_1 = p_2 - p_4).$$

It is easy to accustom oneself to go over freely from one representation to the other.

For the application to the problem of scattering in the field $V(\underset{\sim}{r})$ it is convenient to collect the graphs in (5.33) as follows:

$$\bar{G} = G\{V + VGV + \ldots\}\, G = GAG,$$

where A is the scattering amplitude in the energy normalization:

$$A = V + VGV + VGVGV + \ldots = V + VGA. \tag{5.36}$$

We take for G expression (5.4). Since the field is time-independent, the quantity ϵ which enters G must be taken equal to the energy ϵ_p of the incident particle. The integration over intermediate-state momenta must be carried out with a weight $(2\pi)^{-3}$, corresponding to a summation with weight 1 over all the quasi-discrete states in the normalization volume; just such a summation is implied by the superposition principle. In the momentum representation we have

$$= -iV(p_1 - p_2)\,.$$

Substituting according to the rule we have already found $A \to iA$, $V \to iV$ in (5.36), we get

$$A(p, p') = V(p - p') + \int \frac{V(p - p_1)\, A(p_1, p')}{\varepsilon_p - \varepsilon_{p_1} + i\delta} \frac{d^3 p_1}{(2\pi)^3}, \tag{5.37}$$

which agrees with (5.32) with the replacement $\Gamma \to A$, $U \to V$.

This confirms the relation (5.37).

As an example to illustrate the usefulness of the graphical description and the corresponding symbolic expressions, we shall obtain the integral equation for the amplitude in a form suitable for the investigation of its analytic properties. The graphs for the amplitude may be collected as follows:

$$A = V + V\{G + GVG + \dots\}V = V + V\tilde{G}V. \tag{5.38}$$

Using the recipe already found for reading off the graphs and using for $\tilde{G}$ expression (5.5) with $\epsilon = \epsilon_p$, we obtain

$$A(p, p') = V_{pp'} + \sum_\lambda \frac{A_{p\lambda}A_{\lambda p'}}{\varepsilon_p - \varepsilon_\lambda + i\delta}. \tag{5.39}$$

where the sum over λ implies a summation over the bound states ($\epsilon_\lambda < 0$) and an integration over the continuous spectrum, and $A_{p\lambda} \equiv (e^{-i\underset{\sim}{p}\cdot\underset{\sim}{r}}, V(r)\,\varphi_\lambda(r))$. If we go over from the energy normalization to the usual one (see p. 273) we get

$$f(p, p') = f_B(p - p') - \frac{2\pi}{m}\sum_\lambda \frac{f_{p\lambda}f_{\lambda p'}}{\varepsilon_p - \varepsilon_\lambda + i\delta}, \tag{5.39'}$$

where f_B is the Born scattering amplitude. It follows from this expression that the analytic continuation of the amplitude into the region $\epsilon_p < 0$ has poles at points which coincide with the energies of bound states. However, the converse statement is not correct: not every pole of the scattering amplitude corresponds to a bound state (see p. 237)

Consider the imaginary part of eqn. (5.39') for $\underset{\sim}{p} = \underset{\sim}{p}'$. Assuming that the states φ_λ belonging to the continuous spectrum have the plane-wave normalization, that is have the asymptotic behaviour $\varphi_\lambda \underset{r\to\infty}{\longrightarrow} e^{i\underset{\sim}{p}\cdot\underset{\sim}{r}}$, we get

$$\operatorname{Im} f(p, p) = -\frac{2\pi}{m}(-\pi)\int |f(p, p_1)|^2\, \delta(\varepsilon_p - \varepsilon_{p_1})\frac{d^3p_1}{(2\pi)^3} =$$
$$= \frac{2\pi^2}{(2\pi)^3 m}\frac{p^2}{v}\int |f(\theta, \varphi)|^2\, d\Omega = \frac{p}{4\pi}\sigma,$$

which is just the optical theorem (p. 231).

So far we have learned to read off only graphs describing the motion of one free particle or two interacting particles. Let us now turn to the case of real interest, namely particles moving in a medium.

We begin with the simplest case, that of a single particle, and explain how the Green's function is affected by the identity of particles, i.e. we explain the expressions introduced intuitively on p. 273. Suppose we have a quasiparticle with momentum p (or in a state λ - this does not affect the argument) moving against the background of the other particles, among which is present a quasiparticle with the same momentum with weight n_p. (Here for simplicity we suppress spin indices). Then the transition amplitude for the two particles in question will be represented by the graph

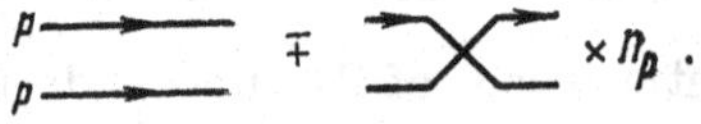

The second graph differs from the first in that the coordinates of the added and background particles are interchanged; the minus sign corresponds to fermions and the plus sign to bosons. The factor n_p takes account of the number of particles with which the coordinates of the initial particle may be interchanged. If both terms are regarded as describing the motion of a single quasiparticle, we get

$$G^{+}(p, \tau) = (1 \mp n_p)\, e^{-i\varepsilon_p \tau}\, \theta(\tau).$$

For hole propagation in a Fermi system we can introduce the number of holes in state $\underset{\sim}{p}$, that is, $\nu_p = 1 - n_p$; then by the same argument we get a factor $(1 - \nu_p) = n_p$. In the case of bosons we must take account of the change in the number of possible permutations n_p arising from the appearance of the hole; this, as we can see, gives a factor n_p. Since these are the same results as are automatically obtained, in the absence of interactions, from our definition of G on p.277 , this strengthens and refines our interpretation of G as the transition amplitude in the medium.

Consider now a quasiparticle in an external field. To second order in the field there appears not only the graph

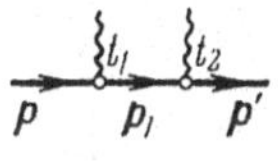

$= a, \; t_2 > t_1$

but also another graph which does not exist in the case of free particles, namely

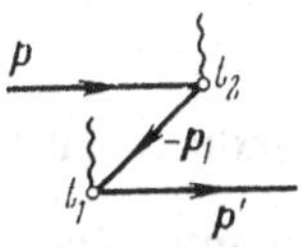

$= b, \; t_2 > t_1$,

This diagram represents the creation of a quasiparticle and quasihole at time t_1 and their subsequent annihilation at time t_2. We would like to know the analytic expression corresponding to graph (b). For this purpose we consider the motion of two particles, our initial particle with momentum $\underset{\sim}{p}$ and a background particle with momentum $\underset{\sim}{p}_1$. Then there corresponds to the sum of the graphs (a) and (b) the processes

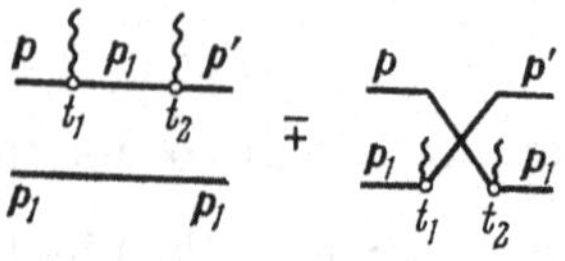

(5. 40)

In the second graph there appear at time t_1 a hole with momentum $-\underset{\sim}{p}_1$ and a second particle with momentum $\underset{\sim}{p}'$; then at time t_2 the original particle occupies the empty place. The graphs

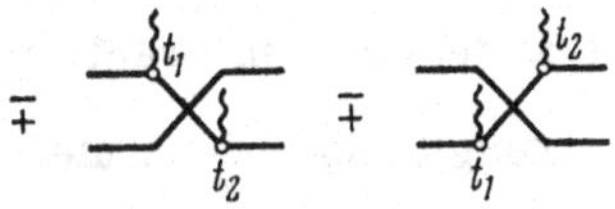

are already taken into account by the factors $(1 - n_{\underset{\sim}{p}})$ and $n_{\underset{\sim}{p}}$ in the particle and hole Green's functions repsectively. The graphs

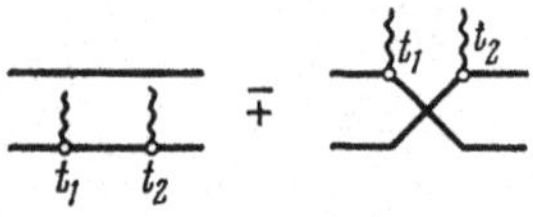

take into account the change of the background in the field and the effect of the added particle on this; these graphs are not related to the motion of the particle in question itself.

Since the second graph in (5. 40) corresponds to interchange of the particles between the interaction events, it must be assigned a minus sign for fermions and a plus sign for bosons. Thus we reach the conclusion that the graph with the reversed arrow is to be interpreted as follows:

$= \mp G^{-}(p, -\tau) = G(p, -\tau)$

Thus, the sum of the graphs (a) and (b) is

$$a + b = G^+ (-iV)\, G^+ (-iV)\, G^+ \mp G^+ (-iV)\, G^- (-iV)\, G^+.$$

Now we see the advantage of the function $G(\mathbf{p}, \tau) = G^+(\mathbf{p}, \tau) \mp G^-(\mathbf{p}, -\tau)$ which we introduced above: both for fermions and for bosons the graphs (a) and (b) can be written as a single graph which combines both processes:

$G^{(2)} =$  $=$

$$= -\int G^+(t_1 - t')\, V G(t' - t'')\, V G^+(t'' - t_2)\, dt'\, dt'',$$

where for $\tau = t' - t'' > 0$ the Green's function is just G^+ and describes scattering, while for $\tau < 0$ it is $\mp G^-$ and describes creation of particle-hole pairs. Here the effect of virtual pair creation on the scattering of a quasiparticle in a field is automatically taken into account. Thus, the expression for the scattering amplitude of a quasiparticle from a scattering centre situated in a medium may be obtained from the expressions found above for a particle by replacing the free-particle Green's function in the intermediate states by the unified Green's function G in the medium. Such a formula may be necessary, for instance, in determining the scattering amplitude of electrons in a metal from impurity atoms.

Similar remarks apply to the scattering amplitude resulting from the interaction of two quasiparticles. For simplicity we shall consider a δ-function type interaction between the particles and represent it by a dot. Consider the graph elements occurring in the scattering amplitude:

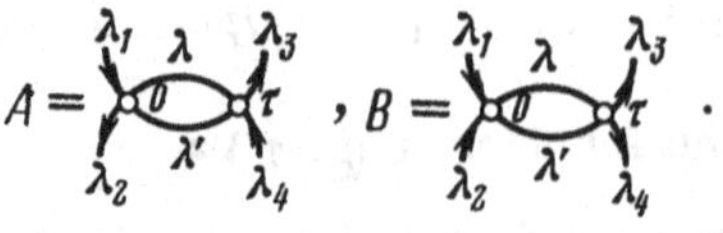

For $\tau > 0$ graph B describes the motion, between the collision events, of two quasiparticles in an N-particle system and contains $G^+_\lambda(\tau)\ G^+_{\lambda'}(\tau)$, while for $\tau < 0$ it corresponds to two quasiholes in an $(N + 2)$-particle system and contains $G^-_\lambda(\tau)\ G^-_{\lambda'}(\tau)$. The two terms can be formally combined into one by introducing $G_\lambda(\tau)\ G_{\lambda'}(\tau)$. The same applies to graph A, which describes the motion of a particle and a hole: there are two different graphs depending on the sign of τ, containing $G^+_\lambda\ G^-_{\lambda'}$ and $G^-_\lambda G^+_{\lambda'}$ respectively. The two can be combined by writing $G_\lambda(\tau)\,G_{\lambda'}(-\tau)$.

As a preparation for later work let us find the diagrams A and B in the (λ, ϵ) representation. Substituting the expression (5.10) for G, we find

$A =$  $$= \int \frac{id\varepsilon}{2\pi} G_\lambda(\varepsilon)\ G_{\lambda'}(\varepsilon - \omega) = \frac{n_\lambda - n_{\lambda'}}{\varepsilon_{\lambda'} - \varepsilon_\lambda + \omega}, \qquad (5.41)$$

$B =$  $$= \int \frac{id\varepsilon}{2\pi} G_\lambda(\varepsilon)\, G_{\lambda'}(E - \varepsilon) = \frac{1 - n_\lambda - n_{\lambda'}}{E - \varepsilon_\lambda - \varepsilon_{\lambda'}}. \qquad (5.42)$$

where E is the total energy of the two particles and ω the total energy of the particle and hole. (In expressions (5.41-2) we have omitted the interaction matrix elements).

We will introduce one further graph element which will be required later: namely the graph

$-i\bar{V}$ =  , (5. 43)

which, as we shall see, describes the interaction of a particular particle with the particles of the medium to first order in the interaction U. The disk is to be interpreted as a Green's function evaluated at equal values of the coordinates; more precisely we associate with the disk the Green's function

$$G(r, r; \tau)_{\tau \to -0} = \mp n(r) \tag{5. 44}$$

(where the upper sign corresponds to fermions and the lower one to bosons, cf. (5.14)), multiplied by a factor γ which we shall now determine. The graph (5.43) describes the action of the external field $\bar{V}$ and at the same time is expressed in terms of the interaction U. We have

$$(-i\bar{V}) = (-iU)\,\gamma G_{\tau \to -0}. \tag{5. 45}$$

In the coordinate representation this reads

$$\bar{V}(r) = \gamma \int U(r - r', t - t')\, G(r', r', -0)\, dr'\, dt' =$$
$$= \gamma \int U(r - r')\, G(r', r')_{\tau \to -0}\, dr',$$

On the other hand, we have the obvious relation

$$\bar{V}(r) = \int U(r - r')\, n(r')\, dr'.$$

Comparing this with (5.44), we get $\gamma = \pm 1$, i.e.

 $= \mp\, G(r, r; \tau = -0)$.

Thus, we must associate with the closed loop the factor $G(r, r, \tau = -0)$ with a minus sign for fermions and a plus sign for bosons, that is, simply the factor $n(\underset{\sim}{r})$.

It will be obvious from the above discussion that the objects to be discussed by the graphical method may be chosen in various ways. We may, if we wish, draw the processes which happen to real particles in the many-body problem; but we can also deal with simpler objects, namely quasiparticles, for which the Green's function has the simple form found on p.283. Once we have established, as in the last section, the connection between particles and quasiparticles and convinced ourselves that for excitation energy corresponding to frequencies $\omega \ll \epsilon_F$ and momenta $k \ll p_F$ the quasiparticles represent the essential features of the excitations to sufficient accuracy, then to describe many processes we need to deal only with quasiparticles. The only exceptions are cases when we have to consider the interaction of the system with particles incident from outside, or when we want to investigate the momentum distribution of real particles, rather than quasiparticles.

In the following we shall choose the objects described by our graphical method to be quasiparticles. To this end we shall introduce a simple expression which determines the interaction of quasiparticles in a system with a δ-function form of interparticle interaction; such an interaction occurs in nuclear matter. We shall further introduce various "charges" which characterize the interaction of the quasiparticles with external fields.

The Interaction Between Quasiparticles

The interaction between quasiparticles differs from the

interaction of two particles in free space. For instance, the interaction between two nucleons in free space is mediated by the exchange of one or more mesons, whereas in the interior of nuclear matter there exists in addition to this mechanism the possibility of interaction by exchange of a particle-hole pair. In the graphs for the scattering amplitude Γ these two mechanisms are represented as follows:

$\Gamma =$ 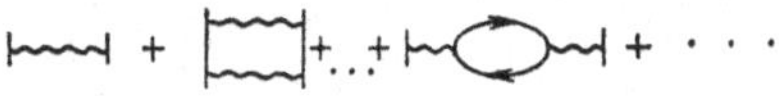

where the wavy line denotes a meson Green's function and the loop corresponds to creation of a particle-hole pair.

Thus, the additional interaction arises from the effect of polarization of the medium. In addition, the Pauli principle implies that even interaction graphs unconnected with polarization are altered, since some of the states are occupied by the other nucleons and so are not available for the interacting particles.

The problem of determining the interaction in matter from the interaction of two particles in free space is extremely complex for strongly interacting particles, since the effect of the medium changes the free-space interaction very substantially. We shall not go into this here; rather, we shall express the interaction between quasiparticles in terms of a few constants which are not calculated but must be found from comparison of theory and experiment.

In the case of the nucleus the range of the interaction between quasiparticles is approximately the same as the range r_o of the interaction potential in free space. This follows because the density of nuclear matter is determined by the condition that the interparticle spacing should be of order r_0; hence the Fermi momentum, which is determined by the density, is related to r_0 by the order-of-magnitude estimate (for $\hbar = m = 1$)

$$p_F r_0 \sim 1.$$

The depth of the effective potential in which the particles composing the nucleus move is of order

$$U = \frac{p_F^2}{2} \sim \frac{1}{r_0^2}.$$

Thus, all the characteristic parameters of nuclear matter, and hence also the effective force range, are determined by the quantity r_0, which is the only characteristic length entering the problem; it therefore characterizes not only the free-space interaction, but also the additional interaction due to the polarizability of nuclear matter.

As we shall see, all problems connected with an external field of frequency ω much less than the Fermi energy ϵ_F, and wave vector small compared with the Fermi momentum p_F, reduce to finding the amplitude for scattering with small momentum transfer ($k \ll p_F$, $\omega \ll \epsilon_F$) in the two-quasiparticle channel, or in other words with small total energy in the quasiparticle-quasihole channel (the horizontal channel in Fig. 47, where $q = (\omega, \underset{\sim}{k})$). In this case, to get a convenient equation, we must classify the graphs occurring in Γ as follows. We separate out and combine into a block diagram $\mathcal{F}$ all diagrams which do not contain parts linked only by two lines corresponding to a quasiparticle and a quasihole. The block diagram $\mathcal{F}$ contains the following graphs:

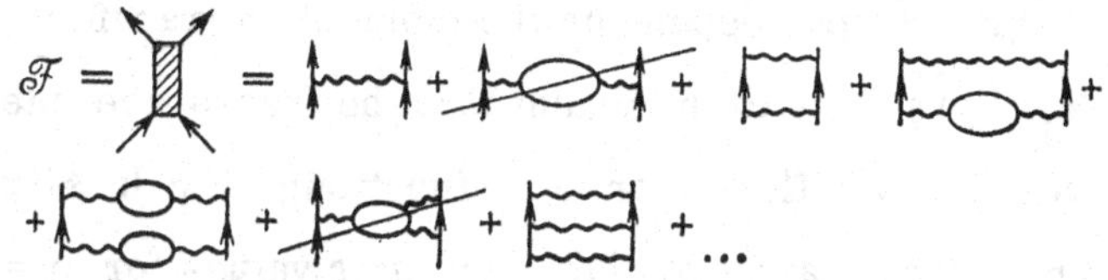

where the deleted graphs are those which by definition do not contribute to $\mathcal{F}$. As we will show, for small momentum transfer these graphs depend critically on the states of the scattered particles; it is for this reason that we exclude them from $\mathcal{F}$.

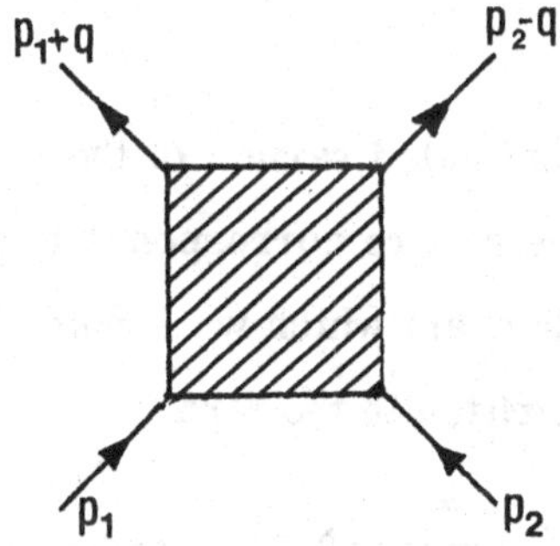

Fig. 47

All the graphs except the deleted ones give a δ-function-type contribution to $\mathcal{F}$ for small momentum transfer. To see this, we use the δ-function nature of the free-space interaction u =  , to represent the first few graphs of $\mathcal{F}$ in the form

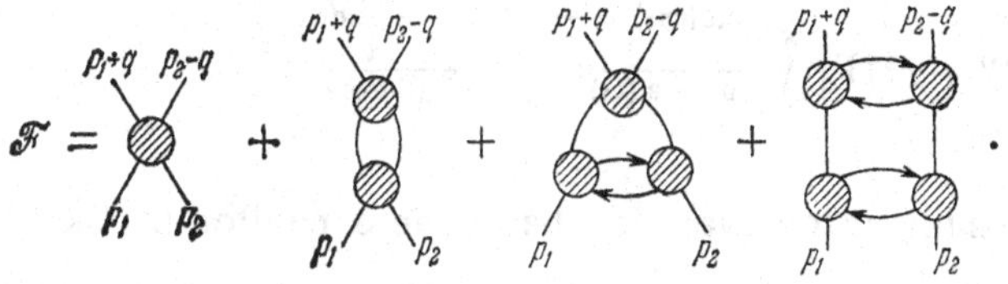

The first three graphs depend on the momentum transfer q ($q_i = (\epsilon_i, k_i)$) only to the extent that the free-space interaction is non-local (i.e. differs from a δ-function); for $k \ll r_0^{-1}$, $\omega \ll c\, r_0^{-1}$ they may be replaced by their values for $q = 0$. To convince ourselves that the more complicated graphs in $\mathcal{F}$ are likewise independent of q, we estimate the graph

$\mathcal{F}^{(4)} =$ 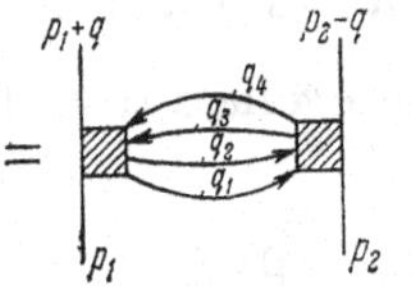 ,

which is a more complicated variant of the fourth graph in $\mathcal{F}$ depicted above. The square corresponds to some "local" (q-independent) block diagram which we denote Γ. The expression for $\mathcal{F}^{(4)}$ may be written in the form

$$\mathcal{F}^{(4)} \sim |\Gamma|^2 \int \prod_i \frac{d^4 q_i}{\omega_i - \varepsilon(k_i)} \delta(q - q_1 - q_2 + q_3 + q_4).$$

Since q_3 and q_4 correspond to holes, k_3 and k_4 are by definition less than p_F. Consider now the region of integration $\epsilon_i > \epsilon_F$ and $k_{1,2} \gg p_F$; then in the denominators corresponding to holes we may neglect $\epsilon(k_{3,4})$ by comparison with $\omega_{3,4}$ and carry out the integration over k_3 and k_4. We get

$$\mathcal{F}^{(4)} \sim n^2 |\Gamma|^2 \int \frac{d^3 k_1 \delta(\omega_1 + \omega_2 - \omega_3 - \omega_4 - \omega) \prod_i d\omega_i}{(\omega_1 - \varepsilon(k_1))(\omega_2 - \varepsilon(k_1))\, \omega_3 \omega_4}.$$

where the integration over k_2 has been cancelled by the factor $\delta(k - k_1 - k_2 + k_3 + k_4)$. It is easy to see that the largest contribution to the integral over the ω_i comes from the region

$\omega_1 \sim \omega_2 \sim \omega_3 \sim \omega_4 \sim \epsilon(k_1)$. We finally get

$$\mathcal{F}^{(4)} \sim n^2 |\Gamma|^2 \int \frac{d^3k_1}{\varepsilon(k_1)} \sim n^2 |\Gamma|^2 L,$$

where L denotes the upper limit of integration over k_1. Thus w we have obtained an apparently divergent expression whose value is actually determined by the nonlocality of the block diagram Γ.

Since the important region of the momentum integration is of order L, the graph in question is independent of q provided

$$\frac{k}{L}, \frac{\omega}{\varepsilon(L)} \ll 1.$$

Depending on the character of the graphs contributions to Γ, the characteristic momentum L may be of order either $m_\pi c$ or $m_n c$ (in nuclear matter: m_π, m_n = pion and nucleon mass respectively). In general, graphs containing more than two lines depend only weakly on the momentum transfer, since in the integration over the 4-momenta of the internal lines it is large energies and momenta ($p \gtrsim p_F, \epsilon \gtrsim \epsilon_F$) which are dominant*.

With the help of the block diagram $\mathcal{F}$ we can classify all graphs contributing to Γ into: (1) graphs not containing two lines in the quasiparticle-quasihole channel (these just form the diagram $\mathcal{F}$), and (2) graphs where in the quasiparticle-quasihole channel we have first the diagram $\mathcal{F}$, followed by two lines (representing a quasiparticle and a quasihole), and finally the sum of all graphs which transfer a quasiparticle and a quasihole to a new state (which

---

*This argument is developed in more detail in the books referred to above (p. 269).

is itself just the block diagram Γ). Thus, in graphical language the equation for Γ reads

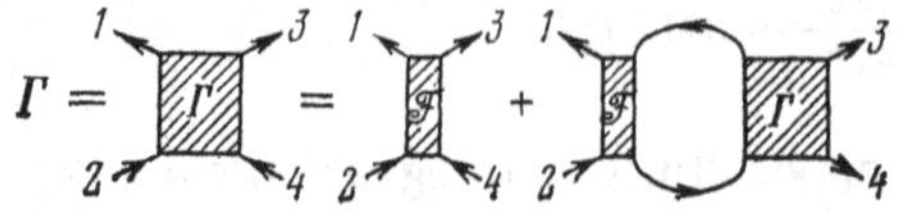

or in symbolic form

$$\Gamma = \mathcal{F} + \mathcal{F} GG\Gamma. \tag{5.46}$$

We write this expression in the (λ, ϵ) representation, omitting for the moment the index λ . Since the block diagram $\mathcal{F}$ is of δ-function form with respect to the difference of the initial times, the expression Γ must also be a δ-function with respect to this difference. Since moreover Γ must be symmetric with respect to the initial and final points (as can be directly verified by collecting the graphs in reverse order) Γ can depend only on the difference of initial and final times and hence in the (λ, ϵ) representation depends on the total energy of the quasiparticle-quasihole pair, which is conserved throughout all sections of this channel. Omitting an overall factor $\delta(\epsilon_1 + \epsilon_2 - \epsilon_3 - \epsilon_4)$ and associating with the graphs $\mathcal{F}$ and Γ the quantities $(-i\mathcal{F}, \; -i\Gamma)$ we obtain

$$\Gamma(\omega) = \mathcal{F} - \mathcal{F} \int G(\varepsilon) G(\omega - \varepsilon) \frac{i\, d\varepsilon}{2\pi} \Gamma(\omega) \equiv \mathcal{F} + \mathcal{F} A\Gamma.$$

Replacing the index λ and using the expression for A found on p. 300 , we get

$$(\lambda_1\lambda_2 | \Gamma | \lambda_3\lambda_4) = (\lambda_1\lambda_2 | \mathcal{F} | \lambda_3\lambda_4) + \\ + \sum_{\lambda\lambda'} (\lambda_1\lambda_2 | \mathcal{F} | \lambda'\lambda) \frac{n_{\lambda'} - n_\lambda}{\varepsilon_{\lambda'} - \varepsilon_\lambda + \omega} (\lambda'\lambda | \Gamma | \lambda_3\lambda_4). \tag{5.47}$$

We will test the correctness of the coefficients in this equation below on a simple example. Since the block diagram $\mathcal{F}$ is of δ-function-like form in the coordinate representation (it is determined by a region of radius $\sim r_0$ near the point in question) we shall call the quantity $\mathcal{F}$ the local interaction amplitude or simply the local interaction.

In cases where pair correlations are important, or where there exist levels near the Fermi surface which "compete" with the one-particle states, the expression for G_λ becomes more complicated and the equation for Γ no longer has the simple form (5.47).

The Local Quasiparticle Interaction

As we have already mentioned, the effective local interaction between quasiparticles will be characterized by a few constants. We shall demonstrate this for the case of the local interaction in a nucleus. We consider first the case of homogeneous nuclear matter; later we shall introduce the corrections due to the finite dimensions of the nucleus. In the momentum representation the local interaction amplitude depends on the two momenta $\underset{\sim}{p}_1$, $\underset{\sim}{p}_2$ and on the momentum transfer $\underset{\sim}{q}$:

$\mathcal{F} =$ 

.

Since $\mathcal{F}$ is only weakly dependent on the momentum variables (it changes appreciably only for large momentum changes $\delta p \sim p_F$, $\delta\epsilon \sim \epsilon_F$), it follows that if we are considering small momentum we may put $q = 0$ in $\mathcal{F}$. (The error so introduced is of the order k/p_F, ω/ϵ_F.) Moreover, to investigate the behaviour of the amplitude Γ near the Fermi surface it is adequate to know $\mathcal{F}$ for $|p_1| = |p_2| = p_F$ and $\epsilon_1 = \epsilon_2 = \epsilon_F$. Then $\mathcal{F}$ depends only on the angle between the initial momenta $\mathbf{p}_1$ and $\mathbf{p}_2$. The interaction between quasiparticles also depends on their spin and isotopic spin quantum numbers. Assuming isotopic invariance, we therefore put

$$\mathcal{F} = C\{f + f'\tau_1\tau_2 + (g + g'\tau_1\tau_2)\,\sigma_1\sigma_2\}, \tag{5.48}$$

where f, f' and g, g' are functions of the angle between $\mathbf{p}_1$ and $\mathbf{p}_2$, and $\boldsymbol{\sigma}$, $\boldsymbol{\tau}$ are, respectively, spin and isotopic spin matrices. We choose the normalization factor C so that

$$C = \frac{1}{dn/d\varepsilon_F} = \frac{\pi^2}{m^* p_F}.$$

Then f, f' and g, g' are dimensionless quantities of order 1.

We have not included in (5) terms of the form $(\mathbf{p}_1 \cdot \boldsymbol{\sigma}_1)(\mathbf{p}_2 \cdot \boldsymbol{\sigma}_2)$, which enter as a relativistic correction and tend to zero for small particle velocities. In fact, since in nuclear matter the velocity at the Fermi surface is not negligibly small compared to the velocity of light ($v/c \sim 1/4$), these terms may be important there. The so-called tensor forces are proportional to k^2 and have, therefore, also not been included in (5.48).

We expand $\mathcal{F}$ in a series in Legendre polynomials in the cosine of the angle between $\mathbf{p}_1$ and $\mathbf{p}_2$:

$$x = \frac{p_1 p_2}{p_F^2}, \quad \mathcal{F} = \sum_l \mathcal{F}_l P_l(x). \tag{5.49}$$

It should be emphasized that this expansion bears no relation to the usual partial-wave expansion of the scattering amplitude; the latter is an expansion in functions P_ℓ of the scattering angle, whereas in $\mathcal{F}$ the scattering angle has been put equal to zero ($k = 0$). The quantities f_ℓ, f_ℓ', g_ℓ, and g_ℓ' must be found by comparing theory and experiment. Such a comparison shows that in nuclei it is the zeroth harmonics in the expansion (5.49) which are most important, that is, the local interquasiparticle interaction is only weakly velocity-dependent.

In the equation for the effective field in an infinite system $\mathcal{F}$ is to be taken for four-momentum q equal to the energy-momentum $(\underset{\sim}{k}, \omega)$ of the external field. Thus, for sufficiently homogeneous external fields we can put $q = 0$ in $\mathcal{F}$, thereby introducing an error of order k/p_F, ω/ϵ_F. In a finite system, however, even if the external field V_0 is homogeneous the effective field V will not be, but will change appreciably over distances of the order of the radius R. We must, therefore, find an expression for $\mathcal{F}$ for $k \sim 1/R$; we can still put $\omega = 0$ provided $\omega \ll \epsilon_F$. Since in a finite system $k/p_F \sim 1/p_F R \sim A^{-1/3}$, we need only keep in $\mathcal{F}$ terms independent of k and linear in k; we may neglect terms of order k^2.

3. THE SOLUTION OF PROBLEMS BY THE GREEN'S FUNCTION METHOD

Dyson's Equation. The Basis of the Shell Model

As we know, the so-called nuclear shell model starts

from the assumption that there exist in the nucleus energy levels similar to those of a system of noninteracting particles in a potential well. Until the advent of the Green's function method the fact that in many cases this model gave very good results was a complete mystery. (A similar difficulty occurred in the theory of metals, where many phenomena were well explained by the free-electron model, in spite of the strong interactions between the electrons.) The Green's function method allows us to explain how one-particle levels can exist in a Fermi system with strong interactions.

We shall denote the exact Green's function by a heavy line:

$G(1,2) =$  ,

and the free-particle Green's function by a thin line:

$G_0(1,2) =$   .

The total transition amplitude is equal to the sum of all possible amplitudes, so that G can be expressed as a sum of graphs:

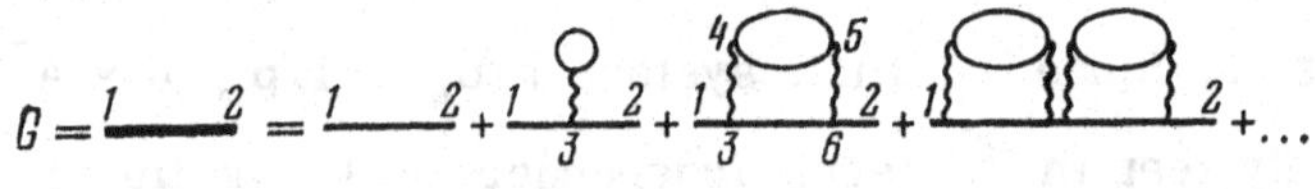

The first graph in the sum corresponds to the transition amplitude when the particle under consideration does not interact with the particles of the medium, that is, to G_0. The second graph represents the elastic scattering of the particle in question by the particles of the medium. The third graph describes free motion up to the point 3, where an inelastic collision occurs; as a result of the latter the particle in question creates a particle-hole pair at the point 4. Subsequently, the pair is annihilated at the point 5, and from points 6 to 2 the particle propagates freely. The meaning of all other graphs can be seen in an analogous way.

We shall classify the various graphs as follows. First of all we take out the single graph corresponding to free propagation. All the other graphs now have the following form: up to a certain point the particle performs free motion. Then there occurs a collision, as a result of which some other particles and holes are formed and annihilated; thereupon free motion occurs once more until a second collision takes place. Therefore, all these graphs have the following structure: first we have free motion, then the sum of all graphs which do not contain parts joined by a single line, and finally the total transition amplitude of the particle from the intermediate to the final state. Thus, the equation for G(1, 2) can be written graphically:

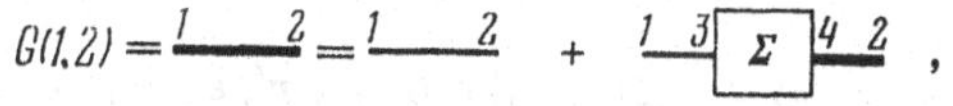

$$G(1,2) = \underset{1 \qquad 2}{\rule{1cm}{0.4pt}} = \underset{1 \qquad 2}{\rule{1cm}{0.4pt}} + \underset{1 \quad 3}{\rule{0.6cm}{0.4pt}}\boxed{\Sigma}\underset{4 \quad 2}{\rule{0.6cm}{0.4pt}} \, , \tag{5.50}$$

where Σ denotes the sum of graphs[*] which do not contain parts joined by a single line. The analytical form of (5.50) is

$$G(1, 2) = G_0(1,2) + \int G_0(1,3) \, \Sigma(3,4) \, G(4,2) d\tau_3 d\tau_4. \tag{5.51}$$

Equation (5.51) is called the Dyson equation. The introduction of the block Σ has the advantage that it is of δ-function-like shape in the coordinate representation and can therefore be characterized by only a few constants. This is vecause Σ contains only graphs with three or more lines, which as we can verify (cf. p.306), are

---

[*] We shall continue to call such a sum a "block diagram".

sharply peaked with respect to the coordinate and time differences, the "spread" of the peak being of order $\delta r \sim p_F^{-1}$, $\delta t \sim \epsilon_F^{-1}$.

In the absence of external fields, when the Hamiltonian of the system is not explicitly time-dependent, all the quantities occurring in (5.51) depend only on the differences of the corresponding time coordinates. It is therefore convenient to go over to a Fourier transform with respect to the time. We then get

$$G(\boldsymbol{r}_1, \boldsymbol{r}_2, \varepsilon) = G_0(\boldsymbol{r}_1, \boldsymbol{r}_2, \varepsilon) + \\ + \int G_0(\boldsymbol{r}_1, \boldsymbol{r}_3, \varepsilon)\,\Sigma(\boldsymbol{r}_3, \boldsymbol{r}_4, \varepsilon)\,G(\boldsymbol{r}_4, \boldsymbol{r}_2, \varepsilon)\,d^3r_3\,d^3r_4. \tag{5.52}$$

In coordinate representation the equation for G_0 has the form (cf. eqn. (5.1), p. 270)

$$(\varepsilon - p^2/2m)\,G_0(\boldsymbol{r}, \boldsymbol{r}', \varepsilon) = \delta(\boldsymbol{r} - \boldsymbol{r}'),$$

where $\underset{\sim}{p}$ is the momentum operator, which acts on the coordinate $\underset{\sim}{r}$. Multiplying (5.52) by $(\epsilon - p^2/2m)$ from the left, we get

$$\left(\varepsilon - \frac{p^2}{2m}\right) G(\boldsymbol{r}, \boldsymbol{r}', \varepsilon) - \int \Sigma(\boldsymbol{r}, \boldsymbol{r}_1, \varepsilon)\,G(\boldsymbol{r}_1, \boldsymbol{r}', \varepsilon)\,d^3r_1 = \\ = \delta(\boldsymbol{r} - \boldsymbol{r}')$$

We have omitted the spin indices for simplicity. Since the block diagram Σ is of δ-function-like form we can write the second term in the form

$$\int \Sigma G\,d^3r_1 = \alpha G(\boldsymbol{r}, \boldsymbol{r}', \varepsilon) + \beta_{\alpha\beta}\frac{\partial^2 G(\boldsymbol{r}, \boldsymbol{r}', \varepsilon)}{\partial r_\alpha \partial r_\beta},$$

where

$$\alpha(\boldsymbol{r}, \varepsilon) = \int \Sigma(\boldsymbol{r}, \boldsymbol{r}_1, \varepsilon)\,d^3r_1,$$
$$\beta_{\alpha\beta}(\boldsymbol{r}, \varepsilon) = \frac{1}{2}\int \Sigma(\boldsymbol{r}, \boldsymbol{r}_1, \varepsilon)(\boldsymbol{r}_1 - \boldsymbol{r})_\alpha(\boldsymbol{r}_1 - \boldsymbol{r})_\beta\,d^3r_1.$$

In a homogeneous system we have $\beta_{\alpha\beta} = \beta\,\delta_{\alpha\beta}$, where β is independent of $\underset{\sim}{r}$. In a finite system this relation is valid to zeroth order in r_0/R, where r_0 is the interparticle spacing and R the radius of the system. Consequently, for $r_0 \ll R$ the equation for G takes the form

$$\left[\varepsilon - \frac{p^2}{2m}(1 + 2m\beta) - \alpha(r, \varepsilon)\right] G(r\ r', \varepsilon) = \delta(r - r'). \tag{5. 53}$$

Let us consider eqn. (5.53) for values of ϵ lying close to the Fermi energy ϵ_F, and measure ϵ from ϵ_F. Then, expanding $\alpha(\underset{\sim}{r}, \epsilon)$ in powers of ϵ and keeping only the first two terms, we get

$$\left(\varepsilon - \frac{p^2}{2m^*} - U(r)\right) G(r, r', \varepsilon) = Z\delta(r - r'), \tag{5. 54}$$

where

$$m^* = \frac{1 - \dfrac{\partial\alpha}{\partial\varepsilon}}{1 + 2m\beta}, \quad Z = 1 - \frac{\partial\alpha}{\partial\varepsilon}, \quad U(r) = \frac{\alpha(r, 0)}{1 - \dfrac{\partial\alpha}{\partial\varepsilon}}.$$

Here m^* is the effective mass, Z the Green's function renormalization constant, and $U(r)$ the effective potential. All these quantities can be expressed in terms of the block diagram Σ, which in turn can be expressed as a perturbation series in the interparticle interaction. Thus, we can, in principle, calculate the quantities m^*, Z, and $U(r)$. For instance, to first order in perturbation theory in the interaction it is easily shown that

$$(m^*)^{(1)} = m, \quad Z^{(1)} = 1, \quad [U(r)]^{(1)} = U_{HF}$$

where U_{HF} is the Hartree-Fock self-consistent potential.

Since the interaction is by no means small in the nucleus, the actual calculation of these quantities is beset by great difficulties. We shall therefore characrerize the potential U(r) by a few constants: the depth of the potential well, its radius, and the width of the d-layer over which the potential drops from its constant value inside the nucleus to its value outside the nucleus, namely zero. Together with m^* and Z, these constants must be found from a comparison of theory and experiment.

We now introduce the system of functions which describe a quasiparticle with energy ϵ_λ close to ϵ_F:

$$\left(\frac{p^2}{2m} + U(r)\right)\varphi_\lambda = \varepsilon_\lambda \varphi_\lambda.$$

Then eqn. (5.54) gives

$$(\varepsilon - \varepsilon_\lambda)\, G_{\lambda\lambda'}(\varepsilon) = Z\delta_{\lambda\lambda'}.$$

This expression differs only by the factor Z from the corresponding equation for quasiparticles which is obtained from (5.10').

Thus, $G_\lambda(\epsilon)$ has a pole at $\epsilon = \epsilon_\lambda$. This result means that the system possesses a branch of single-particle excitations corresponding to the excitations of a gas of quasiparticles.

Instability of the Fermi Distribution in the Case of Attraction. The Occurrence of a Gap in the Energy Spectrum

Let us obtain an equation for the scattering amplitude Γ in the two-particle channel. For this purpose we introduce the block diagram $\mathfrak{U}$, which is defined as not containing, in this channel, parts connected by just two quasiparticle lines. Following a procedure analogous to the one used above to obtain eqn. (5.46), we find

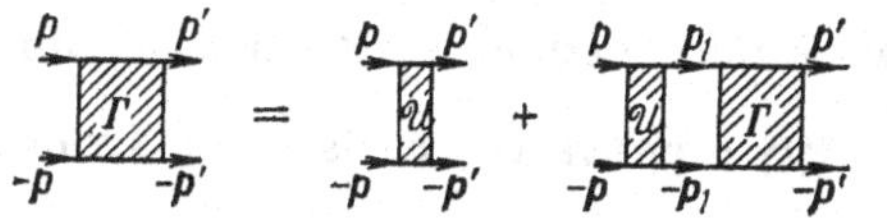

(5.55)

$$\Gamma = \mathcal{U} + \mathcal{U}GG\Gamma.$$

where, in contrast to eqn. (5.46), GG is the two-quasiparticle Green's function. The block diagram $\mathcal{U}$ is δ-function-like with respect to the coordinates and times of the incoming lines for the same reasons as Σ is.

Consider two particles with total energy E and total momentum zero. We calculate the element GG occurring in eqn. (5.55). With the aid of (5.42) we find:

$$\int_{-\infty}^{+\infty} G(p_1, \tau) G(p_1, \tau) e^{iE\tau} d\tau = i \frac{1 - 2n_{p_1}}{E - 2E(p_1) + i\delta} \equiv iB.$$

Substituting this in (5.55) and using the rules for reading off diagrams, we get

$$\Gamma(p, p') = \mathcal{U}_0 + \mathcal{U}_0 \int \frac{1 - 2n_{p_1}}{E - 2E(p_1) + i\delta} \Gamma(p_1, p') \frac{d^3p_1}{(2\pi)^3}. \tag{5.56}$$

We will look for a pole in the expression for Γ. The first term on the right-hand side may then be neglected, and we get a homogeneous equation. To help our intuition we give this equation a form reminiscent of Schrödinger's equation for two particles. Introducing $\Psi(p) = B\Gamma$, we get for Ψ the equation

$$(E - 2E(p)) \Psi(p) = (1 - 2n_p) \mathcal{U}_0 \int \Psi(p') \frac{d^3p'}{(2\pi)^3}.$$

In empty space, where $n_{\underset{\sim}{p}} = 0$, this equation is just the Schrödinger equation in the momentum representation for two particles with a δ-function form interaction potential $\mathcal{U}_0 = \int \mathcal{U}(\underset{\sim}{r} - \underset{\sim}{r}')\, d\underset{\sim}{r}'$. The factor $(1 - 2n_p)$ takes into account the change of this equation in the medium. The energy eigenvalue E is found from the relation

$$1 = \mathcal{U}_0 \int \frac{1 - 2n_p}{E - 2E(p)} \frac{d^3p}{(2\pi)^3} =$$
$$= \mathcal{U}_0 \frac{dn}{d\varepsilon_F} \left\{ - \int_0^{\varepsilon_F} \frac{dE'}{E - 2E'} + \int_{\varepsilon_F}^{\varepsilon_1} \frac{dE'}{E - 2E'} \right\}.$$

We have cut off the integration over E' in the second integral at a value ϵ_1 which is determined by the inaccuracy of the assumption of a δ-function form for $\mathcal{U}$. According to our discussion above, we have $\epsilon_1 - \epsilon_F \sim \epsilon_F$; for small values of $(E - 2\epsilon_F)$ we will see that any error in the definition of ϵ_1 does not affect the result. Denoting the quantity $E - 2\epsilon_F$ by $\epsilon\ (\ll \epsilon_F)$ we find up to logarithmic accuracy

$$1 \simeq \mathcal{U}_0 \frac{dn}{d\varepsilon_F} \ln \frac{-\varepsilon^2}{\varepsilon_F^2} \equiv \gamma \ln \frac{-\varepsilon^2}{\varepsilon_F^2}. \tag{5.57}$$

It follows from (5.57) that a solution exists only for negative $\mathcal{U}_0$:

$$-\varepsilon^2 \simeq \varepsilon_F^2 e^{-1/|\gamma|}.$$

Thus the square of the energy ϵ is negative; this indicates the existence of solutions of the form $\Psi(t) = \Psi_0 \exp\{|\epsilon| t\}$ which increase exponentially in time, that is, that the system is unstable*. The number of such correlated pairs of particles will

---

* This phenomenon was discovered by Cooper in 1956.

grow until the Fermi distribution, and hence the quasiparticle energy, changes in such a way that a stable state results. As we shall see, this phenomenon leads to the appearance of superfluidity, or in the case of charged particles to superconductivity.

We will assume that the instability considered above corresponds to two quasiparticles with opposite spins (this is the case which is realized for electrons in a superconductor). The interaction of pairs of particles with parallel and with antiparallel spins is quite different; as a result of the Pauli principle, two particles with identical spins cannot exist at the same point in space, which weakens the interaction at small distances and can actually change the sign of the quantity γ introduced above.

As a result of the instability there is formed a "condensate" of correlated pairs of particles with total spin zero ("Cooper pairs"). The existence of this condensate drastically changes the properties of particles with energy near the Fermi surface. This can be understood by noticing that a particle can turn itself into a hole with the formation of a Cooper pair. Let us denote the amplitude for such a transition by Δ:

$i\Delta =$  .

Then the amplitude of the inverse transition will be its complex conjugate $(\langle 1|V|2\rangle = \langle 2|V|1\rangle^*)$

$-i\Delta^* =$  .

Then the particle self-energy acquires the extra term

$\Sigma_k =$  .

The hole Green's function between the blocks Δ contains no transitions into Cooper pairs, since such a transition would involve the appearance of a line going rightwards, while the quantity Σ by definition does not contain parts connected by such a line.

If we introduce the Green's function G which includes all graphs other than the one just considered, we get the relation

$$G_s^{-1} = G_0^{-1} - \Sigma = G_0^{-1} - \Sigma_0 - \Sigma_k = G^{-1} - \Sigma_k, \tag{5.58}$$

where $G^{-1} = G_0^{-1} - \Sigma_0$.

Including diagrams of the form Σ_k in the internal lines of Σ_0 does not change this quantity appreciably, since the singularity contained in Σ_k is integrated over in this case. In fact, as we shall see, Σ_k distorts the function G in an energy interval of width Δ, which, if we assume that $\Delta \ll \epsilon_F$, is small compared to the region $(\sim \epsilon_F)$ important in the integration over ϵ and $\epsilon(\underline{p})$. Let us write the equation in the $(\underline{p}, \epsilon)$ representation and measure all energies from the Fermi level. Then we have

$$G_s(p, \varepsilon) = [G^{-1}(p, \varepsilon) - \Sigma_k(p, \varepsilon)]^{-1},$$

where, according to (5.10') and (5.24), $G(\underline{p}, \epsilon)$ is given for small $\epsilon(\underline{p})$ and ϵ by

$$G(p, \varepsilon) \simeq \frac{Z}{\varepsilon - \varepsilon(p) + i\delta \operatorname{sign} \varepsilon}.$$

For $\Sigma_k(\underline{p}, \epsilon)$ we have

$$\Sigma_k(p, \varepsilon) = (i\Delta)(-G(-p, -\varepsilon))(-i\Delta) =$$
$$= |\Delta(p, \varepsilon)|^2 \frac{Z}{\varepsilon + \varepsilon(p) + i\delta \operatorname{sign} \varepsilon}. \tag{5.59}$$

(The change of direction of the arrow on the line between the block diagrams Δ gives, as was shown on p. 298, $-G(-\underline{p}, -\tau)$

in the $(\underset{\sim}{p}, \tau)$ representation or $-G(-\underset{\sim}{p}, -\epsilon)$ in the $(\underset{\sim}{p}, \epsilon)$ representation).

We will obtain (5.59) by a second, much more intuitive method. As is obvious from the expression for G_s, the quantity $Z\,\Sigma_k$ is a correction to the quasiparticle energy. This correction may be found from the usual rules of quantum mechanics: it is just the square of the matrix element for the transition into the intermediate state divided by the energy difference of the initial and intermediate states. In our case there is only one intermediate state, which corresponds to the formation of a hole and a correlated pair; the transition corresponding to the diagram

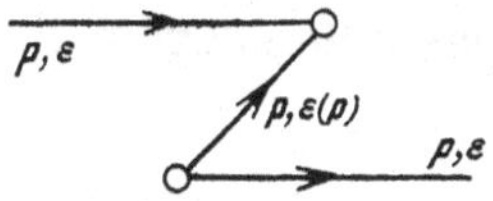

(3 particles in the intermediate state) is strictly forbidden by the Pauli principle (cf. the corresponding discussion for bosons in the next section). Since the initial energy of the particle is just ϵ, and that of the hole in the intermediate state $-\epsilon(\underset{\sim}{p})$, we arrive directly at eqn. (5.59). The quantity $Z\Delta$ plays the role of the corresponding transition amplitude for quasiparticles.

Since the block diagram $\Delta(\underset{\sim}{p}, \epsilon)$ by definition contains no parts joined by a single line, it cannot depend strongly on $\underset{\sim}{p}$ and ϵ and may be evaluated on the Fermi surface:

$$\Delta(p, \varepsilon) \simeq \Delta(p_F, \varepsilon_F) \equiv \Delta e^{i\varphi}.$$

As a result we obtain for G the expression

$$G_s(p, \varepsilon) = \frac{Z}{\varepsilon - \varepsilon(p) - \dfrac{\widetilde{\Delta}^2}{\varepsilon + \varepsilon(p)}}. \tag{5.60}$$

where for brevity we have omitted the term $i\delta$ sign ϵ and have written $\tilde{\Delta} = Z\Delta$.

The poles of G_s correspond to the energies of a new quasiparticle and a new quasihole:

$$(E - \varepsilon(p))(E + \varepsilon(p)) = \tilde{\Delta}^2,$$

whence

$$E = \pm\sqrt{\tilde{\Delta}^2 + (\varepsilon(p))^2}.$$

Near the Fermi surface we may put

$$\varepsilon(p) = \frac{p^2}{2m^*} - \varepsilon_F \simeq v_F(p - p_F).$$

For $\epsilon(p) \gg \tilde{\Delta}$ these expressions go over into the original expression for the quasiparticle and quasihole energies:

$$E(p) \to \pm\varepsilon(p).$$

Thus, a gap has appeared in the energy spectrum of the system. The minimum excitation energy, corresponding to formation of a quasiparticle and a quasihole near the Fermi surface, is given by $[E^+(p_2) - E^-(p_1)]_{min} = 2\tilde{\Delta}$. Expression (5.60) can be written in a form reminiscent of expression (5.10):

$$G_s(p, \varepsilon) = Z\left\{\frac{1 - \nu(p)}{\varepsilon - E(p) + i\delta} + \frac{\nu(p)}{\varepsilon + E(p) - i\delta}\right\},$$

where

$$\nu(p) = \frac{E(p) - \varepsilon(p)}{2E(p)}.$$

The correctness of the signs in front of $i\delta$ follows from the fact that the first term in the curly brackets describes a particle and the second a hole, and that when we go over to the τ-representation $G^+(\tau) = G^-(\tau) = 0$ for $\tau < 0$. Moreover it is obvious that for p far from p_F the expression G_s goes over into G both above and below the Fermi surface.

From expression (5.31), which relates G to the particle distribution, we can convince ourselves that $\nu(p)$ is the new distribution of quasiparticles which replaces the original Fermi distribution. Far from the Fermi surface $\nu(p)$ turns back into the original distribution. Thus, taking account of pair correlations has led to the appearance of a gap in the quasiparticle energy spectrum and to the smoothing out of the jump in the Fermi distribution. As we shall see below, the appearance of a gap in the energy spectrum leads to superfluidity, or in the case of charged particles to superconductivity.

An equation may be obtained for the quantity Δ which relates it to the interaction $\mathcal{U}_0$; this equation corresponds to the graphical equation

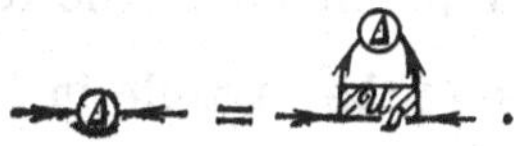

Note that on the right-hand side of this equation the two lines entering Δ are different: one is thin and one heavy. We leave it to the reader to obtain this graphical equation and its analytical form, and to verify that Δ is determined by the relation

$$1 = |\gamma| \int_0^{E_1} \frac{d\varepsilon}{\sqrt{\tilde{\Delta}^2 + \varepsilon^2}}, \quad \tilde{\Delta} \simeq E_1 e^{-1/|\gamma|},$$

where $E_1 \sim \epsilon_F$ and γ was defined in (5.57).

The Energy Spectrum of a Bose System. Superfluidity

The groundstate of an ideal gas of Bose particles is obtained by putting all the particles in the lowest state, which corresponds to $p = 0$ (a "Bose condensate" is formed). When interactions are taken into account, the particles are smeared out in momentum, but a finite fraction of the particles stay in the original $p = 0$ state, just as in the case of a Fermi system the jump at the Fermi surface remains finite (p. 287).

The Green's function of a particle will contain not only diagrams corresponding to interactions with uncondensed particles but also diagrams of the form

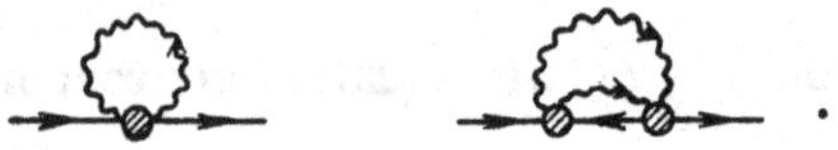

where the wavy lines denote condensate particles. The first diagram corresponds to the interaction of the particle in question with the condensate without a change in the number of condensate particles; diagrams of this type describe potential scattering in the field of the condensate. The second diagram refers to a more complicated process: the particle turns into a hole in the particle distribution and thereby creates two condensate particles, which subsequently recombine with the hole to give back the original particle.

Let us introduce a quantity G defined as the sum of all diagrams which do not contain a transition into a hole and two condensate particles (this is analogous to the G of the previous section):

$$G = G_0 + G_0\Sigma_0 G. \tag{5.61}$$

Then the Dyson equation gives

$$G_s = G + G\Sigma_k G_s. \tag{5.62}$$

Since $\Sigma_0(\underset{\sim}{p}, \epsilon)$ contains no parts linked only by a single line going in either direction, it may be expanded in a series for small $\underset{\sim}{p}$ and ϵ. It follows from (5.61) that $G(\underline{p}, \epsilon)$ is given by

$$G(p, \varepsilon) = [\varepsilon - \varepsilon^0(p) - \Sigma_0(p, \varepsilon)]^{-1}, \quad \varepsilon^0(p) = \frac{p^2}{2m},$$

whence we find for small $\underset{\sim}{p}$ and ϵ

$$G(p, \varepsilon) = \frac{Z}{\varepsilon - \varepsilon(p) + i\gamma} + G_{\text{Reg}} = ZG_Q + G_{\text{Reg}},$$

where

$$Z = \left[1 - \frac{\partial \Sigma_0}{\partial \varepsilon}\right]^{-1}, \quad \varepsilon(p) = \left(\varepsilon^0(p) + \Sigma_0(0, 0) + \frac{\partial \Sigma_0}{\partial p^2} p^2\right) Z.$$

The quantity $G_Q = (\epsilon - \epsilon(p) + i\gamma)^{-1}$ is the propagation function of the "zeroth-approximation" quasiparticles, that is, those which do not include transitions into a quasihole plus two condensate quasiparticles.

This technique of introducing auxiliary Green's functions which correspond to the exclusion of any desired process can be widely used in various branches of theoretical physics. In particular, in nuclear theory one can take pair correlations into account by using auxiliary quasiparticles which exclude the effect of the pairing interaction, and with their help calculate the diagrams which lead to the pair correlations.

Let us find the quantity

$\Sigma_K =$ 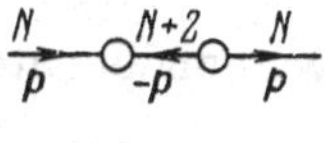

.

where the number of particles in the system is indicated above the lines. The hole with momentum $-\mathbf{p}$ corresponds to a state of the $(N+2)$-particle system with momentum $\mathbf{p}$. Thus, as a result of the existence of a condensate, there takes place a mixing of states:

$$\text{particle in system } (N) \leftrightarrows \text{hole in system } (N+2).$$

In the line connecting the two blocks we must not include transitions into a hole, since Σ_k by definition does not contain parts linked by a single line going rightwards. In the above diagram we have omitted the condensate lines for simplicity.

We will measure all energies from the chemical potential of the system, $\mu = E_0(N+1) - E_0(N)$; this is equal to the energy of a condensate particle, since to go from the groundstate of the N-particle system to the groundstate of the $(N+1)$-particle one we must add a single condensate particle. With energies measured from the chemical potential, then, the energies of the condensate particles are zero and the expression for Σ_k takes a form analogous to the formula (5.59) of the previous section, but with a change of sign (cf. p. 320).

$$\Sigma_k(p, \varepsilon) = (i\Delta)\, G(-p, -\varepsilon)(-i\Delta) = -\frac{Z\,|\Delta|^2}{\varepsilon + \varepsilon(p)}\,.$$

To check the sign in this expression we use ordinary perturbation theory in the same way as we did for the case of pair correlations in Fermi system (p. 321). In the case of Bose systems there are two possible intermediate states: one corresponding to a transition of the quasiparticle into a quasihole and two condensate quasi-particles, and the other describing a transition of the original quasiparticle plus two condensate quasiparticles into two uncondensed quasiparticles with momentum $\underline{p}$ and one with momentum $-\underline{p}$. If we omit for simplicity the (positive) factor Z, then the effect of the first intermediate state is to give a contribution to Σ_k of $|\Delta|^2/(\epsilon + \epsilon(p))$, with the same sign as for a Fermi system. For the second intermediate state, which in the Fermi case does not exist (see the diagram on p. 321) the energy is $2\epsilon + \epsilon(p)$ and the corresponding term in Σ_k is

$$\frac{2|\Delta|^2}{\varepsilon - 2\varepsilon - \varepsilon(p)} = -\frac{2|\Delta|^2}{\varepsilon + \varepsilon(p)}.$$

where the factor 2 takes account of the fact that there are two bosons with momentum $\underline{p}$. The net result is that $\Sigma_k = -|\Delta|^2/\epsilon + \epsilon(p)$, which proves that the expression obtained above for Σ_k is indeed correct; comparison with that expression shows that the quasiparticle transition amplitude is just $Z\Delta$.

The Green's function of interest, G_s, is given by

$$G_s = \frac{1}{G^{-1} - \Sigma_k} = Z\frac{\varepsilon + \varepsilon(p)}{\varepsilon^2 - \varepsilon^2(p) + Z^2|\Delta(p, \varepsilon)|^2} + G_{\text{Reg}}. \tag{5.63}$$

Consider the pole part of this expression for small $\underline{p}$ and ϵ. Among the $p \to 0$ excitations there must necessarily be some which in the limit of small p go over into a displacement of the system as a whole. Such excitations must have a frequency which tends to zero for $p \to 0$ (Goldstone's theorem). (An example of such an excitation is a sound wave). The quantity G_s must therefore have a pole for $p = 0$, $\epsilon = 0$. But the energy $\epsilon(p)$ is given by $\epsilon(p) = Z(\Sigma(0, 0) - \mu + O(p^2))$ (p. 325), and so according to (5.63) we have the condition

$$[\mu - \Sigma_0(0, 0)]^2 = |\Delta(0, 0)|^2,$$

whence

$$\mu = \Sigma_0 \pm \Delta_0, \quad \Sigma_0 \equiv \Sigma\,(0, 0), \quad \Delta_0 \equiv |\Delta\,(0, 0)|. \qquad (5.64)$$

We shall see that the analyticity properties of the Green's function (the condition that the residue at the pole should be positive) force us to choose the lower sign in (5.64).

Let us expand the expressions occurring in G_s in a series around the point $p = 0$, $\epsilon = 0$. In the numerator we can put p, $\epsilon = 0$. Consequently we have

$$\varepsilon + \varepsilon\,(p) \simeq \Sigma_0 - \mu = \mp \Delta_0.$$

As we shall now see, near the pole of G_s it is values of ϵ of order p which are important. Writing

$$Z\Delta_0 = \frac{\Delta_0}{1 - \frac{\partial \Sigma_0}{\partial \varepsilon}} = \widetilde{\Delta}_0, \quad Z\Delta\,(p, \varepsilon) = \widetilde{\Delta}\,(p, \varepsilon),$$

$$\left(\frac{p^2}{2m} + \frac{\partial \Sigma_0}{\partial p^2}\, p^2\right) Z \equiv \frac{p^2}{2m^*},$$

we obtain

$$G_s = Z \frac{\mp \widetilde{\Delta}_0}{\varepsilon^2 - \left(\frac{p^2}{2m^*} + \widetilde{\Delta}_0\right)^2 + |\widetilde{\Delta}\,(p, \varepsilon)|^2}.$$

It is obvious from the analytic properties of the boson Green's function (p. 279) that for $\langle a^+_{\underset{\sim}{p}}\, a_{\underset{\sim}{p}} \rangle = n(\underset{\sim}{p}) \gg 1$ (a condition which as we shall see, is fulfilled in our case) the Green's function must be an even function of ϵ. Hence $|\widetilde{\Delta}\,(\underset{\sim}{p}, \epsilon)|^2$ is also an even function of ϵ:

$$|\widetilde{\Delta}\,(p, \varepsilon)|^2 = \widetilde{\Delta}_0^2 + \frac{\partial |\widetilde{\Delta}|^2}{\partial p^2}\, p^2 + \frac{\partial |\widetilde{\Delta}|^2}{\partial \varepsilon^2}\, \varepsilon^2.$$

As a result, the expression for G_s takes the form:

$$G_s = Z\frac{\mp\tilde{\Delta}_0}{\varepsilon^2 - E^2(p)},$$
$$E^2(p) = c^2p^2 + \left(\frac{p^2}{2m^*}\right)^2 \frac{1}{1 - \frac{\partial|\tilde{\Delta}|^2}{\partial\varepsilon^2}}, \tag{5.65}$$
$$c^2 = \frac{1}{1 - \frac{\partial|\tilde{\Delta}|^2}{\partial\varepsilon^2}}\left(\frac{\tilde{\Delta}_0}{m^*} - \frac{\partial|\tilde{\Delta}|^2}{\partial p^2}\right).$$

Thus the poles of G_s for $p \to 0$ correspond to sound-wave excitations*.

In the denominator of (5.65) we have kept the term $(p^2/2m^*)^2$, which at first sight seems illegitimate, since we have omitted other terms of order p^4 and ϵ^4. However, this term gives the correct transition to the case of large p. (For sufficiently large p we must replace m^* by m and put $\partial|\tilde{\Delta}|^2/\partial\epsilon^2 = 0$, so that E(p) goes over into the free-particle energy). In a system with weak interactions it is legitimate to keep this term, and expansion (5.65) is correct for all momenta.

Let us determine the sign in the numerator of G_s. Consider a system with weak interactions; then $\partial\Sigma/\partial\epsilon$, $\partial|\tilde{\Delta}|^2/\partial\epsilon^2 \ll 1$. The fact (see (5.18)) that the residue at the pole must be positive shows that we must take the lower (plus) sign in (5.65), which corresponds to the minus sign in (5.64). But once the sign is fixed the positiveness of the residue implies

---

* The sound-wave-like character of the spectrum of a nonideal Bose gas was discovered by Bogolyubov (1947). The method used in his work (the method of canonical transformations) has had an important influence on the application of quantum theory to the many-body problem.

that for any form of interaction

$$1 - \frac{\partial |\tilde{\Delta}|^2}{\partial \varepsilon^2} > 0.$$

We leave it to the reader to show that the positiveness of the residue of G implies also $1 - \partial\Sigma/\partial\epsilon > 0$.

From the expression for the velocity of sound we also find

$$\frac{\tilde{\Delta}_0}{m^*} - \frac{\partial |\tilde{\Delta}|^2}{\partial p^2} > 0.$$

and so, finally,

$$G_s = \frac{a}{\varepsilon^2 - E^2(p) + i\delta}, \tag{5. 66}$$

where

$$a = \frac{Z|\tilde{\Delta}_0|}{\left(1 - \frac{\partial |\tilde{\Delta}|^2}{\partial \varepsilon^2}\right)} > 0.$$

We have chosen the sign of the imaginary term so that the particle Green's function $G_s^+(E(p), t) = G_s(E(p), t)_{t>0}$ should correspond to a damped rather than a growing solution.

Let us write G_s in the form

$$G_s = \frac{a}{2E(p)}\left(\frac{1}{\varepsilon - E(p) + i\delta} - \frac{1}{\varepsilon + E(p) - i\delta}\right)$$

and find $G_s(\underline{p}, t)$ for $t < 0$. On integration over ϵ only the second term is left:

$$G_s(p, t) = \int G_s(p, \varepsilon)\, e^{-i\varepsilon t} \frac{i d\varepsilon}{2\pi} \underset{t \to -0}{=} \frac{a}{2E(p)}.$$

On the other hand, we had for a Bose system (see p. 279) the relation $G(\underline{p}, t = -0) = \langle a^+(\underline{p})\, a(\underline{p}) \rangle \equiv n(\underline{p})$. Thus we get for the momentum distribution of the particles

$$n(p) = \frac{a}{2E(p)} \underset{p \to 0}{=} \frac{a}{2cp} \gg 1. \tag{5.67}$$

Thus $n(p)$ has a pole for $p \to 0$.

At large p and ϵ the function G_s (cf. (5.63)) goes over into the free-particle Green's function, that is $E(p)$ goes over into $\epsilon_p^0 = \frac{p^2}{2m} - \mu$. Thus the dependence of $E(p)$ on p begins with a linear region and must go over to a quadratic form for large p. How does $E(p)$ behave in the intermediate region ? A calculation of the specific heat of liquid helium which would correspond to sound waves gives a result smaller than the experimentally observed specific heat. It follows that besides sound waves there must also exist other low-energy excitations (Migdal, 1940). Landau put forward the hypothesis that the $E(p)$ curve has a minimum (see fig. 48). If so, excitations with momentum p near p_0 will give

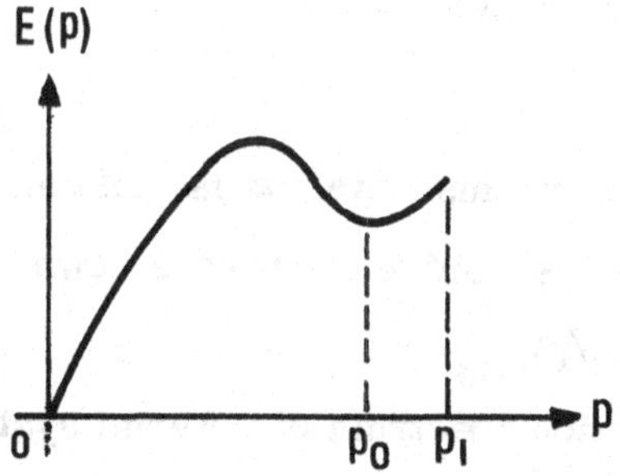

Fig. 48

an extra contribution to the specific heat. This hypothesis has been brilliantly confirmed by experiment.

The spectrum to the right of p_0 disappears at the point p_1 (that is, the pole in the Green's function disappears). The reason for this is the decay of an excitation with momentum p into excitations of lower momentum (Pitaevskii, 1959).

A liquid with the spectrum $E(p) = cp$, like a liquid with the spectrum $E(p) = \sqrt{\Delta^2 + (p - p_F)^2 v_F^2}$ obtained in the last section, will possess the property of superfluidity, that is, it will flow through narrow tubes without friction. To see this, we go over to a coordinate system moving with the liquid. Then superfluidity means that the walls of the tube, or indeed any arbitrary body moving in the liquid, is not slowed down. At low temperatures there is only one possible way of slowing down, namely by transferring energy to the elementary excitations. Suppose the body reduces its momentum by an amount $\mathbf{p}$; then its energy will change by an amount $\mathbf{p}\cdot\mathbf{v}$, where $\mathbf{v}$ is its velocity. Since pv is the largest possible change in the energy of the body for the given transfer of momentum p, the condition for superfluidity to occur (the Landau criterion) is

$$pv < E(p), \tag{5.68}$$

For large velocities v this criterion is violated; the critical velocity v_c at which superfluidity breaks down is given by the condition $v_c = (E(p)/p)_{min}$.

The excitation spectrum of a Fermi system without pair correlations does not satisfy the criterion (5.68) even when the velocity v tends to zero. In such a system the excitation energy $E(\mathbf{p}_1, \mathbf{p}_2)$ is $\epsilon(p_1) - \epsilon(p_2)$, where $p_1 > p_F$, $p_2 < p_F$. At small excitation momenta $E = \mathbf{v}_F \cdot \mathbf{p}$, and so the criterion for superfluidity is violated at extremely small v. However, in the case of a spectrum with pair correlation the criterion for superfluidity is violated only beyond some velocity v. Hence the electrons in a superconductor move through the atomic lattice without being decelerated, if the current is sufficiently weak (the phenomenon of superconductivity). In the case of a spectrum $E(p) = cp$ the critical velocity v_c is equal to c.

The phenomenon of superfluidity in liquid helium was observed experimentally by Kapitza in 1937. As was shown by Landau, all the effects occurring at finite temperature may be explained by a peculiar hydrodynamics of two interpenetrating liquids - a "superfluid" and a "normal" liquid. The theory of liquid helium is one of Landau's most brilliant achievements.

4. A SYSTEM IN AN EXTERNAL FIELD

Many properties of the system, such as static moments, transition probabilities, energies of the first excited states, and so on, are easily determined once we know the way in which the quasiparticle density matrix changes in an external field and also when we add particles to the system. As we saw, (p. 284), the density matrix is simply related to the Green's function. To find it we proceed as follows. We first determine the change in the Green's function in the effective field which is produced in the system under the action of the external field. The effective field is the sum of the external field and the "polarization" field resulting from the redistribution of particles. This redistribution can in turn be expressed in terms of the change of the Green's function in the effective field, and as a result we get a closed system of equations to determine the effective field.

Let us determine, then, the change of the quasiparticle Green's function in an external field, confining ourselves for simplicity to first-order perturbation theory in the field (while, of course, continuing to treat the interparticle interactions exactly). Some of the graphs occurring in the quasiparticle Green's function in an external field, G', are the following

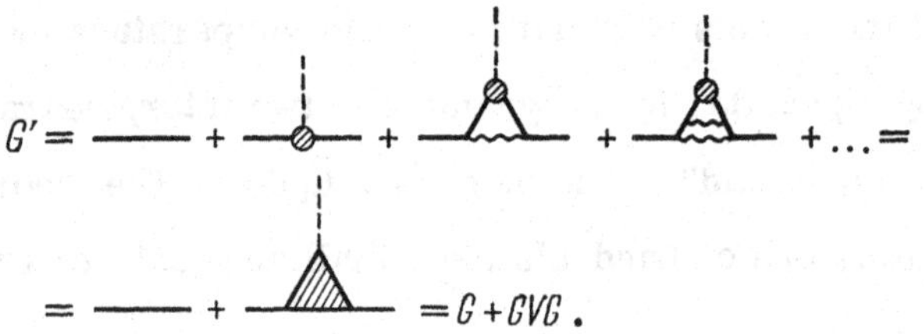

$$G' = \underline{\qquad} + \underline{\quad\circledcirc\quad} + \underline{\quad\bigwedge\quad} + \underline{\quad\bigwedge\quad} + \ldots = \qquad (5.69)$$

$$= \underline{\qquad} + \underline{\quad\bigtriangleup\quad} = G + GVG .$$

where the blob denotes the direct interaction of the quasiparticle with the external field:

$$\underline{\quad\circledcirc\quad} = e_q V^0\!/i ,$$

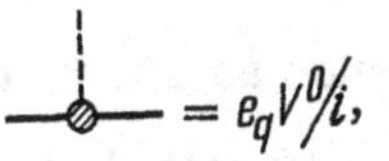

e_q is the "charge" of the quasiparticle. We shall see that for some types of field e_q is not equal to 1.

For noninteracting particles we should have

$$G'_0 = \underline{\qquad} + \underline{\quad\circledcirc\quad} = G_0 + G_0 e_q V^0 G_0 .$$

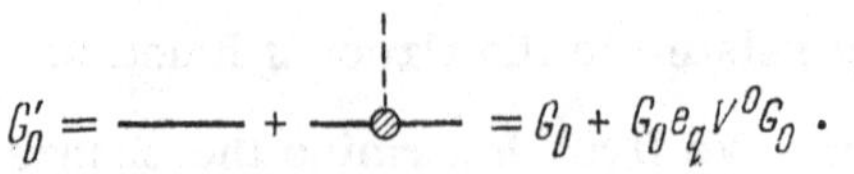

Thus the shaded triangle in eq. (5.69) replaces the dot in the above graph and represents the effective field V acting on a quasiparticle. We now derive an equation for this field.

Among the graphs contributing to V there is one which does not contain interactions between quasiparticles (this is just $e_q V^0$). All the other graphs have the following structure: if we move upwards from the base of the triangle towards its vertex, all graphs begin with an interaction, then we get two lines corresponding to free motion, and then the collection of diagrams which represents the effective field. If we introduce the block diagram $\mathcal{F}$ which contains no parts joined by only two lines, the effective field is determined by the equation

$$V = e_q V^0 + \mathcal{F} GGV, \tag{5.70}$$

or in graphical form

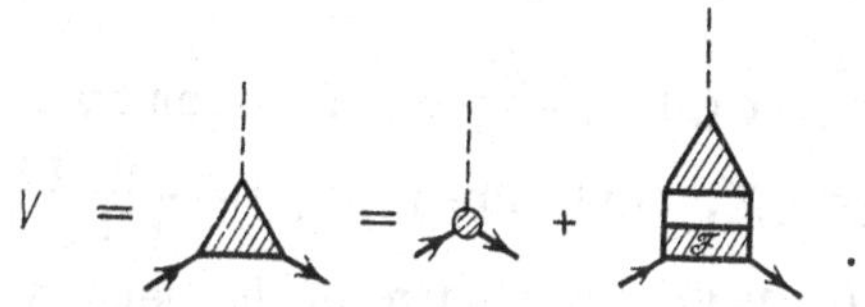

The first term in V describes the direct effect of the external field on the quasiparticle; the second gives the additional field which arises from the polarization of the medium, that is, which is caused by the effect of the redistribution of the nucleons (in the case of a nucleus).

In the λ-representation we get (cf. p. 308).

$$V_{\lambda_1\lambda_2} = e_q V^0_{\lambda_1\lambda_2} - \sum (\lambda_1\lambda_2 \,|\, \mathcal{F} \,|\, \lambda\lambda') \, A_{\lambda\lambda'} V_{\lambda'\lambda}, \tag{5.71}$$

where

$$A_{\lambda\lambda'} = \int G_\lambda(\varepsilon) \, G_{\lambda'}(\varepsilon - \omega) \, \frac{id\varepsilon}{2\pi}. \tag{5.72}$$

As we saw (p. 300)

$$A_{\lambda\lambda'} = \frac{n_\lambda - n_{\lambda'}}{\varepsilon_{\lambda'} - \varepsilon_\lambda + \omega}. \tag{5.72'}$$

Change of the Particle Distribution in a Field

The results we have just obtained have an extremely simple physical meaning, which we shall explain by using the example of a large system. In this case it is convenient to use the momentum representation, which means taking $\varphi_\lambda = e^{i\mathbf{p}\cdot\mathbf{r}}$. Then we can write $A_{\lambda\lambda'}$ in the form

$$A_{\lambda\lambda'} = A_{p,\,p-k} = \frac{n(p) - n(p-k)}{\varepsilon(p-k) - \varepsilon(p) + \omega}.$$

For $k \ll p$ the difference $\epsilon(p) - \epsilon(\mathbf{p}-\mathbf{k})$ is given by $\epsilon(\mathbf{p}) - \epsilon(\mathbf{p}-\mathbf{k}) \approx \frac{d\epsilon}{dp}\cdot\mathbf{k} = \mathbf{k}\cdot\mathbf{v}$, and $n(\mathbf{p}-\mathbf{k}) - n(\mathbf{p}) = -\frac{dn(p)}{d\epsilon}\,\mathbf{k}\cdot\mathbf{v}$. According to eqn. (5.28) (p. 284), the change of the density matrix can be expressed in terms of the change of the Green's function in the field, that is, in symbolic form,

$$\delta\rho = (GGV)_{\tau\to -0} = AV. \tag{5.73}$$

Writing $\delta\rho = f_{\mathbf{k}}(\mathbf{p})$, where $f_{\mathbf{k}}(\mathbf{p})$ is the $\mathbf{k}$-th Fourier component of the distribution function $f(\mathbf{r}, \mathbf{p})$, we obtain

$$(\omega - kv) f_k(p) - \frac{dn(p)}{d\varepsilon} vkV_k = 0, \tag{5.74}$$

or in the coordinate representation

$$\frac{\partial f(\mathbf{r}, \mathbf{p})}{\partial t} + v\nabla f(\mathbf{r}, \mathbf{p}) - \frac{\partial f^0}{\partial p}\nabla V = 0, \tag{5.74'}$$

where $f^0(\mathbf{r}, \mathbf{p}) = n(p)$.

Thus the relation $\delta\rho = AV$ is equivalent to the usual equation for the change in the distribution function in an external field. For values of k comparable to p we would obtain the so-called quantum equation for the distribution function. (quantum kinetic equation). Expression (5.72') leads to this more general equation in the λ -representation.

Eqn. (5.74) allows us not only to find the reaction of the system to the external field, but also to determine the frequencies of those characteristic vibrations of the system whose quantum numbers correspond to the symmetry of the field. Thus, for instance, in the case of a vector field there are excited in the system vibrations of vector type, while for an external field corresponding to a perturbation of the form $H' = \mu \sigma_{\lambda\lambda'} a^{+}_{\lambda} a_{\lambda'} \mathcal{H}$, where $\mathcal{H}$ is the magnetic field, it is spinor-type vibrations which occur. The eigenfrequencies can be determined as the poles of $V(\omega)$ (cf. below).

Spin Polarizability and Quasiparticle Magnetic Moments

For a static field eqn. (5.70) may be simply written in the coordinate representation:

$$V(r) = e_q V^0(r) - \int \mathcal{F}(r - r') \frac{dn}{d\varepsilon_F} V(r')\, dr'. \qquad (5.75)$$

To derive this equation we used the relation

$$\frac{n_\lambda - n_{\lambda'}}{\varepsilon_\lambda - \varepsilon_{\lambda'}} \underset{k\to 0}{=} \frac{dn_\lambda}{d\varepsilon_\lambda} = -\frac{dn_\lambda}{d\varepsilon_F}$$

and also the completeness of the set of functions φ_λ. In view of the δ-function-like form of $\mathcal{F}$, we can write $\mathcal{F} = \mathcal{F}_0 \delta(\underset{\sim}{r} - r')$. We assume, moreover, that the interaction does not depend on the velocities or spin variables of the particles. There we have (cf. p. 310)

$$V(r) = \frac{e_q V^0(r)}{1 + f}, \qquad (5.76)$$

where $f = \mathcal{F}_0(dn/d\epsilon_F)$ characterizes the polarizability of the medium in the external field, and the quantity $(1 + f)/e_q$ is the analog of the dielectric constant.

In the case of a magnetic field acting on a nucleus, the perturbation H' is of the form $H' \simeq a^+_\lambda a_{\lambda'} \sigma_{\lambda\lambda'}$, that is, we have a spin field. If the nucleus were a sufficiently large system, then we would just need to replace f in (5.76) by g (cf. eqn. (5.48)) to obtain the change of the spin part of the single-particle magnetic moment as a result of the spin polarization of nuclear matter (in a nucleus $g \simeq 1$); that is, the quasiparticle spin magnetic moment would be given by

$$\mu^s \simeq \frac{\mu_0^s}{1+g}, \tag{5.77}$$

where μ_0^s is the spin part of the free-nucleon magnetic moment. In practice, the calculation of the spin polarizability of a spherical nucleus requires a numerical solution of eqn. (5.71) in the λ -representation. In deformed nuclei, where the one-particle levels are more densely spaced because of the removal of the degeneracy with respect to the projection of angular momentum, the situation is closer to that in an infinite system and formula (5.77) is approximately correct.

Sound Waves in a Fermi System ("Zero Sound")

In the case of an infinite system we get from eqn. (5.71)

$$V_k \underset{k \ll p_F}{=} e_q V_k^0 + \mathcal{F}_0 \int \frac{n(p) - n(p+k)}{\varepsilon(p) - \varepsilon(p+k) + \omega} 2 \frac{d^3p}{(2\pi)^3} V_k. \tag{5.78}$$

where we took into account the δ-function-like form of $\mathcal{F}$, and also the fact that for small k only values of $\underset{\sim}{p}$ such that $|\underset{\sim}{p}| \simeq p_F$ are important in the integration over $\underset{\sim}{p}$. To calculate the integral in (5.78), we use the smallness of k in the integrand and the formula $dn(p)/d\epsilon = -\delta(\epsilon - \epsilon_F)$; then we get

$$\int \frac{n(p) - n(p+k)}{\varepsilon(p) - \varepsilon(p+k) + \omega} 2 \frac{d^3p}{(2\pi)^3} =$$

$$= -\frac{dn}{d\varepsilon_F}\frac{1}{2}\int_{-1}^{1} \frac{kv_F x}{\omega - kv_F x} dx \equiv -\frac{dn}{d\varepsilon_F}\Phi(k, \omega), \qquad (5.79)$$

where $\Phi(k, \omega)$ is given (for $k^2/4p_F^2 \ll 1$) by the expression

$$\Phi(k, \omega) = 1 - \frac{\omega}{2kv_F}\ln\left|\frac{\omega + kv_F}{\omega - kv_F}\right| + i\pi\frac{\omega}{2kv_F}\theta(kv_F - \omega)$$

where $\theta(x)$ is the usual theta-function (step function). We see that the function Φ determines the dispersion of the characteristic vibrations of the system, that is, the dependence of $\omega(k)$ on k. The appearance of an imaginary part in the equation for V corresponds to the fact that for $\omega < kv_F$ the field can create pairs (when $\omega = \underset{\sim}{k}\cdot\underset{\sim}{v}$) while for $\omega > kv_F$ the creation of pairs is impossible ($\omega > \underset{\sim}{k}\cdot\underset{\sim}{v}$ for all $v < v_F$).

The equation for the effective field takes the form

$$V_k = e_q V_k^0 - \mathcal{F}_0 \frac{dn}{d\varepsilon_F} V_k \Phi(k, \omega). \qquad (5.80)$$

Depending on the symmetry of the field V different terms in the expression for $\mathcal{F}$ (p. 310) may come into this equation. Thus, in the case of a scalar field it is the constant f which comes in, while in the case of a spinor field $V \sim \sigma$ the symbolic product $\mathcal{F}V$ involves the spin-spin interaction constant g (above we have omitted the spin suffices on the matrix element of $\mathcal{F}$ for brevity). Thus, the equation for a scalar field V takes the form

$$V_k = \frac{e_q V_k^0}{1 + f\Phi(k, \omega)}. \tag{5.81}$$

while in the case of a spinor field f is replaced by g.

The characteristic vibrations of the system are determined by the pole in the expression for V_k:

$$1 = -f\Phi(k, \omega). \tag{5.82}$$

Let us assume that $\omega = \gamma k v_F$. Then eqn. (5.82) takes the form

$$1 + f = f\frac{\gamma}{2}\ln\frac{\gamma+1}{\gamma-1}.$$

An undamped solution corresponds to $\gamma > 1$. Such a solution exists if $f > 0$; in that case,

$$\gamma \ln\frac{\gamma+1}{\gamma-1} = 2\frac{1+f}{f}.$$

Since γ does not depend on k, ω has the same dependence on k as for sound waves. These vibrations are therefore called zero sound. In the case $V \sim \sigma$, the constant f is replaced by g; for $g > 0$ there can exist vibrations which might naturally be called "spin sound". Finally, when $V \propto \sigma\tau$, the constant g' (see eqn. (5.48)) enters the problem and we get "spin-isospin-sound".

Plasma Oscillations. Screening of a Charge in a Plasma

As another application of the result obtained above we consider the equation for propagation of an electric wave in a plasma. In this case we should leave in the expression for the interaction $\mathcal{F}$ at small k only the single diagram

$\mathcal{F} =$ 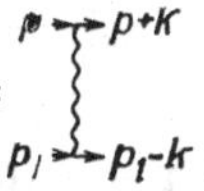

$= \frac{4\pi e^2}{k^2}$, (5.83)

corresponding to the Coulomb interaction. By definition the diagram

does not contribute to $\mathcal{F}$; and the diagram

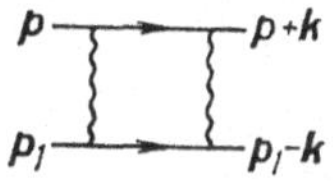

,

may easily be seen to remain finite in the limit $k \to 0$. The same applies to the other possible diagrams in $\mathcal{F}$. Eqn. (5.75) therefore takes the form (as we shall see below, in this case e_q is equal to 1)

$$V_k = \frac{V_k^0}{1 + \frac{4\pi e^2}{k^2} \frac{dn}{d\varepsilon_F} \Phi(k, \omega)}. \tag{5.84}$$

Let us consider the two limiting cases $\omega \ll k v_F$ and $\omega \gg k v_F$. In the first case we find the charge distribution around an external point charge Q situated in a plasma; in the k-representation this corresponds to

$$V_k^0 = \frac{4\pi Q}{k^2}.$$

From (5.84) we get for V_k (since $\Phi(k, 0) = 1$)

$$V_k = \frac{4\pi Q}{k^2 + \varkappa^2}, \quad \varkappa^2 = 4\pi e^2 \frac{dn}{d\varepsilon_F}. \tag{5.85}$$

or going over to the coordinate representation,

$$V(r) = \frac{Q}{r} e^{-\varkappa r}. \tag{5.85'}$$

Our expressions have been obtained for $k \ll p_F$; in the coordinate representation this corresponds to $r \gg 1/p_F$. Thus, at large distances the external charge is entirely screened out by the charges of the plasma.

In the case $\omega \gg kv_F$ it is easy to show that $\Phi(k, \omega) = -k^2 v_F^2/3\omega^2$, and substituting this in (5.84), we get

$$V_k = \frac{V_k^0}{1 - \frac{4\pi e^2 n}{m\omega^2}}. \tag{5.86}$$

This expression has a pole for

$$\omega^2 = \frac{4\pi n e^2}{m}.$$

Since there is no damping at this point, undamped waves can exist in the system. The quantity

$$\omega_p = \sqrt{\frac{4\pi n e^2}{m}} \tag{5.87}$$

is called the plasma frequency.

It is important to note that the above results do not assume that the interaction is small and are exact in the limit of small k.

The expression in the denominator of (5.86) is by definition just the dielectric constant:

$$\varepsilon(\omega) = 1 - \frac{4\pi n e^2}{m\omega^2},$$

This agrees with the result obtained on p. 227.

Conservation Laws and Quasiparticle Charges for Different Fields

The conservation laws impose strong restrictions on the quasiparticle charge e_q. A formal derivation of the results quoted below is given elsewhere*. Here we confine ourselves to physical plausibility arguments.

First of all we consider the consequences implied by the requirement of gauge invariance. This requirement means that an external vector field of the form

$$\frac{\partial \Phi}{\partial x_i} = \left(\frac{\partial \Phi}{\partial r_\alpha}, \frac{\partial \Phi}{\partial t}\right),$$

acting on the protons or the neutrons cannot lead to any physical change in the system. We recall that in the case of the electromagnetic field, a vector potential of the form $A_i = \partial \Phi / \partial x_i$ corresponds to zero electric and magnetic fields, and therefore has no physical effect on the particles.

We first consider a scalar field of the form

$$V(t) = V^0 e^{i\omega t},$$

where V^0 is a constant. This field is a particular case of the fictitious field

$$\frac{\partial \Phi}{\partial x_i}\left(\frac{\partial \Phi}{\partial r_\alpha} = 0, \frac{\partial \Phi}{\partial t} = V^0 e^{i\omega t}\right).$$

Since, in such a field, no physical change of the system whatever can occur, the effective field must be equal to the external field:

$$V = V^0.$$

Now we had

$$V = e_q V^0 + (\mathcal{F} A V).$$

---

* A. B. Migdal, Theory of Finite Fermi Systems and the Atomic Nucleus (Interscience, New York, 1967).

For $V = V^o$ = constant the second term is zero and the condition $V = V^o$ gives $e_q = 1$; if the field V^o acts on the protons (neutrons), then the field V must also act only on the protons (neutrons), so that we find for a scalar field

$$e_q^{pp} = e_q^{nn} = 1, \quad e_q^{np} = e_q^{pn} = 0. \tag{5.88}$$

Since e_q is determined by graphs corresponding to large energies (or, in the coordinate representation, to small distances around the point in question) this result remains true for all scalar fields sufficiently homogeneous in space and time $(k \ll p_F, \omega \ll \epsilon_F)$.

Further information on the quasiparticle charges may be obtained by using the fact that some fields, even though they are not fictitious, may nevertheless not be able to effect a redistribution of the particles and, consequently, the corresponding effective field will be equal to the external one. For instance, if we apply a homogeneous field which acts identically on the two types of particles, the system will oscillate as a whole without any internal change. In this case the perturbation is proportional to the total momentum of the system

$$P = \sum_{n+p} p_i.$$

where sum runs over both neutron and proton states. In our notation this corresponds to a field $V^o_\alpha = p^n_\alpha + p^p_\alpha$, where $\alpha = (x, y, z)$. From the fact that no internal changes take place in the system under these conditions, we easily find

$$V^p_{\lambda\lambda'} [p^p_\alpha + p^n_\alpha] = p_\alpha = (e_q^{pp} + e_q^{np})\, p_\alpha \tag{5.89}$$

where the quantity in the square brackets next to V is the bare perturbation V^o. Thus,

$$e_q^{\mathrm{pp}} + e_q^{\mathrm{np}} = e_q^{\mathrm{nn}} + e_q^{\mathrm{np}} = 1. \tag{5.90}$$

(In a nucleus with $N \simeq Z$ isotopic invariance implies that the effective charge of the neutron field in the case of a bare field acting on the protons, that is e_q^{np}, is equal to the charge e_q^{pn} of the proton field caused by a bare field acting on the neutrons: $e_q^{pn} = e_q^{np}$. Moreover, $e_q^{nn} = e_q^{pp}$). We conclude by analogy that the sum $e_q^{pp} + e_q^{np}$ is one for any perturbation which commutes with the Hamiltonian and which has only diagonal matrix elements in the λ-representation, that is, any perturbation of the form

$$H' = \sum a_\lambda^+ a_\lambda Q_{\lambda\lambda}, \tag{5.91}$$

if the operator H' commutes with H. Perturbations of this type will be called "diagonal". It is easy to see that a diagonal perturbation simply shifts the particle energies and does not lead to any redistribution of the particles, provided only that the new energy levels

$$\tilde{\varepsilon}_\lambda = \varepsilon_\lambda^0 + Q_{\lambda\lambda}$$

have the same order as the original ones (this is always true for small enough $Q_{\lambda\lambda}$).

We now obtain an expression for the effective charge in the case of a spinor field (i.e. a perturbation of the form $\underset{\sim}{\sigma} \cdot \underset{\sim}{H}$). Since the charge e_q is determined by the local interparticle interaction, its value in the nucleus is only slightly different from the corresponding value in infinite nuclear matter of the same density. If we assume that the spin-orbit interaction in nuclear matter is small, the total spin operator of the system commutes with the Hamiltonian. Moreover, in a sufficiently large system the spin-orbit correction to the quasiparticle Hamiltonian is not important and may be neglected. Then the quasiparticle eigenfunctions φ_λ will be eigenfunctions of the operator σ_z. Thus the perturbation is diagonal, and consequently

$$e_q^{pp} + e_q^{np} = 1. \tag{5.92}$$

We write this condition in the form

$$e_q^{pp} = 1 - \xi_s, \quad e_q^{np} = \xi_s. \tag{5.92'}$$

The quantity ξ_s cannot be calculated and must be found from experiment. We shall show that this same constant occurs in the renormalization of the axial nuclear β-decay constant. For allowed Gamow-Teller transitions the interaction with the electron-neutrino field adds to the nucleon Hamiltonian a perturbation proportional to $(\tau_x + i\tau_y)\sigma_z$ (pseudovector type interaction). Let us find the effective charge for such an external field.

We consider first of all the field $\tau_z\sigma_z$:

$$\tau_z\sigma_z = \frac{1+\tau_z}{2}\sigma_z + \frac{\tau_z - 1}{2}\sigma_z = \sigma_z^p - \sigma_z^n.$$

In this case the term $e_q V^o$ in the equation for V is given by

$$e_q[\tau_z\sigma_z]\,\tau_z\sigma_z = e_q[\sigma_z^p]\,\sigma_z - e_q[\sigma_z^n]\,\sigma_z =$$
$$= \begin{vmatrix} e_q^{pp} - e_q^{pn} \\ e_q^{np} - e_q^{nn} \end{vmatrix} \sigma_z = \begin{pmatrix} 1 - 2\xi_s \\ -(1 - 2\xi_s) \end{pmatrix} \sigma_z = (1 - 2\xi_s)\,\sigma_z\tau_z.$$

(The quantities in square brackets indicate the form of the field V^o for which the charge is $e_q[V^o]$.) Thus we get

$$e_q[\tau_z\sigma_z] = (1 - 2\xi_s). \tag{5.93}$$

By virtue of isotopic invariance the same charge will occur for fields of the form $(\tau_x + i\tau_y)\sigma_z$. Thus the factor $e_q = (1 - 2\xi_s)$ gives the renormalization of the pseudovector β-decay constant in nuclear matter. For fields $\tau_x + i\tau_y$ corresponding to Fermi transitions (vector type interaction) we get (most easily by again first considering a field τ_z) $e_q = 1$, that is, the vector interaction is not renormalized.

CHAPTER 6

QUALITATIVE METHODS IN QUANTUM FIELD THEORY

Quantum field theory describes the interaction of elementary particles moving in the vacuum. The number of particles participating in the interaction is not always conserved; particles can be created and disappear in the process of interaction. Even when the number of particles at the beginning and end of the process is the same, the general principles of quantum mechanics tell us that formation of virtual particles in intermediate states will exert an effect on the course of the process. Hence the vacuum must be considered not just as empty space, but as a medium with complex properties. An electron or a nucleon moving in the vacuum will be surrounded by a cloud of virtual particles created by its motion, just like a quasiparticle in the many-body problem.

The diagrams which describe the interaction of elementary particles are no different in form from those in the many-body problem. However, there are some very important differences in their physical meaning. First of all, since the processes take place in the vacuum, they must be independent of the choice of

Lorentz coordinate system; this, as we shall see, imposes extremely strong constraints on the nature of the interaction and on the Green's functions of the particles. Moreover, in quantum field theory we really deal only with "quasiparticles"; "bare" or "naked" particles, like bare interactions, are in this case unobservable. The reason is that in this case, unlike in the many-body problem, we we cannot extract the particles and study their properties outside the "medium". Nevertheless, in developing quantum field theory it is necessary to start by introducing bare particles and bare interactions. An important feature of such a formulation of the theory is the appearance of divergent integrals in all the expressions which connect the bare masses or bare interactions with observable quantities. As we shall see, these divergences are a consequence of one of the fundamental assumptions of quantum field theory, namely the assumption of locality of the interactions. Locality means that particles interact only when their 4-dimensional coordinates are the same; all interactions at a distance are supposed to be a secondary effect arising as a result of exchange of one or more virtual particles emitted and absorbed in local events. Since the bare quantities are unobservable, divergences in the expressions containing them are no argument against the assumption of local interactions, which in fact has been confirmed to high accuracy in the case of quantum electrodynamics (cf. below).

One way out of this difficulty is the following. It is possible to formulate the theory in such a way that from the very outset it contains only observable quantities. Then divergent integrals do not appear in the calculations, despite the locality of the interactions. This program can be carried out, in principle, with the help of dispersion relations. We will explain the idea behind this method with a simple example, which will also help us to trace the origin of the divergences which occur for a local interaction.

Consider the scattering amplitude of two nonrelativistic particles with a local (δ-function) interaction. Suppose the interaction is repulsive, so that there are no bound states. For a δ-function interaction the scattering amplitude is independent of angle (we get only S-wave scattering). The equation for the scattering amplitude in the centre-of-mass system may be written in the form (p.295: we put m = 1)

$$f(p, p') = \lambda - 4\pi \int \frac{f(p, p_1) f(p_1, p')}{p^2 - p_1^2 + i\delta} \frac{d^3 p_1}{(2\pi)^3}.$$

where λ is the Born amplitude, which is independent of $\underset{\sim}{p}$ and $\underset{\sim}{p}'$ in our case and plays the role of "bare" interaction.

Let us assume that λ is sufficiently small and try to obtain f as a power series in λ. In second order in λ we obtain an integral which diverges linearly at the upper limit:

$$f^{(1)} = \lambda, \quad f^{(2)} = \lambda^2 \int \frac{d^3 p_1}{p^2 - p_1^2 + i\delta} \frac{1}{(2\pi)^3}.$$

This divergence is of the same nature as the divergences in quantum field theory. As we shall see, the transition to the relativistic formula changes only the character of the divergence; for bosons the linear dependence on the upper limit is replaced by a logarithmic one and for fermions by a quadratic one. From a mathematical point of view the reason for the divergence is that an integral equation for the amplitude which has a kernel which is singular for $p \to \infty$ does not permit expansion in powers of λ. Thus, the divergences of field theory are nothing mysterious; they are a natural consequence of locality, which leads to integration over an infinite momentum space. If the interaction were nonlocal, then we could not have taken λ out from under the integral sign and the momentum integration would have been cut off at some value $1/r_0$, where r_0 is the range of the interaction.

Let us now attempt to obtain a series not in powers of λ, but in powers of the amplitude corresponding to some arbitrary fixed energy, e.g. zero energy. We express the amplitude in terms of its imaginary part (see p. 232 : we put $\underset{\sim}{p} = \underset{\sim}{p}'$)

$$f(p) = \frac{1}{\pi} \int_0^\infty \frac{\operatorname{Im} f(p_1)\, dp_1^2}{p^2 - p_1^2}.$$

As is easily seen from the optical theorem (cf. formula (b) below) to obtain convergence of the series in powers of f(0) it is necessary to improve the convergence of the integrand. For this purpose we subtract the quantity f(0) from both sides of the above equation. Then we get

$$f(p) = f(0) + \frac{p^2}{\pi} \int_0^\infty \frac{\operatorname{Im} f(p_1)\, dp_1^2}{(p^2 - p_1^2)\, p_1^2}.$$

Since the imaginary part of f(p), according to the optical theorem

$$\operatorname{Im} f(p) = p\, |f(p)|^2$$

depends quadratically on f, formulae (a) and (b) together allow us to obtain an iteration in powers of f(0). Putting f = f(0) on the right-hand side of (b), we get

$$f(p) = f(0) + ip\, [f(0)]^2 + \dots$$

For the case under consideration it is actually possible, as we saw above (p. 233) to obtain a closed equation. According to (a) we have

$$f(p)=\frac{f(0)}{1-ipf(0)}.$$

Note that to get a series in powers of $f(0)$ we had to rewrite the equation for the amplitude in the form (a), that is, subtract a term $f(0)$.

When we go over to the relativistic problem the stronger divergence in the Fermi case means that to get our series in powers of $f(0)$ to converge we would have to subtract not only the term $f(0)$ itself but also a term $(df/dp^2)_0\, p^2$. In higher-order approximations the degree of the divergence increases and the number of subtraction constants grows.

In spite of the attractiveness of the dispersion approach, it can be applied in practice only to the simplest problems. In higher-order approximations in the relativistic problem the number of intermediate-state particles increases and so we get involved with many-particle amplitudes, which lead to immense mathematical difficulties.

It is therefore more convenient to proceed as follows. We cut the divergent integrals off at some large momentum L (or, if the calculation is carried out in coordinate space, at a short distance r_0). Then, if the interaction λ is assumed to be sufficiently weak, we can obtain a perturbation series in powers of λ. It then turns out that in each order of perturbation theory the divergent parts of the integrals (those depending on L) can be taken out in such a way that after a redefinition of the coupling constant (interaction constant) and the constants occurring in the Green's function the remaining expressions are now insensitive to the value of the cutoff L. This is the idea behind the renormalizations which we met in the context of the many-body problem; on going over from a particle to a quasi-particle we had to redefine the mass and coupling constants. The transition from the bare coupling constant λ to the amplitude f(0) in the problem discussed above is an example of renormalization of the interaction.

If the theory is to be physically meaningful, it is necessary that the elimination of divergences by redefinition of the constants can be carried out in all orders of perturbation theory. It turns out that this is not possible for all field theories, that is, not all field theories have the property of renormalizability. The criterion for renormalizability is set by dimensional considerations: if the theory is to be renormalizable then the coupling constant must either be dimensionless or contain a length to some negative power. If the constant has, say, the dimensions of length squared, then the degree of divergence of the integrals will increase with the order of perturbation theory. To see this, note that at large momenta the mass of the particles cannot enter the problem, so that the only possible dimensionless combination at the n-th order is $(\lambda L^2)^n$. Consequently the degree of divergence increases with increasing n, and and, as we shall see, the number of constants which must be redefined or introduced anew to eliminate the divergence increases without limit. We will show that this implies the loss of locality in the interaction.

Thus, renormalizability is the criterion for a physically admissible local interaction. The question of renormalizability leads of its essence to the possibility of formulating the theory in such a way that interactions at momenta much less than the momentum cutoff L should be determined by virtual particles in states with momenta also much less than L; or, in coordinate space, that interactions at distances $r \gg r_0$ should be determined by interactions of virtual particles also at $r \gg r_0$. It is not surprising that not all theories satisfy this requirement.

To illustrate this point we consider an example of a similar situation in classical physics. Suppose we are given the problem of constructing a theory to describe the macroscopic motions of a liquid or gas. The question arises whether these motions will remain macroscopic, or disintegrate into motion on a finer and finer scale until we reach the atomic level. In cases where hydrodynamics is applicable we have an example of a "renormalizable" theory - the effect of small-scale phenomena reduces to the appearance in the theory of the (macroscopic) viscosity and mean density constants. However, for some systems, e.g. for a Fermi gas, it is impossible to construct a theory which describes macroscopic motions; the motions do not remain macroscopic, but decay into motion on an atomic scale.

In discussing the properties of field theory at small distances (or large momenta) another important question arises: is a renormalizable theory logically self-contained, and are its results applicable at arbitrarily short distances ?

In theories of renormalizable type the interaction between two particles is screened by the field of the virtual particles (for a possible exception to this rule see p. 430). As a result, the effective interaction falls off with distance. If we assume that the bare interaction is sufficiently weak, then it is possible to use perturbation theory to discuss the interaction at arbitrary distances. It then turns out that at large distances the interaction is totally screened out by the virtual particles, just as in the screening of an external charge by the electrons in a metal. Thus, particles cannot interact at large distances. This paradox, which has been investigated in detail by I. Ya. Pomeranchuk, has become known as the "zero charge" paradox. In reality this phenomenon is no argument against the view that the theory is logically self-contained, for the following reason.

In the case of quantum electrodynamics, if we decrease the distance between two charges, then in spite of the smallness of the observable charge we will necessarily come to a region where the charge is of order unity and perturbation theory, on which the paradox was based, is no longer valid. Will the charge then continue to increase at even smaller distances ? If it does, we will find an infinite bare charge, which means that the very concept of bare charge is physically meaningless. As we shall see, this is still an open question.

The distances at which the charge $e^2(r)$ becomes of order 1 are extremely small, and it is quite probable that the effect of weak or gravitational interactions will change the theory in such a way that the growth of the charge stops before we reach the region $e^2(r) \sim 1$. In this case the question of "zero charge" in electrodynamics loses any physical interest and remains only of theoretical value for the clarification of the structure of field theories at small distances.

The discussion of this chapter is arranged as follows.

Essentially, we shall consider only one of the problems of quantum field theory, namely the questions connected with the structure of field theories at small distances. This group of questions gives many opportunities for the use of qualitative methods. The concrete problems of quantum field theory (e. g. the calculation of radiative corrections) are bound up with tedious calculations and moreover are in any case excellently explained in other books*. In the first few sections we construct the formal apparatus necessary for our purpose. We find the equations and Green's functions describing free particles with spin 0, $\frac{1}{2}$ and 1; these are obtained as a relativistically invariant form of the two Schrödinger's equations and corresponding Green's functions for particle and anti-particle. The general properties of field theories at small distances are first discussed using a model 4-boson interaction. By considering the simplest diagrams for the two-particle transition amplitude we explain the nature of the divergences and the idea of renormalization; we also find the condition that the theory should be renormalizable. We next carry out a calculation of the scattering amplitude up to third order in the coupling constant. An analysis of the first few terms in the expansion leads to the idea that it is possible to obtain an exact expression for the amplitude from the requirement that it should be independent of the cutoff, that is, from the requirement of renormalizability. The expression thus obtained has the "zero charge" property: for sufficiently small bare charge the charge at large distances tends to zero.

---

*A. I. Akhiezer and V. B. Berestetskii, Elements of Quantum Electrodynamics, (London, Oldbourne Press, 1962).
N. N. Bogolyubov and D. V. Shirkov, Introduction to the Theory of Quantized Fields (New York, Interscience, 1958).

After that we consider a realistic theory, quantum electrodynamics. We explain the physical meaning of the corrections to the photon Green's function by a comparison with classical electrodynamics, and determine, to second order in e^2, the correction to Coulomb's law induced by polarization charges. After these preparatory problems we use simple physical considerations, without any expansion in e^2, to find a corrected Coulomb's law, from which we derive a formula for the interaction of particles at arbitrarily short distances (the Gell-Mann-Low formula). We discuss the possible limits of applicability of quantum electrodynamics*.

1. CONSTRUCTION OF RELATIVISTIC EQUATIONS

When we want to obtain equations (or Green's functions) to describe elementary particles, a decisive role is played by symmetry requirements, that is, by the requirement of invariance relative to general Lorentz transformations, including rotations in space-time and reflections (the requirements of reflection symmetry and reversibility). Another important condition which places constraints

---

* An exposition of other problems in field theory similar to the one given in this chapter may be found in the book by R.P.Feynman "Theory of Fundamental Processes" (W.A. Benjamin, 1961) and in V.N.Gribov's lectures "Quantum Electrodynamics" (Proceedings of the 9th Winter School of the Leningrad Nuclear Physics Institute, part 1, Leningrad 1974) (in Russian).
An extremely clear account of concrete calculations for various processes is given in the book by L.B.Okun' , "Weak Interactions of Elementary Particles", Israel Program for Scientific Translations, Jerusalem, 1965.

on the equations is the stipulation of simplicity; for instance, we try to choose equations involving only the lowest-order derivatives or, in the case of an external field, with the smallest possible number of parameters describing the interactions of the particle with the field. The requirement that the equations be simple, or, better, beautiful, is not absolute, but has played in the past an extremely important role in the discovery of the laws of nature.

Lorentz Invariance

The equations which describe the motion of particles in empty space must be Lorentz covariant, that is, they must preserve their form when we go over to a moving coordinate system by the Lorentz transformation

$$t' = \frac{t + vr}{\sqrt{1 - v^2}}, \quad r' = \frac{r + vt}{\sqrt{1 - v^2}}.$$

It is sufficient to consider an infinitesimal Lorentz transformation, and we shall do this in what follows:

$$t' = t + vr, \quad r' = r + vt, \quad v \to 0.$$

Lorentz transformations leave invariant the interval

$$(x_1 - x_2)^2 = (t_1 - t_2)^2 - (r_1 - r_2)^2$$

between two points in space-time: we have

$$(x_1 - x_2)^2 = (x_1' - x_2')^2 = \text{inv}.$$

In classical theory all quantities are either scalars or four-vectors or 4-tensors, that is, quantities transforming as a product of four-vector components $x_\mu = (t, \underset{\sim}{r})$. Some examples are the energy-momentum four-vector $p_\mu = (E, \underset{\sim}{p})$, the current four-vector $J_\mu = (\rho, \underset{\sim}{j})$, the electromagnetic field intensity tensor $F_{\mu\nu} = \partial_\mu A_\nu - \partial_\nu A_\mu$ associated with the four-vector potential $A_\mu = (\varphi, \underset{\sim}{A})$, and so on. All four-vectors transform according to the law

$$A_0' = A_0 + vA, \quad A' = A + vA_0.$$

so the scalar product of any two four-vectors is on invariant:

$$(AB) \equiv A_0B_0 - AB = (A'B') = \text{inv}.$$

It is possible to preserve the usual rules of vector multiplication if we introduce the metric tensor

$$g_{\mu\nu} = \begin{pmatrix} 1 & 0 & 0 & 0 \\ 0 & -1 & 0 & 0 \\ 0 & 0 & -1 & 0 \\ 0 & 0 & 0 & -1 \end{pmatrix}.$$

Then $AB = g_{\mu\nu} A_\mu A_\nu$. Instead of using $g_{\mu\nu}$ we shall interpret summation over (repeated) Greek indices in an invariant sense, that is,

$$AB \equiv A_\mu B_\mu \equiv A_0 B_0 - A_i B_i.$$

In some cases we shall go over to a Euclidean metric by introducing $A_4 = iA_0$; such vectors will be denoted by a tilde:

$$\tilde{A}\tilde{B} = \tilde{A}_\nu \tilde{B}_\nu = -AB.$$

In quantum mechanics observable quantities are represented by operators $\hat{V}$, which are characterized by a set of matrix elements $V_{ij} = \langle \Psi_i^* \, V \Psi_j \rangle$. The matrix elements must transform like the corresponding classical quantities, i.e. like vector or tensor components. However, the wave functions Ψ themselves may transform according to a more complicated law, since they enter the matrix elements bilinearly. An analogous situation is already met in nonrelativistic quantum mechanics: when we analyse invariance relative to 3-dimensional rotations we introduce not only tensors but also spinors. Tensors transform like products of 3-dimensional coordinate vectors, but spinors transform according to a more complicated law, since bilinear combinations of spinors transform like vectors and tensors.

For instance, for a particle of spin $\frac{1}{2}$ the wave function $\varphi = \binom{\varphi_1}{\varphi_2}$ transforms as follows when the coordinate system is rotated by a small angle θ around the axis $\underline{n}$:

$$\varphi' = \varphi + i\frac{\sigma n}{2}\theta\varphi \tag{6.1}$$

where the components of $\underset{\sim}{\sigma}$ are the Pauli matrices. The bilinear combinations $\varphi^{+}\varphi$ and $\varphi^{+}\underset{\sim}{\sigma}\varphi$ thus transform as a scalar and a 3-dimensional pseudovector respectively (this was indeed what led to the discovery of the spinor transformation law).

Let us now try to find the Lorentz transformation law for spinors. Since the Pauli matrices form a complete set, this law must be analogous to (6.1):

$$\varphi' = \varphi + c_1\frac{\sigma v}{2}\varphi. \tag{6.2}$$

Relation (6.1) corresponds to the coordinate transformation

$$x_1' = x_1 + \theta x_2, \quad x_2' = x_2 - \theta x_1. \tag{6.3}$$

where for definiteness we have considered a clockwise rotation around the z-axis. In order that the Lorentz transformation

$$x_3' = x_3 + vt, \quad t' = t + vx_3$$

should coincide with the formula for rotation of coordinates when we introduce the Euclidean metric it = x_4, it is necessary to choose $i\theta = \pm v$ (where the choice of sign depends on whether we choose a right-handed (+) or left-handed (-) system of space coordinates). Hence to obtain (6.2) we must replace $i\theta$ in (6.1) by $\pm v$. Thus, $c_1 = \pm 1$.

There are two possible types of spinor, with different transformation properties:

$$\varphi_+' = \varphi_+ + \frac{\sigma v}{2}\varphi_+, \quad \varphi_-' = \varphi_- - \frac{\sigma v}{2}\varphi_-. \tag{6.4}$$

Neither of these spinor fields has a definite parity, since the quantity $\underset{\sim}{\sigma}\cdot\underset{\sim}{v}$ changes sign under reflection. Under reflection φ_+ transforms into φ_- and vice versa. These fields describe two 2-component particles (e.g. neutrino and antineutrino). From such fields one can form two linear combinations with definite parity:

$$\varphi = \varphi_+ + \varphi_-, \; \chi = \varphi_+ - \varphi_-.$$

where the field φ has the opposite parity to χ. The transformation laws for these quantities are obtained from (6.4):

$$\varphi' = \varphi + \frac{\sigma v}{2}\chi, \quad \chi' = \chi + \frac{\sigma v}{2}\varphi. \tag{6.5}$$

These transformation laws preserve their form on reflection.

Thus, the requirements of Lorentz invariance and reflection symmetry lead to the result that a particle with spin $\frac{1}{2}$ must be described by a 4-component spinor $\Psi = \{ {\varphi \atop \chi} \}$. We shall see below that this fact is connected with the existence of antiparticles; the four components of a relativistic spinor correspond to the two possible spin projections of the particle and two of the antiparticle.

Without specifying for the moment the physical meaning of the spinors φ and χ, we can form from them bilinear combinations transforming like scalars, vectors and tensors under Lorentz transformations. We will restrict ourselves to the scalar and vector cases, which are important for subsequent work. The scalar has the form

$$S = \varphi^+\varphi - \chi^+\chi.$$

since under a Lorentz transformation each of the terms $\varphi^+\varphi$, $\chi^+\chi$ acquires the same additional term $\frac{1}{2}(\chi^+ \boldsymbol{\sigma}\cdot\mathbf{v}\,\varphi + \varphi^+ \boldsymbol{\sigma}\cdot\mathbf{v}\,\chi)$. which therefore cancels in S. If instead of the difference we form the sum

$$V_0 = \varphi^+\varphi + \chi^+\chi,$$

then the additional terms do not cancel but add, and so

$$V_0' = V_0 + \mathbf{v}\mathbf{V}.$$

where

$$V_i = \chi^+\sigma_i\varphi + \varphi^+\sigma_i\chi. \tag{6.6}$$

Thus, V_0 transforms like the fourth component of a four-vector $V_\mu = (V_0, \mathbf{V})$. It is easy to verify that the quantities V_i transform like the space components of a four-vector V_μ: substituting (6.5) in (6.6), we get

$$V_i' = V_i + \frac{1}{2} v_k \chi^+ (\sigma_i\sigma_k + \sigma_k\sigma_i)\,\chi + \frac{1}{2} v_k \varphi^+ (\sigma_i\sigma_k + \sigma_k\sigma_i)\,\varphi.$$

Using the anticommutation relation $\sigma_i\sigma_k + \sigma_k\sigma_i = 2\delta_{ik}$, we arrive at the correct transformation law

$$V_i' = V_i + v_i V_0.$$

The bilinear combinations we have just found are usually written in terms of the bispinor $\Psi = \binom{\varphi}{\chi}$ and the Dirac matrices

$$\gamma_0' = \begin{pmatrix} I & 0 \\ 0 & -I \end{pmatrix}, \quad \gamma_i = \begin{pmatrix} 0 & \sigma_i \\ -\sigma_i & 0 \end{pmatrix},$$
$$S = \Psi^+\gamma_0\Psi \equiv \overline{\Psi}\Psi, \quad V_\mu = \Psi^+\gamma_0\gamma_\mu\Psi \equiv \overline{\Psi}\gamma_\mu\Psi.$$

The Dirac matrices obey the anticommutation relations

$$\gamma_\mu\gamma_\nu + \gamma_\nu\gamma_\mu = 2g_{\mu\nu}.$$

Maxwell's Equations

As an illustration of the use of symmetry properties, we shall show how a contemporary theoretician might have deduced Maxwell's equations in free space had they not been known. Our task is to find relations between the electric field $\underset{\sim}{\mathcal{E}}(\underset{\sim}{r}, t)$ and the magnetic field $\underset{\sim}{\mathcal{H}}(\underset{\sim}{r}, t)$. These relations must be linear up to the very large values of the fields $(\mathcal{E}, \mathcal{H} \sim \mathcal{E}_c \sim 10^{16}\,\mathrm{V/cm})$ where vacuum polarization becomes important (cf. the estimate on p. 5). First we have to find out the symmetry properties of the fields $\underset{\sim}{\mathcal{E}}$ and $\underset{\sim}{\mathcal{H}}$; this can be demonstrated by very simple experiments, e.g. on the deflection of an electron beam in electric and magnetic fields. In such an experiment we find that the force acting on an electron is given by $\underset{\sim}{\mathcal{F}} = e\,\underset{\sim}{\mathcal{E}} + \frac{e}{c}\,\underset{\sim}{v} \times \underset{\sim}{\mathcal{H}}$, from which it follows that $\underset{\sim}{\mathcal{H}}$ is a pseudovector (axial vector); that is, in contrast to a vector quantity $\underset{\sim}{\mathcal{H}}$ is unchanged by the operation of reflection. Moreover, since $\underset{\sim}{\mathcal{H}}$ is produced by a current, that is by a quantity proportional to the velocity of charged particles, the magnetic field must change sign under the operation of time reversal. On the other hand the electric field is a (polar) vector, and since it can be produced by stationary charges, must be invariant under time reversal. Then the lowest-order equations which can relate $\underset{\sim}{\mathcal{E}}$ and $\underset{\sim}{\mathcal{H}}$ will be

$$\frac{\partial \mathcal{E}}{\partial t} = a \underset{\sim}{\nabla} \times \underset{\sim}{\mathcal{H}}, \quad \frac{\partial \mathcal{H}}{\partial t} = b \underset{\sim}{\nabla} \times \underset{\sim}{\mathcal{E}}.$$

The argument is that curl $\underset{\sim}{\mathcal{H}}$ is the only quantity which does not change sign on reflection but does change sign when t is replaced by -t. (A term of the form $\underset{\sim}{r} \times \underset{\sim}{\mathcal{H}}$ would violate the translational symmetry of space: all other possibilities are similarly excluded).

We have not included in the equation derivatives of higher than first order; their inclusion would lead to the introduction of additional constants and would spoil the beauty of the theory. Moreover, the introduction of higher-order space derivatives would force us to introduce also higher-order time derivatives (otherwise the symmetry of space and time which is dictated by relativistic invariance would be destroyed); then the values of the fields $\underset{\sim}{\mathcal{E}}$ and $\underset{\sim}{\mathcal{H}}$ at time t would be determined not only by their values at the initial time, but also by the values of their time derivatives.

Both the constants introduced above have the dimensions of velocity; one of them may be chosen arbitrarily, thereby defining the units of $\underset{\sim}{\mathcal{E}}$ and $\underset{\sim}{\mathcal{H}}$ relative to one another. We will put $b = c$ (where c is the velocity of light). Then, eliminating $\underset{\sim}{\mathcal{H}}$ from the equations and using the fact that curl curl $\underset{\sim}{\mathcal{E}} = -\nabla^2 \underset{\sim}{\mathcal{E}} + \underset{\sim}{\nabla} \operatorname{div} \underset{\sim}{\mathcal{E}} = -\nabla^2 \underset{\sim}{\mathcal{E}}$, we find

$$\frac{\partial^2 \mathcal{E}}{\partial t^2} = -ac\nabla^2 \mathcal{E}.$$

If the velocity of propagation of the waves is to be equal to c, we must put a = - c, when our equations just become Maxwell's equations. We could use the requirement of relativistic invariance to find the transformation law of $\underset{\sim}{\mathcal{E}}$ and $\underset{\sim}{\mathcal{H}}$ which compensates a Lorentz transformation of the coordinates in such a way as to preserve the form of the equations. It is more convenient, however, to obtain the result by introducing the four-vector A_ν :

$$E_i = \dot{A}_i - \partial_i A_0, \quad \mathcal{H} = \underset{\sim}{\nabla} \times A, \quad (c \equiv 1),$$

$$\Box A_\nu - \partial_\nu \partial_\mu A_\mu = 0, \quad \Box = \frac{\partial^2}{\partial t^2} - \nabla^2,$$

whence the four-vector nature of A_ν is evident. Thus, $\underset{\sim}{\mathcal{E}}$ and $\underset{\sim}{\mathcal{H}}$ transform as the components of the four-dimensional tensor $F_{\mu\nu} = \partial_\mu A_\nu - \partial_\nu A_\mu$.

The Klein-Gordon-Fock Equation

Let us obtain the Lorentz-invariant equation describing a particle with spin zero. The number of components of a wave function with spin j in the rest frame is determined by the number of possible projections of j on some fixed axis, i.e. it is 2j + 1. So in our case the wave function must have just one component. As we shall see in our discussion of the Green's function in a field (cf. next section), it is impossible to construct a relativistically invariant theory for particles alone; the theory must inevitably contain also antiparticles with the same mass, which are described by a single equation along with the particles. Let us obtain this equation.

The wave functions of the particles and antiparticles in the $(\underset{\sim}{p}, t)$ representation obey the equations

$$i \frac{\partial \Psi_+(P)}{\partial t} = E(p)\,\Psi_+(p), \quad i \frac{\partial \Psi_-(P)}{\partial t} = E(p)\,\Psi_-(p), \tag{6.7}$$

where

$$E(p) = \sqrt{p^2 + m^2}.$$

We introduce the functions

$$\Psi = \Psi_+ + \Psi_-^*, \quad \Psi_1 = \Psi_+ - \Psi_-^*,$$

Then from (6.7) we get

$$i \frac{\partial \Psi}{\partial t} = E(p)\,\Psi_1, \quad i \frac{\partial \Psi_1}{\partial t} = E(p)\,\Psi.$$

Elimination of Ψ_1 gives

$$-\frac{\partial^2 \Psi}{\partial t^2} = (p^2 + m^2)\,\Psi.$$

In the coordinate representation we get the Klein-Gordon-Fock equation

$$(\Box + m^2)\,\Psi = 0. \tag{6.8}$$

It is obvious from the definition of $\Psi(\underline{p}, t)$ that the terms in $\Psi(\underline{r}, t)$ with negative frequencies correspond to particles, while terms with positive frequencies describe antiparticles. Equation (6.8) is relativistically invariant, as is immediately obvious in the p-representation: $(p_0^2 - \underline{p}^2 - m^2)\Psi = 0$. Multiplying (6.8) by Ψ^* and subtracting the equation for Ψ^* multiplied by Ψ, we obtain the continuity equation

$$\frac{\partial \rho}{\partial t} + \operatorname{div} j = 0,$$

where ρ and j_α form the four-dimensional current:

$$j_\nu = \frac{1}{i}\left(\Psi^* \frac{\partial \Psi}{\partial x_\nu} - \Psi \frac{\partial \Psi^*}{\partial x_\nu}\right).$$

The density ρ is given by the expression

$$j_0 = \rho(r, t) = \frac{1}{i}\left(\Psi^* \frac{\partial \Psi}{\partial t} - \Psi \frac{\partial \Psi^*}{\partial t}\right),$$

or in the (p, t) representation

$$\rho(p) = 2E(p)(\Psi_-^*(p)\Psi_-(p) - \Psi_+^*(p)\Psi_+(p)).$$

Thus the quantity ρ is the difference in the density of particles and antiparticles, and the continuity equation expresses the conservation of the difference in the numbers of particles and antiparticles. (Of course, in the absence of a field each of these numbers is separately conserved.) For charged particles the continuity equation must obviously correspond to the conservation of charge, and hence the charge of an antiparticle can differ only by a sign from that of a particle.

Thus eqn. (6.8) is a relativistically invariant way of writing the two Schrödinger equations (6.7); this will be useful when we introduce fields later on.

The Dirac Equation

Our problem is to obtain a relativistically invariant equation describing a particle with spin $\frac{1}{2}$. As we already remarked in the context of the Klein-Gordon-Fock equation, this is impossible to do if we use only the field of a single particle; it is necessary to construct a single equation which will describe both particle and antiparticle. For definiteness we shall talk in terms of an electron and a positron.

Consider first the equation in the rest frame. Then we obtain two independent equations, one for the electron and one for the positron:

$$i\frac{\partial \Psi_+}{\partial t} = m\Psi_+, \quad i\frac{\partial \Psi_-}{\partial t} = m\Psi_-.$$

each of these functions has two components corresponding to the two possible spin projections. We write $\Psi_+ = \varphi$, $\Psi^*_- = \chi$ and introduce the 4-component function $\Psi = \{ {\varphi \atop \chi} \}$. Then the equation for Ψ takes the form

$$i\gamma_0 \frac{\partial \Psi}{\partial t} = m\Psi. \tag{6.9}$$

We must now write a Lorentz-invariant equation which contains the components of the momentum operator $p_\mu = -i \frac{\partial}{\partial x_\mu}$ and goes over for $p_i \Psi = 0$ into eqn. (6.9). We already know that the quantity $\bar{\Psi} \gamma_\mu \Psi$ is a four-vector, while $\bar{\Psi} \Psi$ is a scalar. It therefore follows that if A_μ is a four-vector, the quantity $A_\mu \gamma_\mu \Psi$ transforms like Ψ. (Multiplication of this quantity from the left by $\bar{\Psi}$ $(\equiv \Psi^+ \gamma_0)$ gives a scalar, just as does multiplication of $\bar{\Psi}$ by Ψ). In the present case we have at our disposal only the four-vector $ip_\mu = (\frac{\partial}{\partial t}, \frac{\partial}{\partial x_\mu})$. Thus we arrive at the Dirac equation

$$i\gamma_\mu \frac{\partial}{\partial x_\mu} \Psi = m\Psi. \tag{6.10}$$

Multiplying the equation which is the Hermitian conjugate of (6.10) from the right by γ_0 and using the properties of the γ_μ, we obtain an equation for $\bar{\Psi}$:

$$-i \frac{\partial \bar{\Psi}}{\partial x_\mu} \gamma_\mu = m\bar{\Psi}. \tag{6.11}$$

Multiplying (6.10) from the left by $\bar{\Psi}$ and (6.11) from the right by Ψ and subtracting one from the other, we get the continuity equation

$$\frac{\partial}{\partial x_\mu} \bar{\Psi}\gamma_\mu\Psi = 0. \tag{6.12}$$

We shall see below that $\bar{\Psi} \gamma_\mu \Psi$ represents a four-current.

In the momentum representation we have

$$\hat{p}\Psi \equiv \gamma_\mu p_\mu \Psi = m\Psi. \tag{6.13}$$

Applying this equation twice, we get

$$\hat{p}^2\Psi = p^2\Psi = m^2\Psi,$$

and hence

$$p^2 = \omega^2 - \mathbf{p}^2 = m^2.$$

This equation has two solutions: $\omega = \pm (\mathbf{p}^2 + m^2)^{\frac{1}{2}}$. The positive frequencies correspond to particles, the negative ones to antiparticles. Of course, this does not mean that the energy E of an antiparticle is negative - it simply means that it is the complex conjugate of the antiparticle wave function which comes into Ψ, so that instead of the factor e^{-iEt} there appears the factor e^{iEt}, i.e. a negative frequency. We had to introduce the complex conjugate in order to obtain in (6.9) the matrix γ_0 which allowed us to write a covariant equation.

The Dirac equation establishes a connection between the first and second components of Ψ. From (6.13) we have

$$(E(\mathbf{p}) - m)\varphi = \sigma\mathbf{p}\chi, \quad (E(\mathbf{p}) + m)\chi = \sigma\mathbf{p}\varphi. \tag{6.14}$$

We write $\Psi(x)$ for a plane wave in the form

$$\Psi_p^{(\sigma)}(\alpha, x) = u_\alpha^{(\sigma)} e^{ipx}. \tag{6.15}$$

where the index σ defines the sign of the energy and the sign of the spin projection, that is it "labels" the function, while the index α is a spinor variable numbering the component. From (6.14) we find the $u_\alpha^{(\sigma)}$, which we normalize so that $\bar{u}u = 1$:

$$u^{(+)s} = \sqrt{\frac{E+m}{2m}} \left\{ \begin{array}{c} \varphi^{(s)} \\ \frac{\sigma p}{E+m} \varphi^{(s)} \end{array} \right\},$$

$$u^{(-)s} = \sqrt{\frac{|E|+m}{2m}} \left\{ \begin{array}{c} -\frac{\sigma p}{|E|+m} \chi^{(s)} \\ \chi^{(s)} \end{array} \right\}, \tag{6.16}$$

where $\varphi^{(s)}$ is a spinor corresponding to the two spin projections: $(\varphi^{(s)}, \varphi^{(s')}) = \delta_{ss'}$. In the rest frame $\Psi^{(\sigma)}(x)$ goes over for $E(p) > 0$ into $\Psi^{(+)s} = \{ {\Psi_+^{(s)} \atop 0} \}$, and for $E(p) < 0$ into $\Psi^{(-)s} = \{ {0 \atop \Psi^{(s)*}} \}$. It follows from (6.16) that the function $u^{(-)s}(p, \alpha)$ describes a positron with momentum $-p$.

The most important consequence of the Dirac equation is the prediction of antiparticles with the same mass as the particle but with opposite parity. So long as there are no external fields or interactions, particles and antiparticles propagate independently and the Dirac equation is simply a compact way of writing the Schrödinger equation and the condition of Lorentz invariance. The advantages of the Dirac equation will become clear below, when we introduce interactions.

The Green's Function of a Spinless Particle

The Green's function of a free particle in the (p, τ) representation has the form (see p. 271)

$$G^+(p, \tau) = e^{-iE^+(p)\tau}\theta(\tau),$$

where $E^+(p) = \sqrt{m_+^2 + p^2}$. Fourier-transforming with respect to τ, we have (p. 272)

$$G^+(p, p_0) = \frac{1}{p_0 - E^+(p) + i\delta}.$$

This expression is not relativistically covariant. In fact, the Green's function $G(\underset{\sim}{p}, p_0)$ must be covariant under the Lorentz transformation of the four-vector $(\underset{\sim}{p}, p_0)$ given by

$$p' = p + vp_0, \quad p_0' = p_0 + vp.$$

However, the quantity $E^+(\underset{\sim}{p})$ standing after p_0 changes according to the law

$$E'(p) = E(p') = E(p) + vp\frac{p_0}{E(p)}.$$

To obtain a covariant expression, it is necessary to postulate the existence of another particle with a Green's function

$$G^-(p, \tau) = e^{-iE^-(p)\tau}\theta(\tau), \quad E^- = \sqrt{m_-^2 + p^2}$$

and to introduce, as we did above (p. 274), a Green's function defined for all τ:

$$G_1(p, \tau) = \begin{cases} G^+(p, \tau), & \tau > 0, \\ G^-(p - \tau), & \tau < 0. \end{cases} \tag{6.17}$$

Then we have

$$-G_1(p, p_0) = \frac{1}{E^+(p) - p_0 - i\delta} + \frac{1}{E^-(p) + p_0 - i\delta}.$$

If this expression is to have a covariant form, we must assume that $E^+(\underset{\sim}{p}) = E^-(\underset{\sim}{p}) = E(\underset{\sim}{p})$, that is, that the mass of the second particle is equal to the mass of the first $(m_+ = m_- = m)$. Then

$$G_1(p, p_0) = -\frac{2E(p)}{E^2(p) - p_0^2 - i\delta}. \tag{6.18}$$

The denominator contains the invariant $\underset{\sim}{p}^2 + m^2 - p_0^2$.

We introduce the invariant quantity $G(p) = G_1(p)/2E(\underset{\sim}{p})$. Note that $G(p)$ is just the Green's function of the Klein-Gordon-

Fock (KGF) equation, which describes a particle with zero spin:

$$\begin{aligned} &\Box \Psi + m^2\Psi = 0, \\ &(\Box + m^2)\, G(x, x') = \\ &= -i\delta(x - x'),\quad x = (\boldsymbol{r}, t). \end{aligned} \tag{6.19}$$

In fact, in the four-momentum representation we have

$$G(p) = \frac{1}{p^2 - m^2 + i\delta}, \tag{6.19'}$$

which is just $G_1/2E(\mathbf{p})$.

In some cases it is convenient to use the Green's function in the coordinate representation. Writing $x_1 - x_2 = x$, we find

$$G(x_1 - x_2) = G(x) = i\int \frac{e^{-ipx}}{p^2 - m^2 + i\delta}\cdot\frac{d^4p}{(2\pi)^4}.$$

To evaluate this integral it is convenient to go over to the Euclidean variables $p_4 = ip_0$, $x_4 = ix_0$. It is clear from Fig. 49 that we can make the indicated deformation of the integration contour in the

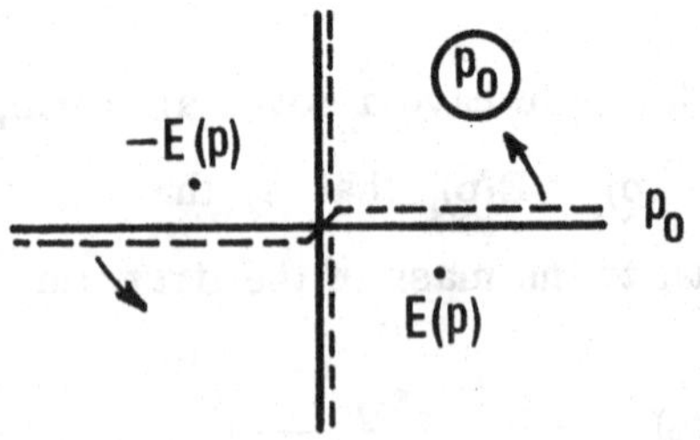

Fig. 49

complex plane without crossing any singularities of the integrand (we shall consider this deformation in more detail below). Using the relation

$$\frac{1}{\bar{p}^2+m^2}=\int_0^\infty e^{-\alpha(\bar{p}^2+m^2)}\,d\alpha,\quad \bar{p}^2=p^2+p_4^2=-p^2,$$

we obtain

$$G(x)=\int_0^\infty e^{-\alpha m^2}d\alpha\prod_{i=1}^{4}\left(\int \exp\left(-\alpha p_i^2+ip_ix_i\right)\left(\frac{dp_i}{2\pi}\right)\right)=$$

$$=\frac{1}{(4\pi)^2}\int_0^\infty du\exp\left\{-\frac{\tilde{x}^2}{4}u-\frac{m^2}{u}\right\}. \qquad (6.20)$$

At large distances we can expand the argument of the exponential and use the method of steepest descents to get ($\tilde{x}=\sqrt{x_i^2}$)

$$-\frac{\tilde{x}^2}{4}u-\frac{m^2}{u}=-m\tilde{x}-\frac{\tilde{x}^3}{8m}(u-u_1)^2,\quad u_1=2m/\tilde{x},$$

whence

$$G(x)\simeq\frac{1}{8\pi^2}\sqrt{\frac{2\pi m}{\tilde{x}^3}}\,e^{-m\tilde{x}},\quad \tilde{x}\gg 1/m. \qquad (6.21)$$

In the case of small x we can neglect the m^2 in the argument, and (6.20) gives

$$G(x)=\frac{1}{4\pi^2\tilde{x}^2}=-\frac{1}{4\pi^2x^2}. \qquad (6.22)$$

The Green's function has a very simple form in the mixed ($\mathbf{r}$, p_0) representation. From eqn. (6.19) we get

$$\nabla^2 G(p_0,\mathbf{r})+(p_0^2-m^2)\,G(p_0,\mathbf{r})=\delta(\mathbf{r}),$$

whence

$$G(p_0,\mathbf{r})=-\frac{1}{4\pi r}\exp\left[i(p_0^2-m^2)^{1/2}r\right]. \qquad (6.23)$$

The plus sign in the exponent replaces the rule for going around the pole in the momentum representation.

The Green's Function of a Particle With Spin $\frac{1}{2}$

We first obtain the Green's function in a frame in which the particle and antiparticle are at rest; then their Green's functions in the τ-representation have an identical form:

$$G^+_{ss'}(\tau) = e^{-im\tau}\theta(\tau)\delta_{ss'}, \quad G^-_{ss'}(\tau) = e^{-im\tau}\theta(\tau)\delta_{ss'},$$

where s, s' are spin indices. We introduce as above (p. 271) the unified function

$$G_{ss'}(\tau) = \begin{cases} G^+_{ss'}(\tau), & \tau > 0, \\ -G^-_{ss'}(-\tau), & \tau < 0. \end{cases}$$

Going over from the τ- to the p_0-representation, we can write G in the form of a matrix element of a 4 x 4 matrix:

$$G(p_0) = \begin{pmatrix} I\dfrac{p_0 + m}{p_0^2 - m^2 + i\delta} & 0 \\ 0 & -I\dfrac{p_0 - m}{p_0^2 - m^2 + i\delta} \end{pmatrix},$$

$$I = \begin{pmatrix} 1 & 0 \\ 0 & 1 \end{pmatrix}.$$

The expression in the upper left-hand corner corresponds to a particle, and the one in the bottom right-hand one to an antiparticle. In other words, we represent $G_{ss'}$ as $\bar{\Psi}_s G \Psi_{s'}$, where $\Psi_s = \begin{pmatrix} \varphi_s \\ \chi_s \end{pmatrix}$ is a bispinor describing a particle and antiparticle with spin s, and $\bar{\Psi} = \Psi^+\gamma_0$. We can rewrite $G(p_0)$ in the form

$$G(p_0) = \frac{\gamma_0 p_0 + m}{p_0^2 - m^2 + i\delta}.$$

To obtain G(p) in an arbitrary coordinate system we must write G in invariant form. To do this we have to replace the p_o^2 in the denominator (which goes with the scalar m^2) by the 4-dimensional scalar $p^2 = p_o^2 - \underset{\sim}{p}^2$, and $\gamma_o p_o$ in the numerator by the scalar $\gamma_\nu p_\nu$. (The scalar character of this product follows from the fact that the matrix elements $\bar{\Psi} \gamma_\nu \Psi$ form a four-vector.) Thus we get

$$G(p) = \frac{\hat{p} + m}{p^2 - m^2 + i\delta}, \tag{6.24}$$

where $\underline{p} = \gamma_\nu p_\nu$. The matrix elements of G must be understood as $\bar{\Psi}_1 G \Psi_2$. Expression (6.24) is just the Green's function of the Dirac equation, defined by the relation

$$(\hat{p} - m) G = I, \ G = \frac{1}{\hat{p} - m} = \frac{p + m}{p^2 - m^2}.$$

Let us find G in the coordinate representation:

$$G(x) = i \int e^{-ipx} \frac{p + m}{p^2 - m^2} \frac{d^4 p}{(2\pi)^4} = (m + i\gamma_\nu \partial_\nu) G_s \tag{6.25}$$

where G_s is the Green's function of a scalar particle, given by (6.20). For $m^2 x^2 \ll 1$ we get

$$G(x) = \frac{i}{2\pi^2} \frac{\gamma_\nu x_\nu}{x^4}. \tag{6.26}$$

Thus, G(p) is a matrix $G_{\alpha\beta}(p)$ in the space of the spinor variables α. The transition to the σ-representation can be made with the help of the functions $u^{(\sigma)}(\alpha)$ introduced on p. 369. For the internal lines of diagrams it is more convenient to use the α-representation (cf. below).

The Photon Green's Function

To describe scalar and spinor particles we started from Schrödinger's equation and used the requirement of Lorentz invariance to find the relativistic equations and the corresponding Green's functions. In the case of the quanta of the electromagnetic field - photons - our starting point is the Lorentz-invariant Maxwell equations, and what we have to find is the quantum-mechanical interpretation of these equations.

We apply to the vector potential A_μ the Lorentz condition

$$\partial_\mu A_\mu = 0. \tag{6.27}$$

Then the Maxwell equation reduces to four KGF equations, one for each component of A_μ:

$$\Box A_\mu = 0. \tag{6.28}$$

Thus we can interpret $A_\mu(x)$ as the wave function of a Bose particle with zero mass; the vector index μ corresponds to the spin projection of this particle.

In the case of a massive particle, we can determine the spin of the particle by going over to the rest frame; in this frame the number of components of the wave function is $(2j + 1)$. For a photon, however, no rest frame exists. In this case we can determine the spin by using gauge invariance: fields A_μ and $A'_\mu = A_\mu + \partial_\mu f$, where f is an arbitrary function of coordinates and time, are physically indistinguishable and must describe one and the same particle. In any fixed coordinate system it is possible to choose the function f so that A_0 is zero at all space-time points; in that case A_μ becomes a three-dimensional vector and hence describes a particle with spin 1. However, the Lorentz condition leaves only two independent spin projections, corresponding to the two polarizations of electromagnetic waves:

$$A_\mu(x) = e_\mu^{(1,2)}(ae^{ikx} + a^* e^{-ikx}). \tag{6.29}$$

Here the first term corresponds to the wave function of a photon, and the second is the complex conjugate wave function of an antiphoton. Since the field A_μ is real, the antiphoton is identical to the photon, and the current which corresponds to the difference in number of particles and antiparticles is zero. The wave functions describing a single photon in unit volume corresponds to putting $a = (2k_0)^{-\frac{1}{2}}$ in (6.29) (cf. the remarks on p. 412).

The vectors $e_\mu^{(\lambda)}$ may be chosen orthonormal:

$$e_\mu^{(\lambda)} e_\mu^{(\lambda')} = \delta_{\lambda\lambda'}.$$

Moreover, the $e_\mu^{(\lambda)}$ must satisfy the "transversality" condition:

$$k_\mu e_\mu^{(\lambda)} = 0.$$

Thus, for instance, taking A_0 to be zero and choosing the z-axis along $\underset{\sim}{k}$, we get

$$e_\mu^{(1)} = (0, 1, 0, 0), \quad e_\mu^{(2)} = (0, 0, 1, 0).$$

Now we can construct the photon Green's function as the Green's function of a particle satisfying the KGF equation. In the $(\lambda, \underset{\sim}{k})$ representation we have

$$D_{\lambda\lambda'}(k) = \frac{\delta_{\lambda\lambda'}}{k^2 + i\delta}. \tag{6.30}$$

To write this equation in covariant form, we proceed as we did to obtain the Green's function of a spinor particle; that is, we introduce a 4 x 4 matrix such that its matrix elements give the quantity

$D_{\lambda\lambda'}$:

$$D_{\lambda\lambda'} = e_\mu^{(\lambda)} D_{\mu\nu} e_\nu^{(\lambda')}. \tag{6.31}$$

The condition (6.31) does not determine $D_{\mu\nu}$ uniquely, but only up to a "longitudinal" term $(d(k^2) - 1)\, k_\mu k_\nu / k^2$:

$$D_{\mu\nu}(k) = \frac{g_{\mu\nu} + (d(k^2) - 1)\, k_\mu k_\nu / k^2}{k^2 + i\delta}. \tag{6.32}$$

The longitudinal term gives no contribution to (6.31) because of the transversality condition; the term in $g_{\mu\nu}$ gives (6.30) because of the orthomality of the functions $e_\mu^{(\lambda)}$. The gauge function $d(k^2)$ in (6.32) may be chosen with an eye to convenience in the intermediate calculations; in the final result it cancels out because of the gauge invariance of the theory. The usual choice is $d(k^2) = 1$.

We could have obtained this result in another way: we could have introduced a vector particle with mass m and subsequently gone over to the case of the photon by letting the mass tend to zero. The Green's function of a vector particle in the rest frame has a form identical to (6.30), but in contrast to the photon there are three possible polarizations and hence three basis vectors $e_\mu^{(\lambda)}$. We have to find the four-tensor expression $D_{\mu\nu}^{(m)}$ whose matrix elements in the rest frame give $D_{\lambda\lambda'}^{(m)}$

$$e_\mu^{(\lambda)} D_{\mu\nu}^{(m)} e_\nu^{(\lambda')} = \frac{\delta_{\lambda\lambda'}}{k_0^2 - m^2}.$$

Hence it follows, similarly to (6.32), that

$$D_{\mu\nu}^{(m)} = \frac{1}{k^2 - m^2}\left(g_{\mu\nu} - \frac{k_\mu k_\nu}{m^2}\right). \tag{6.33}$$

To go over to the case $m \to 0$, we have to assume that the longitudinal component of $D_{\mu\nu}$ (the second term in the bracket) does not enter observable expressions. This is just the expression of gauge invariance. Thus, gauge invariance is an inevitable consequence of the zero mass of the photon.

Since the Green's function $D_{\mu\nu}$ which we have introduced is the Green's function of the d'Alembert equation

$$\Box D_{\mu\nu} = -ig_{\mu\nu}\delta(x-x'),$$

we can use $D_{\mu\nu}$ to determine the classical field A_{μ}^{ex} induced by a current j_{μ}^{ex}:

$$\Box A_{\mu}^{ex} = j_{\mu}^{ex},$$

whence we get

$$A_{\mu}^{ex}(x) = i\int D_{\mu\nu}(x-x')\, j_{\nu}^{ex}(x')\, d^4x'. \tag{6.34}$$

In particular, for the field of a stationary charge e we have $j_0 = e\,\delta(\mathbf{r})$, $A_0^{ex} = ie\int D_{00}(\mathbf{r}, t-t')\,dt'$. Hence, using the expression for the Green's function of spinless particles with mass zero, we get

$$-i\int D(\tau, \mathbf{r})\,d\tau = D_{00}(\omega=0, \mathbf{r}) = -\frac{1}{4\pi r}, \quad A_0^{ex} = \frac{e}{4\pi r}. \tag{6.35}$$

2. DIVERGENCES AND RENORMALIZABILITY

Using the Green's functions found in the proceding section and introducing interactions between the particles, we can now proceed to explain the nature of the divergences of field theory and to eliminate them by redefining the constants entering the theory. We will first explain these questions in the context of the simplest model of quantum field theory, the model of a four-boson interaction.

The Local Interaction Between Particles

In this section we shall consider the possible types of interaction between particles and select a very simple model, which will be used below to explore the properties of quantum field theory at small distances.

Quantum field theory starts out from the assumption of a local inter-particle interaction, which means that particles interact only when their space and time coordinates are identical*. The experimentally observed interaction at a distance is conceived to be a secondary process arising from the emission and absorption by the interacting particles of other virtual particles, the emission and absorption events being themselves local. If the interaction is such that the local event creates a single particle, then the process of particle-particle scattering is represented by the diagram of Fig. 50(a).

In Fig. 50(a) the particle 1 moves freely from x_1 to y_1; at point y_1 it emits some particle which is then absorbed by particle 2 at the point y_2. In the case of electrodynamics the

---

*We will not discuss here attempts to construct nonlocal theories; so far these have produced no solid results.

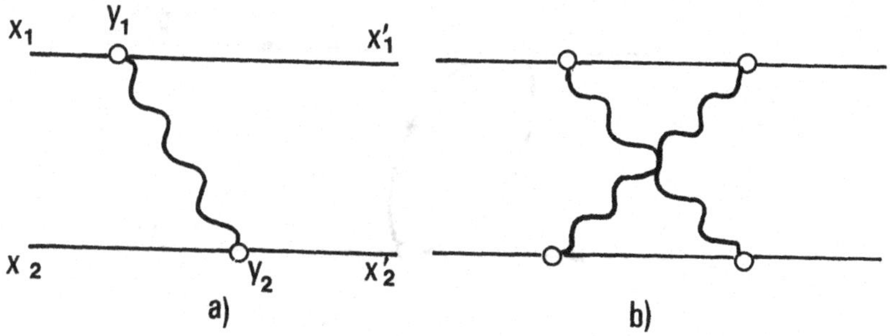

Fig. 50

emitted particle is a photon, so that the Coulomb interaction is the result of the exchange of virtual photons. In the case of the nuclear (strong) interaction of two nucleons the virtual particle is a π-meson or any other particle which interacts strongly with nucleons (following a suggestion of L.B. Okun', such particles are collectively known as "hadrons"). The interaction at a distance can also be mediated by more complicated processes, such as that shown in Fig. 50(b). In the case of electrodynamics this process leads only to a small correction to the Coulomb interaction, since the interaction of electrons with the electromagnetic field is characterized by the small dimensionless parameter $\alpha = e^2/\hbar c = 1/137$. In the case of the strong interactions, on the other hand, the corresponding dimensionless parameter is not small, and diagrams of the type (50(b)) must be taken into account, along with other more complicated diagrams in which a large number of virtual particles participate.

When two particles are emitted in the local event, the interaction at a distance is described by the diagram of Fig. 51. Such a case occurs for the weak interactions; in fact, Fig. 51

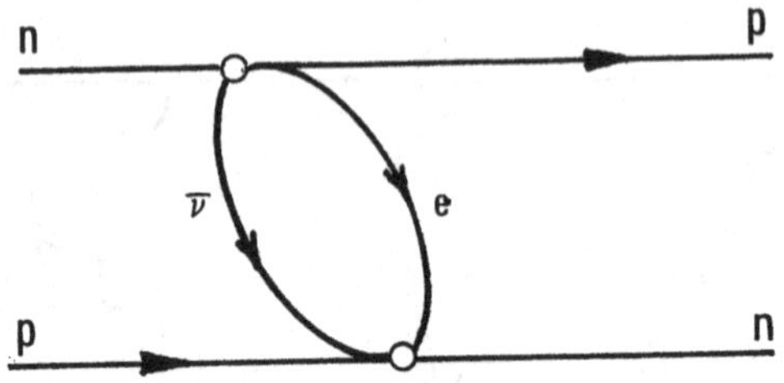

Fig. 51

represents the process of weak interaction of two nucleons. This interaction is mediated by the exchange of two particles, an electron and an antineutrino.

Thus, the requirement of locality means that between the interaction events the particles move freely, and the size of the (four-dimensional) region in which the interaction takes place is assumed to be zero. This simple and beautiful assumption leads to serious difficulties: the integrals over space which describe some processes turn out to involve divergences coming from integration over the region of small distances between interaction events, or equivalently, if the calculation is carried out in the momentum representation, from high momenta of the virtual particles. This indicates that our methods of description of quantum systems are inapplicable at short distances. A consistent description of systems in small regions of space-time will quite possibly demand a fundamental revision of our concepts.

However, if we leave aside this unsolved problem, we may attempt to construct a theory which will be applicable when we investigate processes which take place in four-dimensional regions much greater than the four-dimensional interval r_0 which defines the limit of applicability of the basic theory. This is just the way in which all macroscopic theories are constructed: for example, the construction of the electrodynamics of a medium for fields which vary slowly in time and space on the atomic scale does not require a knowledge of atomic physics but only the introduction of the electric and magnetic susceptibilities.

To carry out such a program we must first of all clearly understand the nature and essential features of the divergences occurring in the theory. It is not essential to investigate this fundamental question in the context of actually existing particles and interactions; indeed, it is advisable to consider it first in a simple model of a field theory for a single type of particle. The very simplest type of local interaction corresponds to Fig. 52.

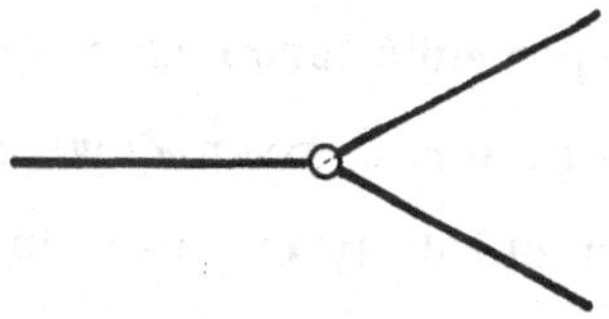

Fig. 52

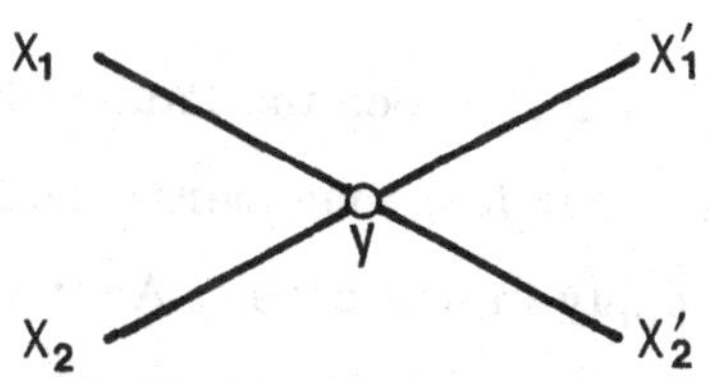

Fig. 53

However, it is easy to see that such a theory is unstable with respect to creation of infinite numbers of particles: for an interaction of this type the energy density is of the form $m^2\varphi^2 + g\varphi^3$, and whatever the sign of g the energy can be decreased indefinitely by letting $g\varphi$ tend to $-\infty$. As a result there occur divergences which have nothing to do with the ones we have to investigate. The simplest sensible theory, therefore, corresponds to scalar particles with the interaction shown in Fig. 53. If such a theory is to be stable, the constants characterizing the interaction must correspond to a repulsion between the particles. The term in the field energy corresponding to Fig. 53 has the form $\lambda\varphi^4$; then stability requires $\lambda > 0$, which indeed corresponds to repulsion.

Notice that in a theory describing two types of particles (fermions and bosons) the interaction represented in Fig. 54 (where the wavy line indicates a boson) does not lead to instability. Such an interaction corresponds to a term in the energy of the form

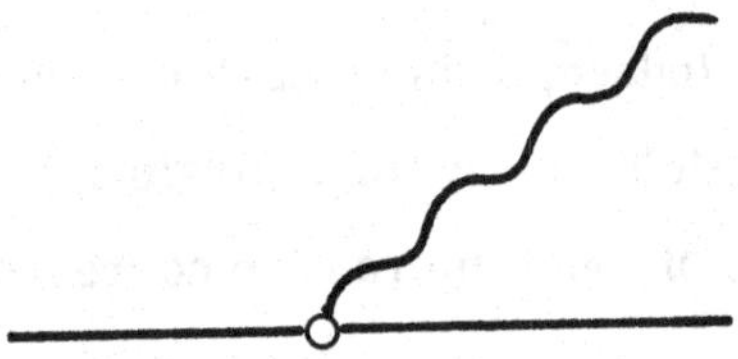

Fig. 54

$g\Psi^+\Psi\varphi$. Since the Pauli principle prevents fermions being created in large numbers, the simple argument given above does not apply in this case. An interaction of this type is used in the theory of the strong interaction of elementary particles. The same remark applies to electrodynamics (cf. p. 409).

Feynman Graphs in a Scalar Theory

We will now explain how the two-particle Green's function must be calculated in the 4-boson interaction model, that is, we find the analytic expressions corresponding to the simplest diagrams occurring in this transition amplitude. By so doing we automatically find the rules for reading off arbitrary diagrams composed of these basic elements.

The amplitude for the simplest process (that shown in Fig. 53) has the form

$$A(x_1, x_2, x_3, x_4) = \int d^4y G(x_1 - y)\, G(x_2 - y)\, (-i\hat{V}_y) G(y - x_3)\, G(y - x_4).$$

The interaction $\hat{V}_y$ cannot depend explicitly on the point y; that would violate the homogeneity of space and lead to non-conservation of energy and momentum, since the conservation laws for these quantities are a direct consequence of the homogeneity of time and space. Generally speaking it is possible to include in $\hat{V}_y$ a dependence on the gradients: $\hat{V}_y = V(\partial/\partial y)$, where $\partial/\partial y$ acts on any of the four propagation functions. We shall however consider the simplest variant of field theory, that is one without gradients; we put therefore $\hat{V} = \lambda$. (The theory then makes sense only if $\lambda > 0$). The diagram of Fig. 53 corresponds to the lowest-order perturbation-theoretic approximation in λ, which is valid for $\lambda \ll 1$.

According to the superposition principle the transition amplitude has contributions not only from Fig. 53 but from all possible diagrams corresponding to different intermediate states. In second order in λ there are three possible graphs; they are shown in Fig. 55. These three diagrams differ by permutations of their external points x_1, x_2, x_3, x_4, so that the sum of the diagrams is symmetric with respect to these four variables, as it must be for scalar particles obeying Bose statistics.

As regards the four-dimensional coordinates of the internal points $y_1 = (\tau_1, \rho_1)$, $y_2 = (\tau_2, \rho_2)$, we must integrate over the

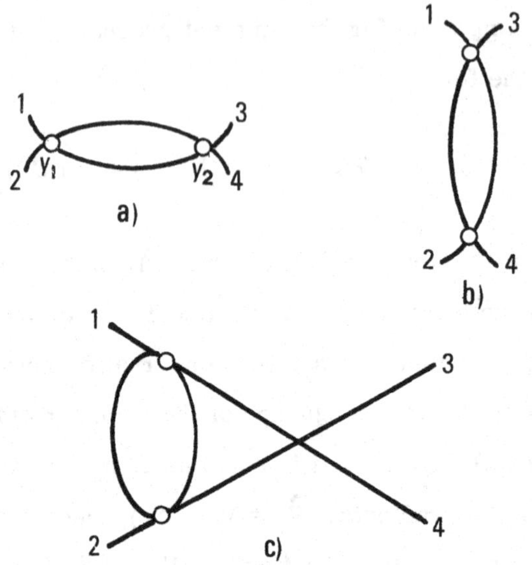

Fig. 55

whole space-time. Different regions of integration over the times τ_1, τ_2 correspond to different intermediate states; for instance, in diagram (a) the region $t_1, t_2 < \tau_1 < \tau_2 < t_3, t_4$ corresponds to the process of repeated scattering of particles 1 and 2, whereas the region $t_1, t_2 < \tau_2 < \tau_1 < t_3, t_4$ corresponds to the creation from the vacuum at time τ_2 of four particles, of which two annihilate at time τ_1 with the initial particles 1, 2 while the other two propagate to the points 3, 4. The existence of annihilation processes is dictated by Lorentz invariance; the integral over the region $\tau_1 < \tau_2$ is not invariant (it is possible to go over to a moving coordinate system where $\tau_1' > \tau_2'$) and we obtain a Lorentz-invariant amplitude only by combining it with the region $\tau_1 > \tau_2$, that is, by taking account of annihilation processes. It is essential to require, moreover, that the constant λ corresponding to scattering should be identical to the analogous constant corresponding to annihilation and creation of particles out of the vacuum. The connection between scattering and annihilation processes ("crossing symmetry") is a characteristic feature of a local relativistic field theory and is reliably confirmed by experiment.

Another important principle used in constructing a field theory is the principle of identity of particles. It is well known from nonrelativistic quantum mechanics that states which differ by permutations of the coordinates (or momenta) of particles of the same type are identical and should not be counted separately in the sum over intermediate states. If we are going to sum over all such states, the correct normalization can be established by attaching to each intermediate state with n particles the factor $1/n!$, corresponding to the number of identical permutations. Thus, diagram (a) must be multiplied by $1/2! = 1/2$ to exclude identical states for $\tau_1 < t < \tau_2$. Then to preserve Lorentz invariance we must also add a factor of $1/2$ in the other regions of integration over τ_1 and τ_2, e.g. for $\tau_2 < \tau_1$. In this region the factor $1/2$ takes account of the indistinguishability of the particles created from the vacuum. The diagrams of Figs. 55(b) and (c) may be analysed similarly.

Estimation of Divergences: The Idea of Renormalization

Having found the rules of reading off diagrams, we may now proceed to a more detailed analysis of the resulting expressions. Here from the very start we encounter divergent integrals. We will show, in fact, that Feynman diagrams containing closed loops diverge when integrated over the internal coordinates or momenta. Consider, for instance, the simplest diagram, Fig. 55(a). The divergence here is associated with the region $y_2 \to y_1$, where each of the Green's functions $G(y_1 - y_2)$, $G(y_2 - y_1)$ behaves like

$$G(y) \to \text{const}/y^2, \quad y^2 \ll m^{-2}.$$

(see p. 373). As a result we get an integral which is logarithmically divergent at the lower limit:

$$\frac{1}{2}\lambda^2 \int \frac{d^4y_1 d^4y_2}{y_{12}^4} G(x_1 - y_1) G(x_2 - y_1) G(x_3 - y_2) G(x_4 - y_2). \quad (6.36)$$

This is particularly obvious if we go over to a Euclidean metric by replacing y_{10}, y_{20} by iy_{14}, iy_{24}; then the part of the integral which is of interest behaves like $\int \frac{y^3 dy}{y^4} \sim \ell n(1/mr_0)$.

The divergent part of this integral may be factorized, giving:

$$\frac{1}{2}\lambda^2 \int \frac{d^4y_{12}}{y_{12}^4} \int d^4y_1 G(x_1 - y_1) G(x_2 - y_1) \times$$
$$\times G(x_3 - y_1) G(x_4 - y_1); \; (|y_{12}| \ll |x_i - y_1|)$$

which has the same form as the first-order diagram (Fig. 53). Just like the first-order diagram, the divergent part of the second-order diagram corresponds to a point interaction, and it makes sense to combine them by redefining the point interaction constant λ . This is the basic idea of renormalisation.

The deletion of the point contributions corresponds to the subtraction from the factor $G(x_1-y_1)\,G(x_2-y_1)\,G(y_2-x_3)\,G(y_2-x_4)$ of its value for $y_2 = y_1$, after which we get a convergent integral over y_1 and y_2. Not all divergences reduce to a renormalization of the coupling constant. For instance, consider the second-order graph for the Green's function shown in Fig. 56. It contains the product of three Green's functions $G(y)$ and diverges for $y \to 0$. Here the divergence is no longer logarithmic, but depends quadratically on the lower limit r_0:

$$\int d^4y G^3(y) \sim \int d^4y \cdot y^{-6} \sim r_0^{-2}.$$

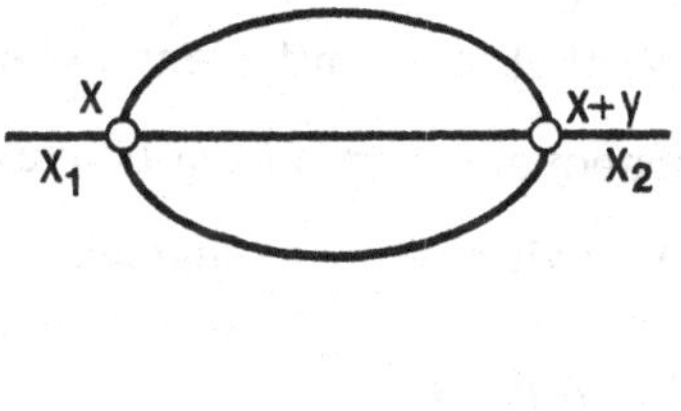

Fig. 56

We shall show that the divergent part of this integral can be included in the renormalization of the particle mass and of the Green's function. For this purpose it is convenient to write the exact equation for the Green's function in the Dyson form, as we did in the many-body problem (see p. 314)

$$(\Box + m_0^2)\bar{G}(x-x') + i\int \Sigma(x-x_1)\bar{G}(x_1-x')\,d^4x_1 = \\ = -i\delta(x-x'). \qquad (6.37)$$

The quantity $\Sigma(x, x')$ is called the self-energy part (in the case of bosons, as here, it is also called the polarization operator); in the absence of external fields the homogeneity of space-time implies that Σ, like the Green's function itself, can depend only on the coordinate difference $(x - x')$. The quantity $\Sigma(y)$ includes all graphs which cannot be decomposed into parts connected by a single line; in second order perturbation theory in λ it is just the internal part of the diagram of Fig. 56. Σ contains no diagrams corresponding to iteration of this graph - they are already taken into account in $\tilde{G}(x_1 - x')$.

Thus, in our theory integrals of $\Sigma(y)$ over d^4y diverge for $y \to 0$. To factor out this divergence we expand the $\tilde{G}(x_1 - x')$ under the integral in (6.37) in a power series in y, assuming that $|x - x'| \gg r_0$ (where r_0 defines the limits of applicability of the theory). We shall see that it is sufficient to keep only two terms in the expansion; subsequent terms give convergent integrals and are not sensitive to the value of r_0. Indeed we have

$$\int \Sigma(y)\tilde{G}(x - x' + y)\, d^4y =$$
$$= \tilde{G}(x - x')\int \Sigma(y)\, d^4y + \partial_\nu \tilde{G}(x - x')\int \Sigma(y)\, y_\nu d^4y +$$
$$+ \frac{1}{2}\partial_\mu \partial_\nu \tilde{G}(x - x')\int \Sigma(y)\, y_\mu y_\nu d^4y + \ldots$$

Because of the isotropy of space-time, the integral containing y_ν linearly is zero, and the one containing it bilinearly can be written

$$\int \Sigma y_\mu y_\nu d^4y = \frac{1}{4}\delta_{\mu\nu}\int \Sigma y^2 d^4y.$$

In second order perturbation theory we have $\Sigma \sim \lambda^2/y^6$. Hence the integrals are of order

$$\int \Sigma d^4y \sim \frac{\lambda^2}{r_0^2}, \quad \int \Sigma y^2 d^4y \sim \lambda^2 \ln \frac{1}{r_0}.$$

All subsequent terms in the expansion lead to convergent integrals. Let us denote this part of Σ by Σ' and introduce the notations m'^2 and C for the two divergent parts, so that

$$i\int \Sigma \bar{G} d^4y = i\int \Sigma' \bar{G} d^4y + m'^2 \bar{G} + C\Box \bar{G}.$$

Substituting this in eqn. (6.37), we get

$$(1-C)\Box \bar{G} + (m_0^2 + m'^2)\bar{G} + i\int \Sigma' \bar{G} d^4y = -i\delta(x-x').$$

where the last term is written in symbolic form.

We introduce the notation

$$\frac{m_0^2 + m'^2}{1-C} = m_1^2, \quad \frac{i\Sigma'}{1-C} = \mu^2, \quad \bar{G}(1-C) = G_R.$$

Then we get for G_R an equation which contains only observable (convergent) quantities in place of the "bare" ones:

$$(\Box + m_1^2 + \mu^2)G_R = -i\delta(x-x').$$

The quantities entering this equation are finite. Let us express m_1^2 in terms of the experimental mass m^2. To do this we go over to the momentum representation. The function G_R must have a pole at $p^2 = m^2$, which corresponds to the expression for the particle energy, $E(\underset{\sim}{p}) = (\underset{\sim}{p}^2 + m^2)^{\frac{1}{2}}$; consequently :

$$m^2 = m_1^2 + \mu^2 \; (p^2 = m^2).$$

Thus the extraction of the divergent parts of the graphs has led to a mass renormalization and to a change in the coefficient

$$\hat{V}_{\text{eff}}(y) = \lambda_F^2 \int G(z)\, G(z)\, d^4z + \lambda_F^2 \int G(z)\, G(z)\, z_\mu\, d^4z \frac{\partial}{\partial y_\mu} + \\ + \frac{\lambda_F^2}{2} \int G(z)\, G(z)\, z_\mu z_\nu d^4z \frac{\partial}{\partial y_\mu} \frac{\partial}{\partial y_\nu} .$$

So the divergences lead to a change in the structure of the original interaction - we are forced to consider an interaction depending on the gradients. When we go on to more complicated diagrams, for instance the one in Fig. 57, there appear terms with higher

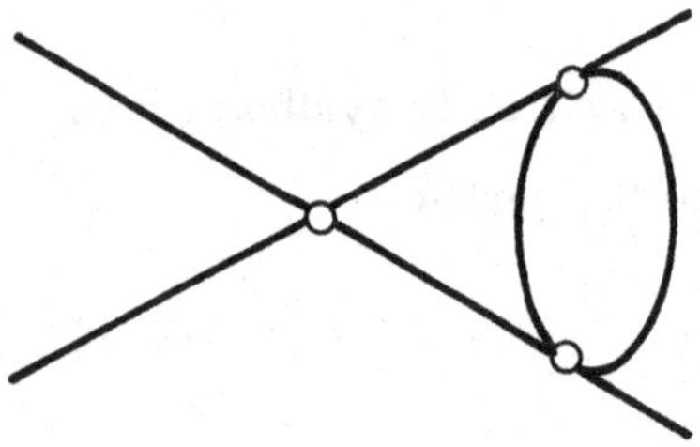

Fig. 57

derivatives ($\partial^3/\partial y^3$, $\partial^4/\partial y^4$ etc.)

It is possible to regard the added terms as renormalizations of the parameters of an initial interaction of the form

$$\hat{V}_y = \lambda_F + C_1 \gamma_\mu \frac{\partial}{\partial y_\mu} + C_2 \frac{\partial^2}{\partial y^2} + \ldots$$

This series does not terminate and contains an infinite number of gradients. As a result, the original interaction must be taken to be nonlocal; it could, for instance, have the structure

$$\hat{V}_y = \left(\frac{g}{m}\right)^2 \sum_n \left(\frac{-\partial^2}{\partial y^2}\right)^n m^{-2n} = \frac{g^2}{m^2 + \Box},$$

corresponding to exchange of a particle with mass m.

In other words, a local 4-fermion interaction is meaningless. A consistent (i.e. renormalizable) interaction between fermions can be constructed if we start out from a local interaction of fermions with bosons, as in Fig. 54. This idea is used in the theory of the strong interaction of elementary particles and in contemporary models of the theory of the weak interaction.

What is the general rule ? How does one tell at a glance whether a theory is renormalizable ?

We can find the criterion from dimensional analysis. The coupling constant λ for four scalar particles is dimensionless ; this is shown by the fact that, as we saw above, the corrections to the amplitude have the form

$$\lambda + \lambda^2 \int \frac{d^4y}{y^4},$$

which proves the statement. The constant corresponding to a fermion-boson interaction (as in Fig. 58) is also dimensionless, as can be seen from a comparison of the two simple graphs shown

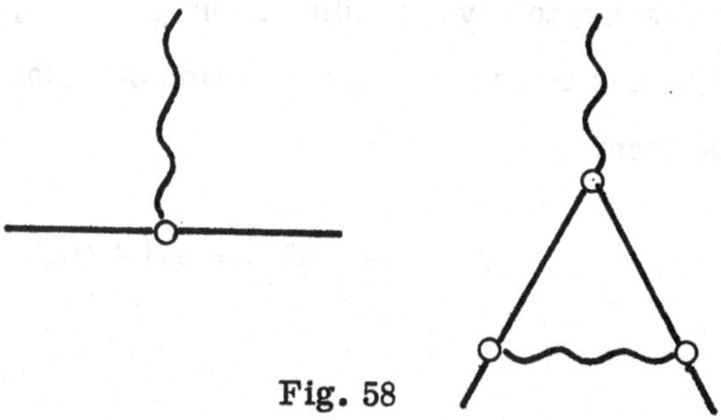

Fig. 58

in Fig. 58. If the first graph is associated with a coupling constant g, the second has the order of magnitude

$$g^3 \int G_F G_F G_B d^4x_1 d^4x_2 \sim g^3 \frac{x^4 x^4}{x^6 x^2},$$

that is, the constant g is dimensionless. A similar estimate with the replacement of g by e leads to the conclusion that the electromagnetic coupling constant e is also dimensionless. On the other hand the coupling constant λ_F for a 4-fermion interaction has the dimensions of length squared; this may be seen from a comparison of two diagrams for the scattering amplitude:

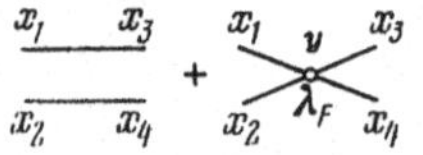

$$= G_F G_F + \lambda_F \int G_F^4 d^4y \,.$$

Hence the dimensionality of λ_F is ℓ^2. The corrections therefore have the form

$$\lambda_F + \lambda_F^2 \int \frac{d^4y}{y^6},$$

that is, they diverge quadratically and, as we have seen, require gradients to be added to the interaction.

The coupling constant g for the three-boson interaction mentioned above may be easily seen from Fig. 58 (by changing the fermions to bosons) to have the dimension $g \sim \ell^{-1}$; consequently, corrections to this constant from the region of distances $\sim r_0$ will have the form

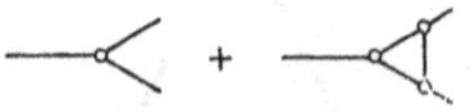

$$= g + g^3 \frac{y^8}{y^6} = g\,(1 + O\,(g^2 r_0^2)),$$

that is, there are no divergences in the interaction in such a theory. The only renormalization which must be carried out is a mass renormalization in second order of perturbation theory, corresponding to the graph

$\Sigma^{(2)} =$ 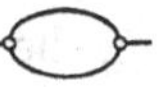 ,

which gives $m'^2 \sim g^2 \ln(1/mr_0)$. In general, for any point interaction with coupling constant γ the n-th order corrections to a dimensionless quantity will have the form

$$(\gamma r_0^{-d})^n,$$

where $[r_0^d]$ is the dimensionality of γ. In the case of a 4-fermion interaction the n-th order correction will contain r_0^{-2n}, which leads to the appearance of high derivatives in the interaction. In cases where the coupling constant γ is dimensionless, the n-th order divergences have the form (cf. below)

$$\gamma^n \left(C_0 \ln^n \frac{1}{mr_0} + C_1 \ln^{n-1} \frac{1}{mr_0} + \ldots \right).$$

We can use dimensional considerations to convince ourselves immediately that in the theory of the 4-boson interaction there do not occur local interactions involving a large number of particles, e.g. of the form

The dimensionality of the coupling constant for such an interaction can be found by comparing the two graphs

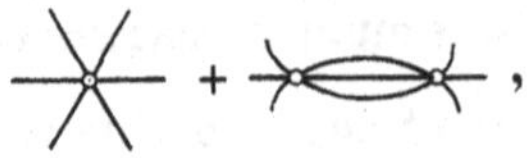

whence the dimensionality of λ_6 is ℓ^2 and hence $\lambda_6 \sim r_0^2 \to 0$. The same result may be obtained by considering the simplest graph, namely

$\lambda_6 =$ 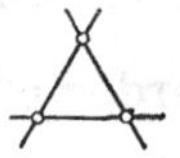 .

We leave it to the reader to convince himself that in the case of a 4-fermion interaction there occurs a local interaction for any even number of particles.

We have reached the important conclusion that not all field theories can be constructed on the assumption of locality; for a local theory to be renormalizable the coupling constant must be dimensionless. This is a necessary but not sufficient condition; for instance, in theories containing vector particles with nonzero mass it can be seen from the expression (6.33) for the Green's function of such a particle that the mass does not drop out of the theory for large 4-momenta $k^2 \gg m^2$, and the dimensional considerations given above do not apply. In fact, we find integrals which diverge according to a power law.

The Logarithmic Approximation and Renormalizability

As we have seen, in theories with a dimensionless coupling constant there occur logarithmically divergent integrals, which can be included in the renormalizations of the coupling constant, the mass and the Green's functions. In each order of perturbation theory the power of the logarithm is increased.

To carry out a program of renormalization in higher orders of perturbation theory, we use the logarithmic approximation, which can be applied to many problems in quantum field theory. The logarithmically divergent integrals which occur in renormalizable theories are cut off at some distance r_0 much smaller than the distances of interest (the distances between the end points of the transition amplitude). As a result there occur large logarithms $\ell \sim \ln (y^2/r_0^2)$, which for $\lambda \ll 1$ allows us to keep in each order of perturbation theory only terms containing the highest power of the logarithm. We shall first carry out this procedure for the first few orders of perturbation theory and subsequently generalize the results to all orders, using a property of renormalizability to be formulated below.

The procedure of cutting off integrals will be meaningful if the answer does not depend on the precise method of cut-off, that is, if the integrals which are left are determined by the region $y^2 \gg r_0^2$. As we shall now see, the logarithmic integrals occurring in the theory do indeed have this property.

It will be convenient to work not in the coordinate but in the momentum representation, in which the Green's functions $G(k)$ have the simple form

$$G(k) = \frac{1}{k^2 - m^2}.$$

In the momentum representation the contribution of the graph of Fig. 53 is simply

$$-i\lambda \frac{1}{k_1^2-m^2}\,\frac{1}{k_2^2-m^2}\,\frac{1}{k_3^2-m^2}\,\frac{1}{k_4^2-m^2}\,\delta\left(\sum_{i=1}^{4} k_i\right)(2\pi)^4.$$

The factor $-i\delta(\Sigma k)/(2\pi)^4$, which expresses the law of conservation of energy-momentum, and the factors $(k_i^2 - m^2)^{-1}$ corresponding to the external Green's functions, will occur in all diagrams, and we shall henceforth omit them. The expression remaining after this is called the particle scattering amplitude.

The second-order graphs correspond to a correction to λ of the form

$$= -\frac{1}{2}\lambda^2 \int \frac{d^4k}{(2\pi)^4 i}\,\frac{1}{[(p_1+p_2-k)^2-m^2+i\varepsilon]\,[k^2-m^2+i\varepsilon]}\,. \qquad (6.38)$$

For simplicity, we shall take the external momenta p_1, p_2, p_3, p_4 to be purely space-like, i.e. put $p_{oi} = 0$. (We can relax this assumption after obtaining the final result). Then the integrals occurring in perturbation theory will be purely real. In fact, the integral over the energy component k_o in (6.38) has the form

$$\int_{-\infty}^{+\infty} \frac{dk_0}{2\pi i}\,\frac{1}{[k_0^2-(p_1+p_2-k)^2-m^2+i\varepsilon]\,[k_0^2-k^2-m^2+i\varepsilon]}\,. \qquad (6.39)$$

The singularities of the integrand are shown in Fig. 59. Without crossing any singularities, we can deform the integration contour C_1 into the imaginary axis C_2, that is, go over to an integral over $k_4 = -ik_o$:

$$\int_{-\infty}^{+\infty} \frac{dk_0}{2\pi i} = \int_{-\infty}^{+\infty} \frac{dk_4}{2\pi}.$$

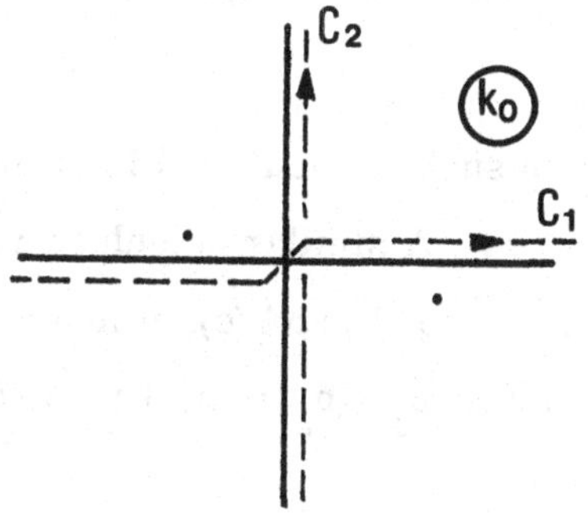

Fig. 59

This trick is called a Wick rotation. After the Wick rotation momentum space is Euclidean : $k^2 = k_o^2 - \underset{\sim}{k}^2 = - k_4^2 - \underset{\sim}{k}^2 = - \underset{\sim}{k}_\nu^2 = - \tilde{k}^2$. Thus the integral (6.38) reduces to an integral over Euclidean space:

$$= -\frac{1}{2}\lambda^2 \int \frac{d^4\tilde{k}}{(2\pi)^4} \frac{1}{[\tilde{k}^2 + m^2]\,[(p_1 + p_2 - \tilde{k})^2 + m^2]}.$$

Such integrals are positive and easily estimated, since the angular integration is carried out over the finite surface area $S = 2\pi^2$ of the unit hypersphere in 4-dimensional space, so that the divergences are associated only with the infinite region of integration over $|\tilde{k}| = \sqrt{\tilde{k}^2}$.

Let us cut off the integration over $|\tilde{k}|$ at some large value $L \gg m, |\tilde{p}_1 + \tilde{p}_2|$. This quantity will correspond to a cut-off at distances of order r_o in the coordinate representation, where $L \sim 1/r_o$. We will calculate the integral assuming that

$|p_1 + p_2|^2 \equiv p_{12}^2 \gg m^2$ (this is the region required below). Deleting m^2 and p_{12}^2 in the integrand, we get

$$\int_{p_{12}}^{L} \frac{d^4\tilde{k}}{(2\pi)^4\tilde{k}^4} = \frac{1}{16\pi^2}\int_{p_{12}^2}^{L^2} \frac{d\tilde{k}^2}{\tilde{k}^2} = \frac{1}{16\pi^2}\ln\frac{L^2}{p_{12}^2}.$$

The relative error in such a calculation is of order $[\ln(L^2/p_{12}^2)]^{-1} \ll 1$. A similar calculation can be carried out for the diagrams of Fig. 55(b) and (c), which differ from the above are by the replacements $p_1 + p_2 \to p_1 + p_3$ and $p_1 + p_2 \to p_1 + p_4$ respectively.

The sum of all the second-order corrections along with the bare coupling constant has the form

$$\lambda + A^{(2)}(p_1, p_2, p_3, p_4) = \\ = \lambda - \frac{1}{2}\frac{\lambda^2}{16\pi^2}\left(\ln\frac{L^2}{p_{12}^2} + \ln\frac{L^2}{p_{13}^2} + \ln\frac{L^2}{p_{14}^2}\right).$$

Let us assume that all the momenta p_1, p_2, p_3, p_4 are of the same order and much less than L. Then we can take all three terms in the brackets equal, and get

$$\lambda + A^{(2)}(p) \simeq \lambda - \frac{3}{2}\frac{\lambda^2}{16\pi^2}\ln\frac{L^2}{p^2} \equiv \lambda\left(1 - \frac{3}{2}\xi\right).$$

Here and below we use the notation $\xi = (\lambda/16\pi^2)\ln(L^2/p^2)$.

The correction just obtained is of the order of the first term when $(\lambda/16\pi^2)\ln(L^2/p^2) \sim 1$. In that case the next-order corrections (of order λ^3) may become important if they are multiplied by a factor $(\ln(L^2/p^2))^2$; corrections of order $\lambda^3 \ln(L^2/p^2)$ and of order λ^3 may be neglected. We now consider the third-order graphs (Fig. 60).

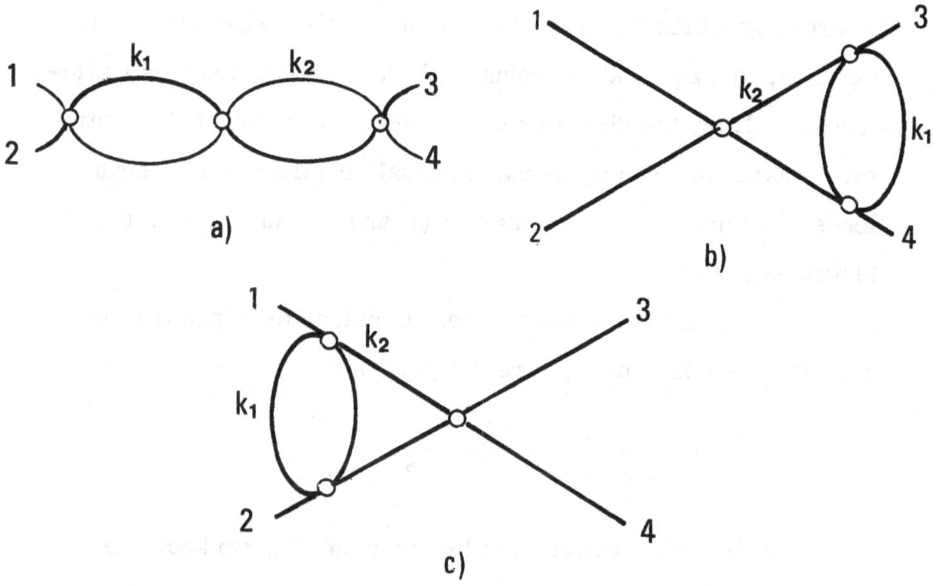

Fig. 60

We do not consider graphs of the form

which correspond to renormalization of the mass of the incoming particles and renormalization of the Green's functions. The first renormalization indicates a redefinition of the mass of the incoming particles; as to the factor $\sqrt{Z}$, which must come into the new coupling constant from each line, it is easy to see that it has the form $1 + C_1 \lambda^2 \ell n(L^2/p^2)$, that is, it should not be kept when we pick out the leading logarithmic terms.

Besides the graphs we have drawn, there are diagrams differing from them by interchanges of the external momenta :

$p_1 \leftrightarrow p_3$, $p_1 \leftrightarrow p_4$. To logarithmic accuracy, for $p_1 \sim p_2 \sim \cdots \sim p$ interchange of the momenta does not affect the contribution of the diagram, and so to take account of these permutations it is sufficient to multiply the diagrams of Fig. 60 by a factor of 3. Moreover, taking the identity of the "internal" particles into account leads to a factor $\frac{1}{2} \cdot \frac{1}{2}$ in diagram (a) and to a factor $\frac{1}{2}$ for diagrams (b) and (c).

Diagram (a) is the simplest to calculate, since the integrations over k_1 and k_2 are independent :

 $=$ (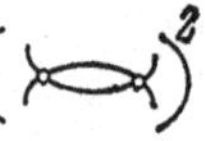 $)^2 = \lambda \frac{1}{4} \xi^2$.

In the double integral corresponding to graph (b), we know the contribution of the integral over k_1 :

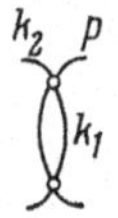

$=$

$$= \frac{1}{2} \lambda \int \frac{d^4 k_1}{(2\pi)^4} \frac{1}{k_1^2 (k_1 - p - k_2)^2} = \frac{1}{2} \lambda \frac{1}{16\pi^2} \ln \frac{L^2}{(p + k_2)^2} \,.$$

In the integration over k_2 the important region is $k_2 \gg p$:

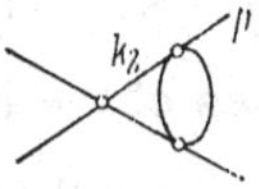

$$= \lambda^2 \int_p^L \frac{d^4 k_2}{(2\pi)^4} \frac{1}{k_2^4} \cdot \frac{1}{2} \lambda \frac{1}{16\pi^2} \ln \frac{L^2}{k_2^2} \,.$$

We first integrate over angles : $\int d^4 k_2 = \pi^2 \int k_2^2 \, dk_2^2$ and introduce the logarithmic variable $\eta = (\lambda / 16\pi^2) \, \ell n(L^2/k^2)$; then we obtain the simple integral

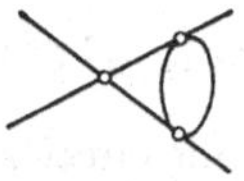

$$= \lambda \frac{1}{2} \int_0^{\xi} d\eta \cdot \eta = \lambda \frac{\xi^2}{4}.$$

Graph (c) differs from (b) by a permutation of the external momenta and hence gives the same contribution:

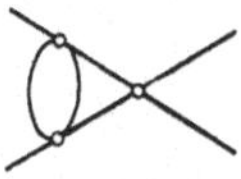

$$= \lambda \frac{\xi^2}{4}.$$

Thus, we have found the contribution of the sum of third-order diagrams

$$A^{(3)}(p) = \lambda \cdot 3 \left(\frac{\xi^2}{4} + \frac{\xi^2}{4} + \frac{\xi^2}{4} \right) = \lambda \cdot \frac{9}{4} \xi^2.$$

At each successive order of perturbation theory we will get an extra factor of λ and one extra integration, that is one extra power of ξ. Thus the total amplitude has the form

$$A \equiv \lambda f(\xi) = \lambda \left(1 - \frac{3}{2} \xi + \frac{9}{4} \xi^2 + \ldots \right). \tag{6.40}$$

A calculation of the fourth-order graphs (which we shall omit), gives a term in $f(\xi)$ equal to $-(3\xi/2)^3$. Thus, the first few terms in the function $f(\xi)$ are powers of the quantity $3\xi/2$. If we assume that subsequent terms also obey this rule, that is that $f(\xi)$ is the sum of a geometric series, then we get

$$A = \frac{\lambda}{1 + \frac{3}{2}\xi} = \frac{\lambda}{1 + \frac{3}{2} \frac{\lambda}{16\pi^2} \ln \frac{L^2}{p^2}}. \tag{6.41}$$

On physical grounds we expect that the cutoff radius L must somehow drop out of the final answer. The analysis of diagrams in coordinate space led us earlier to the idea of renormalization, that is, of redefinition of the bare coupling constant by inclusion in it of the divergent contributions from higher-order diagrams. Formula (6.41) is suited to such a renormalization procedure, as is immediately obvious if we rewrite it in the form

$$A(p) = \frac{1}{\lambda^{-1} + \frac{3}{2}\frac{1}{16\pi^2}\ln L^2 - \frac{3}{2}\frac{1}{16\pi^2}\ln p^2}.$$

If L is changed we can change λ in such a way that the amplitude A(p) remains invariant.

If we fix the value of the amplitude $A(p^2)$ at some point $p^2 = \mu^2$, so that

$$A(\mu) = \frac{1}{\lambda^{-1} + \frac{3}{2}\frac{1}{16\pi^2}\ln\frac{L^2}{\mu^2}} \equiv \lambda_R$$

and call the quantity $A(\mu)$ the renormalized coupling constant λ_R, then the relation between the amplitude A(p) and λ_R will no longer contain the cutoff radius:

$$A(p) = \frac{\lambda_R}{1 - \frac{3}{2}\frac{\lambda_R}{16\pi^2}\ln\frac{p^2}{\mu^2}}. \qquad (6.42)$$

For definiteness we may choose μ to be the renormalized particle mass m.

Thus, the geometrical progression observed in the first few terms of the expression for $A(\xi)$ is no accidental peculiarity of these first terms, but reflects an important property of the theory in question, the property of renormalizability. Indeed, we shall now obtain the expression (6.42) without using perturbation-theoretic expansions in λ, directly from the requirement of renormalizability, that is, that the amplitude A should not depend on the cut off momentum L. In that case the total derivative of the amplitude (expressed in terms of λ and $\ln(L^2/p^2)$) with respect to $\ln L^2$ at constant λ_R should be zero. Using (6.40) and differentiating A with respect to $(16\pi^2)^{-1} \ln L^2 = u$, we find

$$\frac{dA}{du} = 0 = \lambda^2 \frac{df}{d\xi} + \left(\frac{\partial\lambda}{\partial u}\right)_{\lambda_R} \left(f(\xi) + \frac{df}{d\xi}\xi\right).$$

In this equation we may vary ξ for fixed L, i.e. preserving the values of λ and $(d\lambda/du)_R$. Putting $\xi = 0$ (p = L), we find

$$\left(\frac{\partial\lambda}{\partial u}\right)_{\lambda_R} f(0) = -\lambda^2 \left(\frac{\partial f}{\partial\xi}\right)_0.$$

Using the values of f(0) and f'(0), we get from (6.40)

$$\left(\xi + \frac{2}{3}\right) f'(\xi) + f(\xi) = 0.$$

The solution of this equation is

$$f(\xi) = \frac{1}{1 + \frac{3}{2}\xi},$$

which leads to the expression (6.41) assumed above, and hence to (6.42). It follows from (6.42) that A increases with increasing p (which is equivalent to decreasing distance x). When we approach the pole of this expression the formula is no longer valid. The criterion for the applicability of eqn. (6.42) is that the quantity A should be small. To explain this point we go over to the coordinate representation. In the diagrams for the amplitude $A(x)$ for $m x \ll 1$ the dominant distances between scattering events for the virtual particles are also of order x. In fact, all diagram elements corresponding to small distances (i.e. diverging at small x) are absorbed into the renormalizations of the coupling constant, mass and Green's function, while distances greater than x give a small contribution because of the fall-off of the Green's functions. (This can be easily checked on the example of the graphs considered above). Thus the scattering of virtual particles will be determined by the quantity $A_1 \sim A(x)$, which will replace λ_R at short distances. Hence the criterion of applicability of our initial expression (6.41) is not $\lambda \ll 1$, but $A \ll 1$.

These considerations lead to the notion that it may be possible to formulate the idea of renormalizability more generally, without assuming the smallness of the constant λ_R. Below we shall return to this question when we consider the properties of quantum electrodynamics at ultra-small distances. We notice that we could have obtained the same results without introducing the bare coupling constant λ and the corresponding cut-off radius L (or r_0 in the coordinate representation) at all. Instead, we could have introduced some quite arbitrary cut-off radius r_c much larger than the limit of applicability r_0 of the theory, but at the same time much smaller than the distances x in which we are interested. Then we can include the contribution from the regions of integration smaller than r_c in the arbitrary coupling constant λ_c (which will be local to within an error $\sim r_c^2$). For $\lambda_c \ll 1$ the amplitude A is given by $A = A(\lambda_c \ell n(x^2/r_c^2))$, or in the momentum representation by $A = A(\lambda_c \ell n(L_c^2/p^2))$. Since the point $r_c = L_c^{-1}$ is arbitrary, the amplitude cannot depend on the choice made for it. Repeating the above derivation with L replaced by L_c and λ by λ_c, we arrive at the same results.

3. QUANTUM ELECTRODYNAMICS AT SMALL DISTANCES

The scalar field theory discussed above does not describe any real physical system. The scalar particles which exist in real life (the mesons) interact not only among themselves but with spinor particles (baryons) and in addition the coupling constant is large, so that perturbation theory is inapplicable. The theory of a scalar field with small coupling constant is therefore only the simplest possible model we could find to explore the general properties of field theory at small distances. We shall now go over to a realistic theory, quantum electrodynamics, that is, the theory which describes the interactions of electrons, positrons and photons.

The Local Interaction in Quantum Electrodynamics

In section 1 we found the electron (positron) and photon Green's functions, but did not introduce any interaction between these particles. The simplest electromagnetic process is represented in Fig. 61, where the line with the arrow corresponds to the propagation of an electron (positron) and the wavy line to propagation of a photon. At the point x a local interaction takes place; the most general expression for the amplitude of the process in Fig. 61 has the form

$$-i\bar{\Psi}_1 \int d^4x G(x_1 - x)\, \hat{\Gamma}_\mu \left(\frac{\partial}{\partial x}\right) G(x - x_2)\, D_{\mu\nu}(x - x')\, \Psi_2 e_\nu, \quad (6.43)$$

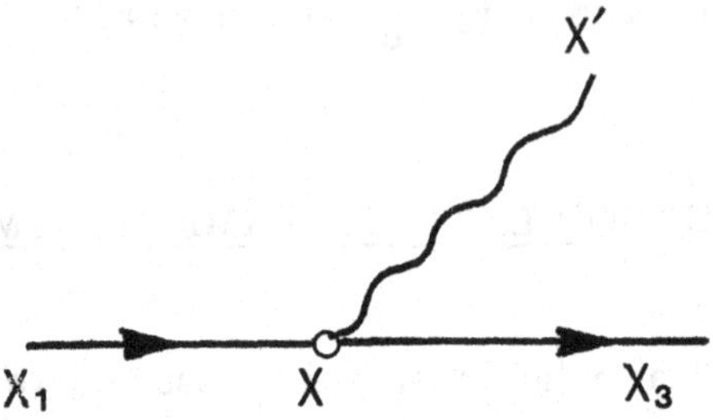

Fig. 61

where Ψ_1, Ψ_2 and e_ν are respectively the wave functions of the electron (or positron) and photon, and $\hat{\Gamma}_\mu(\partial/\partial x)$ is some unknown function. The derivatives $\partial/\partial x$ may act on any of the three Green's functions; explicit dependence of Γ_μ on x is excluded by the requirement of the homogeneity of space. Lorentz invariance of the transition amplitude requires that the quantity $\bar{\Psi}_1 \Gamma_\mu \Psi_2$ should transform like a four-vector, which must be formed from γ_μ and $\partial/\partial x_\mu$. However, any gradient-dependence of Γ_μ would mean the introduction of a coupling constant with nonzero dimension, which would destroy the renormalizability of the theory; to eliminate the divergences we would have to introduce an infinite number of terms of the form $(\partial/\partial x)^n$, that is, lose locality. To illustrate this we estimate the contribution of the third-order graphs for a vertex of the form

$$\Gamma_{1\mu} = i\mu_1 (\gamma_\mu\gamma_\nu - \gamma_\nu\gamma_\mu)\partial/\partial x_\nu.$$

In the momentum representation we have

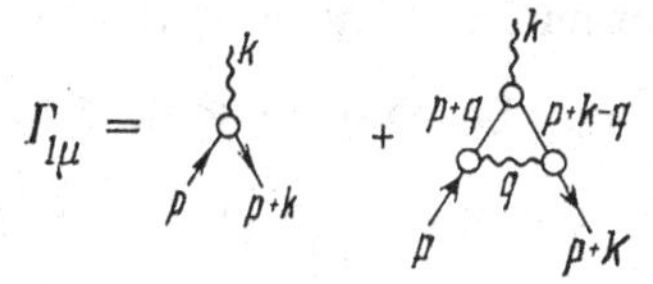

The region of integration q >> p, k gives

$$\Gamma_{1\mu} \sim \Gamma^{(0)}_{1\mu}\left(1 + \mu_1^2 \int q^2 \frac{1}{q^2}\frac{1}{q}\frac{1}{q}\, d^4q\right) = \Gamma^{(0)}_{1\mu}\,(1 + \mu_1^2 L^2).$$

So we have obtained a quadratically divergent correction, corresponding to the dimensionality of μ_1 ($[\mu_1] = 1/m$).

Thus we arrive at the so-called minimal electromagnetic coupling

$$\Gamma_\mu = e\gamma_\mu \tag{6.44}$$

with a dimensionless coupling constant which, as we shall see, is just the charge of the electron in Heaviside (rationalized Gaussian) units. ($e^2/4\pi = 1/137$). Any corrections to this vertex arise only as a secondary effect when we calculate graphs of higher order in e. Thus, for instance, if we put vertices given by (6.44) in the third-order graph considered above, we get a correction to Γ_μ of just the type given above, where μ_1 is the correction to the electron magnetic moment. This correction may be calculated with the aid of (6.44) up to sixth order in e, and the results agree with experiment to great accuracy.

To convince ourselves that the coupling constant e is indeed nothing but the electronic charge, it is sufficient to take the matrix element of expression (6.44) corresponding to a transition of a nonrelativistic electron with emission of a photon, and compare it with the expression obtained on p. 57. The wave function corresponding to a single photon in unit volume is given by $e_\mu^{(\lambda)}/\sqrt{2k_0}$; the normalization may be easily checked by calculating the electromagnetic field energy $\int \frac{1}{2}(\mathcal{E}^2 + \mathcal{H}^2)dV = \omega$. As a result we obtain in the transverse gauge

$$-ie\bar{u}_\alpha^{(\sigma)}(\gamma_i)_{\alpha\beta}u_\beta^{(\sigma')}e_i^{(\lambda)}\frac{1}{\sqrt{2k_0}}. \tag{6.45}$$

Neglecting the change in the electron momentum, we get

$$\bar{u}(p_1)\gamma_\mu u(p_1) = \frac{p_\mu}{m}.$$

That this expression is correct can be checked immediately by multiplying by p_μ and using the Dirac equation. The correction to the electron Hamiltonian corresponds to expression (6.45) without the factor (-i), that is

$$H' = e(ve^{(\lambda)})\frac{1}{\sqrt{2k_0}}, \quad v_i = \frac{p_i}{m},$$

which exactly corresponds to eqn. (1.23) on p. 57 (here the charge e is in Heaviside units). Since $u^{(\sigma)}(\underset{\sim}{p}, \alpha)$ for negative energy corresponds to a positron with momentum $-\underset{\sim}{p}$ (cf. p. 369), it follows that in the case of a positron this expression changes sign, i.e. the charge of the positron is opposite to that of the electron. Since the quantity $e_i^{(\lambda)}/\sqrt{2k_0}$ is the vector potential corresponding to a single photon, the matrix element in an arbitrary electromagnetic field A_μ is obtained by the replacement of $e_\mu^{(\lambda)}/\sqrt{2k_0}$ by A_μ. This defines the rule for introducing the electromagnetic field into the Dirac equation - we simply add to the term $\gamma_\mu p_\mu$ a term $e\gamma_\mu A_\mu$. In fact, the equation for the Green's function then gives $G^{(1)} = G(-ie\gamma_\mu A_\mu)G$ in accordance with eqn. (5.33). Calculation of higher-order terms in e in Γ_μ gives not only a correction to the magnetic moment but also a correction to the interaction of the electron with the field of the nucleus, which leads to the Lamb shift of the atomic energy levels discussed on p. 80.

Consider the transition amplitude corresponding to the scattering of two particles, e.g. an electron and a proton. To lowest order in e this process is described by the graph

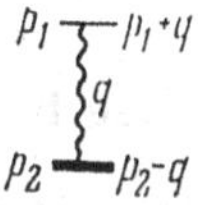

. (6. 46)

where short ends on the electron and proton lines indicate that the propagation functions corresponding to the incoming and outgoing lines are not included in the transition amplitude in question. The transition matrix element corresponding to this graph can be written in the form

$$-e^2(-i)^2\,\overline{\Psi}{}^e_1\gamma_\mu\Psi^e_2 D_{\mu\nu}(q)\,\overline{\Psi}{}^p_1\gamma_\nu\Psi^p_2.$$

For small 4-momenta q ($q \ll M_p$) we can take the motion of the proton as given; then the factor to the right of $D_{\mu\nu}$ is the q-th component of the four-current of a proton moving with momentum p_2, and the expression $D_{\mu\nu}(q)\, j^p_\nu(q)$ is the q-th component of the vector potential

$$A_\mu(x) = i\int D_{\mu\nu}(x-x')\, j^p_\nu(x')\, d^4x', \tag{6. 47}$$

induced by the proton current (see p. 379). In this case the problem of scattering of the electron reduces to the problem of scattering in an external field $A_\mu(x)$. To take the recoil of the proton into account it is sufficient to take $j^p(x)$ to be the current "of the transition", that is the matrix element of the current operator between the initial and final proton states. A consideration of the graphs which give corrections to $D_{\mu\nu}$ leads to the conclusion that $D_{\mu\nu}$ in (6.47) should be replaced by the exact function $\tilde{D}_{\mu\nu}$; we now turn to a consideration of this quantity.

Vacuum Polarization

In the presence of an external field there arise in the vacuum polarization charges and currents associated with the appearance of virtual pairs. As we shall see, the additional charges induced in the vacuum screen out any external charges placed in it. A similar process takes place in a dielectric; thus, to clarify the physical picture of these vacuum processes it is useful to pursue the analogy with the classical electrodynamics of a polarizable medium.

We write down the Dyson equation (see pp. 313, 389) for the exact photon Green's function $\tilde{D}_{\mu\nu}$, which contains all possible virtual processes which take place when a photon propagates in the vacuum. As we shall see, this equation has a simple analog in classical electrodynamics. We introduce the block diagram $\Pi_{\mu\nu}(x-x')$ which contains no parts linked by a single photon line. Then repeating the calculation on p. 313, we find in operator form

$$\tilde{D} = D + D\Pi\tilde{D}.$$

Multiplying this equation from the left by $-iD^{-1}$, we find in the coordinate representation

$$\square \tilde{D}_{\mu\nu}(x-x') + i\int \Pi_{\mu\gamma}(x-x_1)\tilde{D}_{\gamma\nu}(x_1-x')\,d^4x_1 = \\ = -ig_{\mu\nu}\delta(x-x'). \qquad (6.48)$$

We write this equation in operator form and multiply it from the right by the quantity j^{ex}, where j^{ex} is the current due to the external charges placed in the vacuum. Then we have

$$\square \tilde{D}j^{ex} + i\Pi\tilde{D}j^{ex} = -ij^{ex}.$$

Since $\tilde{D}j^{ex} = -iA$, this equation is just the equation for the vector potential in a polarizable medium, $\square A = j + j^{ex}$; the quantity $j = i\Pi A$ is the polarization current, that is $\square A + i\Pi A = j^{ex}$. Thus, the quantity Π defines the polarization current induced by the vector potential, and $\tilde{D}$ is the Green's function of the homogeneous equation for the potential A in the polarizable medium. Putting back the indices and the explicit integration, we find

$$A_\mu(x) = i\int \tilde{D}_{\mu\nu}(x-x')\,j^{ex}_\nu(x')\,d^4x'. \qquad (6.49)$$

This formula generalizes the analogous expression (6.47) to the case of a polarizable medium. The polarization current is given by the expression

$$j_\mu(x) = i\int \Pi_{\mu\nu}(x-x')\,A_\nu(x')\,d^4x'. \qquad (6.50)$$

To clarify these formulae we consider the case of a stationary charge at rest at the origin: $j^{ex}_i = 0$, $j^{ex}_0 = e_0\,\delta(\underset{\sim}{r})$ and $A_i = 0$, $e_0 A_0 = V(\underset{\sim}{r})$. For the field $V(\underset{\sim}{r})$ we get from (6.49)

$$V(r) = ie_0^2\int \tilde{D}_{00}(t, r)\,dt = -e_0^2\tilde{D}_{00}(\omega=0, r), \qquad (6.51)$$

where $\tilde{D}(\omega, \underset{\sim}{r})$ is the exact photon Green's function in the mixed representation. The density of induced charges is given by

$$\rho(r) = -i e_0 \int \Pi_{00}(r - r', \tau) \tilde{D}_{00}(\omega = 0, r')\, d\tau\, dr'. \qquad (6.52)$$

If we substitute in (6.51) the unperturbed Green's function $D_{00}(\omega = 0, \underset{\sim}{r} = 0) = -1/4\pi r$ (see p. 379) we get Coulomb's law. Since the current $j_\mu(x)$ must obey the continuity equation

$$\frac{\partial j_\mu}{\partial x_\mu} = 0,$$

it follows from (6.50) that

$$\frac{\partial \Pi_{\mu\nu}(x - x')}{\partial x_\mu} = 0. \qquad (6.53)$$

Moreover, as can be seen from the graphical definition of Π, we have $\Pi_{\mu\nu} = \Pi_{\nu\mu}$, so that the condition (6.53) automatically guarantees gauge invariance, that is, that the current is unchanged if a term of the form $\partial_\nu f$ is added to A_ν. However, as we shall see, in quantum electrodynamics the quantity $\Pi_{\mu\nu}(x)$ has a strong singularity as $x \to 0$, and the condition (6.53) is violated at the point $x = 0$. This means that for small x quantum electrodynamics must be modified in such a way as to guarantee gauge invariance and current conservation. In the next section we shall use the expression (6.51) to find the corrections to Coulomb's law due to the formation in the vacuum of induced charges with density given by the relation (6.52).

Radiative Corrections to Coulomb's Law

Quantum electrodynamics predicts deviations from Coulomb's law. These so-called radiative corrections are connected with higher-order diagrams. As we shall now see, the physical nature of the corrections to Coulomb's law is determined by the polarization of the vacuum. In spite of the fact that the radiative corrections contain the small parameter $\alpha = e^2/\hbar c = 1/137$, they are important from the point of view of principle, since they allow us to explain the character of the divergences at small distances. Indeed, it was precisely the investigation of the radiative corrections to quantum electrodynamics which gave birth to the idea of renormalization, which lies at the very foundation of the contemporary theory of elementary particle interactions.

Let us then find the first radiative correction to Coulomb's law. Consider an infinitely heavy charged particle at rest at the origin. The field produced by such a particle is given by expression (6.51); in the zeroth approximation this just gives Coulomb's law. The change in Coulomb's law is determined by the appearance of screening (polarization) charges and consequently by the correction to $D_{\mu\nu}$. To lowest order in the bare charge e_0 we have

$D^{(2)}_{\mu\nu} =$ 

. (6.54)

We will carry out the calculation in the mixed representation. Then using the fact that $i\Pi(0, \underset{\sim}{r}) \equiv \int \Pi(\tau, \underset{\sim}{r})\, d\tau$, we get

$$D^{(2)}_{\mu\nu}(0, r) = \\ = -\int D_{\mu\gamma}(0, r - r_1)\, \Pi_{\gamma\rho}(0, r_1 - r_2)\, D_{\rho\nu}(0, r_2)\, dr_1\, dr_2. \qquad (6.54')$$

where $D_{\mu\nu}(0, \underset{\sim}{r})$ is the free Green's function in the mixed representation (see p. 379)

$$D_{\mu\nu}(0, r) = -\frac{1}{4\pi r} g_{\mu\nu}$$

(we have chosen the transverse gauge $\alpha = 1$, cf. p. 378), and $\Pi_{\mu\nu}$ is the polarization operator introduced in the last section, which corresponds to the internal part of the diagram (6.54) :

$\Pi_{\mu\nu}(r,t) =$ 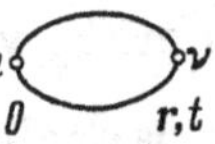

.

It is more convenient to calculate $\Pi_{\mu\nu}(x)$ in the coordinate representation. The diagram for $\Pi_{\mu\nu}(x)$ is read off as follows :

$$\Pi_{\mu\nu}(x) = -(-ie_0)^2 \, \mathrm{Sp}\, \{\gamma_\mu G(x)\gamma_\nu G(-x)\}.$$

The trace (Sp) of the matrix corresponds to summation over all possible spin states of the virtual electron-positron pair. The extra minus sign is connected with the fact that $G^+(-t) = -G^-(t)$, where G^- is the positron Green's function (cf. p. 374). As will become clear below, we need $\Pi_{\mu\nu}$ at distances much less than the Compton wavelength, so in the electron and positron Green's functions we can put $m = 0$; then (cf. p. 375)

$$\Pi_{\mu\nu}(x) = \frac{e_0^2}{4\pi^4} \mathrm{Sp}\left(\gamma_\mu \frac{\hat{x}}{x^4} \gamma_\nu \frac{\hat{x}}{x^4}\right).$$

It is elementary to calculate the trace, using the relations

$\gamma_\nu \hat{x} = - \hat{x}\gamma_\nu + 2x_\nu, \quad x^2 = x^2, \quad Sp(\gamma_\mu \gamma_\nu) = 4g_{\mu\nu}$:

$$\Pi_{\mu\nu}(x) = \frac{e_0^2}{\pi^4}[2x_\mu x_\nu - x^2 g_{\mu\nu}]\, x^{-8}. \qquad (6.55)$$

We notice that for $\tilde{x} \gg 1/m$ we get, from expressions (6.25) and (6.21) for G(x) the result $G(\tilde{x}) \sim e^{-m\tilde{x}}$, so that $\Pi_{\mu\nu}(\tilde{x})$ falls off exponentially for large $\tilde{x}$:

$$\Pi_{\mu\nu}(\tilde{x}) \underset{\tilde{x} \gg 1/m}{\sim} e^{-2m\tilde{x}}.$$

It is easy to check that the expression (6.55) satisfies the condition (6.53) everywhere except at x = 0. For small distances $(x < r_0)$ this expression must be modified, either as a result of fundamental changes in the theory or, if the theory is internally consistent (cf. next section) by taking account of more complicated processes. It is natural to assume that these modifications will preserve gauge invariance.

For subsequent purposes it is sufficient to assume that the addition to A_ν of a constant term does not change the current (cf. (6.50)), i.e. that the corrected expression for $\Pi_{\mu\nu}$ satisfies the condition

$$\int \Pi_{\mu\nu}(x)\, d^4x = 0. \qquad (6.56)$$

Since we are interested in distances $x \gg r_0$, the difficulties connected with the behaviour of $\Pi(x)$ for $x < r_0$ can be avoided, as we shall see.

To find the quantity of interest, namely $\Pi_{00}(\omega = 0, \underset{\sim}{r}) \equiv \Pi(\underset{\sim}{r})$, we must integrate (6.55) over t. The correct method of going round the poles corresponding to $t^2 = \underset{\sim}{r}^2$ is defined by adding to $\underset{\sim}{r}^2$ an infinitesimal negative imaginary term; the sign of this term is determined by the condition that $\Pi(\omega, \underset{\sim}{r})$ corresponds to a diverging wave, $\Pi(\omega, \underset{\sim}{r}) \sim e^{i\omega r}$. Differentiating the integral

$$I(r) = -i \int_{-\infty}^{+\infty} \frac{dt}{t^2 - r^2 - i\delta} = \frac{\pi}{r}$$

with respect to $\underset{\sim}{r}^2$, we easily obtain

$$\Pi_{00}(\omega = 0, r) = -i \frac{e_0^2}{\pi^4} \int \frac{t^2 + r^2}{(t^2 - r^2)^4} dt = A \frac{e_0^2}{r^5}, \quad A = \frac{1}{4\pi^3}. \quad (6.57)$$

For the correction to the Coulomb interaction we find, using (6.51) and (6.54'), the result

$$\delta V(r) = -\ e_0^2 D_{00}^{(2)}(\omega = 0, r) =$$
$$= -\frac{e_0^2}{4\pi} \int \frac{1}{|r - r_1|} \Pi(\rho) \frac{1}{4\pi} \frac{1}{|r_1 + \rho|} dr_1 \, d\rho.$$

The density of induced charges is given by

$$\rho_1(r) = -\ \frac{e_0}{4\pi} \int \Pi(\rho) \frac{1}{|r + \rho|} d\rho.$$

or, using (6.56), by

$$\rho_1(r) = -\frac{e_0}{4\pi} \int \Pi(\rho) \left(\frac{1}{|r + \rho|} - \frac{1}{r} \right) d\rho.$$

In this expression we must expand the first term in the bracket in Legendre polynomials $P_\ell(\underset{\sim}{\rho} \cdot \underset{\sim}{r} / \rho r)$. Using the fact that $\Pi(\rho) = \Pi(|\underset{\sim}{\rho}|)$, we get

$$\rho_1(r) = e_0^3 A \int_r^{\sim 1/m} \frac{1}{\rho^5} \left(\frac{1}{r} - \frac{1}{\rho} \right) \rho^2 d\rho = \frac{e_0^3 A}{6} \frac{1}{r^3}.$$

In δV we cut off the integration over r_1 at the lower end at the limit of applicability of the theory, r_0. Substituting ρ_1 in the expression for δV and expanding $|\underset{\sim}{r} - \underset{\sim}{r}_1|^{-1}$ in Legendre polynomials $P_\ell(\underset{\sim}{r} \cdot \underset{\sim}{r}_1/rr_1)$, we obtain to logarithmic accuracy

$$V(r) = \frac{e_0^2}{4\pi r}\left(1 - \frac{e_0^2}{12\pi^2}\ln\frac{r^2}{r_0^2}\right). \qquad (6.58)$$

This expression is valid for $r^2 \ll 1/m^2$, since we have used the massless Green's functions to obtain it. At large distances the logarithmic integral will be cut off at the Compton wavelength, and we obtain Coulomb's law with a corrected charge :

$$V(r) = \frac{e^2}{4\pi r},$$

where

$$e^2 = e_0^2\left(1 - \frac{e_0^2}{12\pi^2}\ln\frac{1}{m^2 r_0^2}\right); \qquad (6.59)$$

The quantity e^2 is by definition just the observed electron charge. Eliminating the bare charge e_0 from (6.58), we find

$$V(r) = \frac{e^2}{4\pi r}\left(1 + \frac{e^2}{12\pi^2}\ln\frac{1}{m^2 r^2} + O(e^4)\right) = \frac{e^2_{\text{eff}}}{4\pi r}.$$

The effective charge increases for decreasing r, as it should, since the screening action of the polarization charges is then decreased.

The Electromagnetic Interaction at Ultra-Small Distances

Quantum electrodynamics, like the four-boson interaction theory discussed above, is a renormalizable theory (cf. p. 395); it is characterized by a dimensionless constant, the fine structure constant $\alpha = e^2 = 1/137$.

If we introduce a bare charge e_o and a cut-off radius L, then the relation between the observed and the bare charge has a form similar to (6.59):

$$e^2 = e_0^2 f\left(e_0^2 \ln \frac{L}{m}\right). \tag{6.60}$$

This formula assumes that the bare charge e_o is small but the quantity $e_o^2 \ell n\,(L/m)$ is of order unity. Then we need keep, in the diagrams of perturbation theory, only the leading terms, i.e. those in $(e_o^2\ \ell n(L/m))^n$. That such terms do in fact occur is obvious from an analysis of the diagrams with the help of a Wick rotation like that carried out for the scalar theory. The assumption that the bare charge e_o is small has been made only for simplicity - the only thing actually necessary in what follows is that the observed charge e should be small.

The stipulation of renormalizability allows us to find the function $f(\xi)$ up to an unknown constant $f'(0)$. As in the case of the scalar theory, the only function of the form $e_o^2 f(e_o^2\ \ell n\ \frac{L}{m})$ which allows us to compensate a variation of the cut-off radius δL by a variation of the bare charge δe_o has the form

$$e^2 = \frac{e_0^2}{1 - f'(0)\, e_0^2 \ln \frac{L}{m}}.$$

We will obtain this formula by using simple intuitive considerations, from which it will follow that for any type of charged particle we must have

$$f'(0) < 0.$$

Consider the potential $\varphi(\underset{\sim}{r})$ of a set of stationary bare charges e_o distributed with density $n_o(\underset{\sim}{r})$. The potential φ

satisfies Poisson's equation

$$\nabla^2\varphi = -e_0^2\,(n_0 + n_1), \tag{6.61}$$

where n_1 is the charge density arising in the vacuum as a result of its polarization by the field. According to the results obtained above, if we introduce the quantity $\Pi_0(\rho) = \Pi(\rho)/e_0^2$ we get

$$n_1(r) = \int \Pi_0(\rho)\,(\varphi(r+\rho) - \varphi(r))\,d\rho. \tag{6.62}$$

It follows from this relation that the total charge of the vacuum remains zero ($\int n_1(r)\,dr = 0$). At distances $\rho \gg 1/m$, $\Pi_0(\rho)$ decreases exponentially, while at small distances ($\rho \ll 1/m$) we have according to (6.57)

$$\Pi_0(\rho) = A\,|\rho|^{-5}, \quad A = 1/4\pi^3 > 0.$$

Let us separate the integration over ρ in (6.62) into three regions : $\rho \ll r$, $\rho \sim r$ and $\rho \gg r$. In the first region we can expand $\varphi(r + \rho)$ in powers of ρ. After averaging over the direction of ρ, we obtain

$$\overline{\varphi(r+\rho) - \varphi(r)} = \\ = \bar{\rho}_i\partial_i\varphi(r) + \frac{1}{2}\,\overline{\rho_i\rho_k}\,\partial_i\partial_k\varphi(r) + \ldots \simeq \frac{1}{6}\,\rho^2\nabla^2\varphi.$$

The contribution of this region to $n_1(r)$ is therefore

$$n_1(r) \simeq \frac{1}{6}\,A\nabla^2\varphi \int_{r_0}^{r} \frac{d^3\rho}{\rho^3} = \frac{2\pi}{3}\,A\nabla^2\varphi\ln\frac{r}{r_0}.$$

For $r > 1/m$ the upper limit should be replaced by $1/m$. The lower limit is defined by the limit of validity of the theory. As we mentioned above (p. 408), we could perfectly well take as the lower limit an arbitrary point r_c such that $1/m > r_c > r_0$; this would correspond to a charge e_c. Then the renormalizability of the theory implies that the final result, when expressed in terms of the observed charge, will not depend on e_c and r_c. Below we shall confirm this.

The region $\rho \gg r$ gives effectively no contribution to $n_1(\underset{\sim}{r})$ for $r > 1/m$, in view of the fast fall-off of $\Pi(\rho)$ for $\rho > 1/m$. For $r < 1/m$ the contribution of this region is of order $A\varphi(\underset{\sim}{r})/\rho_1^2$, where $\rho_1 \gg r$, i.e. it is small compared to the contribution of the first region.

Finally, by making the replacement $\varphi(\underset{\sim}{r} + \underset{\sim}{\rho}) - \varphi(\underset{\sim}{r}) \to \underset{\sim}{\rho} \cdot \underset{\sim}{\nabla}\varphi \sim \underset{\sim}{r} \cdot \underset{\sim}{\nabla}\varphi$ we get from the region $r \sim \rho$ a term of the form $\delta n_1 \sim A\, \underset{\sim}{r} \cdot \underset{\sim}{\nabla}\varphi/r^2$. The coefficient in front of this term may be found from the following consideration : as we know from ordinary electrodynamics, the density of induced charges can be expressed in terms of the divergence of the polarization vector ; hence, $n_1(\underset{\sim}{r})$ must have the form of the divergence of some vector $\underset{\sim}{P} = f(\underset{\sim}{r})\, \underset{\sim}{\nabla}\varphi$, whence we get $n_1 \sim \text{div}\, \underset{\sim}{P} = f(r)\nabla^2\varphi + \frac{df}{dr}\, \underset{\sim}{r} \cdot \underset{\sim}{\nabla}\varphi$. Thus, the term corresponding to the region $r \sim \rho$ must supplement the term proportional to $\nabla^2\varphi$ in $n_1(\underset{\sim}{r})$ in such a way that we obtain the divergence of a vector. We therefore finally get

$$n_1(r) = \frac{2\pi}{3} A \,\text{div}\left(\ln\left(\frac{r}{r_0}\right) \nabla\varphi\right).$$

Consider first the case when the density $n_0(\underset{\sim}{r})$ is spread out over a region $r > 1/m$. Substituting the expression for $n_1(\underset{\sim}{r})$, in which r must be replaced by $1/m$, in Poisson's equation, we get

$$\nabla^2 \varphi = -e^2 n_0 (r), \tag{6.63}$$

where

$$e^2 = \frac{e_0^2}{1 + e_0^2 \frac{2\pi}{3} A \ln \frac{1}{mr_0}}. \tag{6.64}$$

As is obvious from (6.63), e determines the interaction between charges which are far apart and therefore is just the observed electron charge. Expression (6.64) establishes the relation between e^2 and the square of the bare charge e_o^2. The quantity A, which defines the polarizability of the vacuum, is positive and hence the observed charge is less than the bare one.

Consider now the case in which the bare charges are distributed over a region $r < 1/m$, and in particular when there is a single charge at the origin : $n_o(\underset{\sim}{r}) = \delta(\underset{\sim}{r})$. We write $n_1(\underset{\sim}{r})$ in the form

$$n_1 (r) = \frac{2\pi}{3} A \left[\operatorname{div} (\ln (mr) \nabla\varphi) - \ln (mr_0) \nabla^2 \varphi\right].$$

Transferring the second term to the left-hand side of eqn. (6.61) and going over to the charge e, we get

$$\nabla^2 \varphi = -e^2 (n_0 (r) + n_R (r)). \tag{6.65}$$

where the quantity

$$n_R (r) = \frac{2\pi}{3} A \operatorname{div} (\ln (mr) \nabla\varphi)$$

may be called the renormalized density of vacuum charges.

Let us introduce the quantity

$$\mathcal{D} = -\left(1 - e^2 \frac{2\pi}{3} A \ln \frac{1}{mr}\right) \nabla\varphi ,$$

which is analogous to the electric displacement vector. Eqn. (6.65) then gives

$$\operatorname{div} \mathcal{D} = e^2 n_0(r).$$

Thus, the quantity

$$\varepsilon = 1 - e^2 \frac{2\pi}{3} A \ln \frac{1}{mr} \tag{6.66}$$

is the dielectric susceptibility of the vacuum at small distances from the charges.

Let us find the distribution of vacuum charges around a charge e_0 situated at the origin: $n_0(\mathbf{r}) = \delta(\mathbf{r})$. The charge inside a sphere of radius r, $e^2(r)$, is connected with φ by the obvious relation $-\nabla\varphi = \frac{e^2(r)}{4\pi r^3}\mathbf{r}$. Using the expression $\mathcal{D} = (e^2/4\pi r^3)\mathbf{r}$, we obtain

$$e^2(r) = \frac{e^2}{\varepsilon} = \frac{e^2}{1 - e^2 \frac{2\pi^2}{3} A \ln \frac{1}{mr}}. \tag{6.67}$$

For $r \geq 1/m$, $e^2(r)$ goes over into the observed charge e^2, and for $r = r_0$ into the bare charge. Since the polarization process does not change the total charge of the vacuum, the screening vacuum charge which appears near the charge e_0, is compensated by an equal and opposite charge which goes off to infinity, just as happens when a charge is placed in an infinite dielectric.

It follows formally from (6.67) that for

$$r \sim r_1 \sim \frac{1}{m} \exp\left(-\frac{3}{8\pi^2 A e^2}\right)$$

the charge $e^2(r)$ tends to infinity. In fact, however, formula (6.67) is inapplicable at such small distances. Indeed, when we found $\Pi(\mathbf{r})$ we assumed that the dimensionless charge $e^2(r)$ is small, and did not take into account the possible dependence of

$\Pi(\underset{\sim}{r})$ on $e^2(r)$.

In the region $r \sim r_1$, when $e^2(r) \sim 1$, charges created at relative distances $\sim r$ can in their turn interact, with an interaction characterized by a charge $\sim e^2(r)$. As we shall see below, this natural assumption is another formulation of the property of renormalizability.

Thus the "constant" A in (6.64) can in fact depend on r through $e^2(r)$:

$$A = A\,(e^2\,(r)).$$

If this dependence is taken into account, we get instead of (6.64)

$$e^2 = \frac{e_0^2}{1 + e_0^2 \frac{4\pi}{3} \int\limits_{\ln r_0}^{\ln 1/m} A\,(e^2\,(\rho))\, d\ln\rho}.$$

The corrected formula for $e^2(r)$ will have the form of an integral equation:

$$e^2\,(r) = \frac{e^2}{1 + \frac{4\pi}{3} e^2 \int\limits_{\ln 1/m}^{\ln r} d\ln\rho\, A\,(e^2\,(\rho))}. \qquad (6.68)$$

Differentiating (6.68) with respect to $\ell n\, r$, we can obtain the differential equation first found by Gell-Mann and Low:

$$\frac{de^2\,(r)}{d\ln r} = -\beta\,(e^2\,(r)),$$

where the Gell-Mann-Low function $\beta(e^2)$ is related to the function $A(e^2)$ by the relation

$$\beta\,(e^2) = e^4\,\frac{4\pi}{3} A\,(e^2). \qquad (6.69)$$

The implicit solution of this equation has the form

$$\int_{e^2}^{e^2(r)} \frac{dx}{\beta(x)} = \ln \frac{1}{mr}. \tag{6.70}$$

where we have introduced the observed charge $e^2 = e^2(m^{-1})$. The relation between the observed charge and the bare charge $e_0^2 = e^2(r_0)$ is given by (6.70) for $r \sim r_0$:

$$\int_{e^2}^{e_0^2} \frac{dx}{\beta(x)} = \ln \frac{1}{mr_0}. \tag{6.71}$$

These formulae obviously satisfy the renormalizability relation and indeed were found by Gell-Mann and Low precisely by starting from this requirement.

The qualitative derivation given above allows us to grasp the physical meaning of the Gell-Mann-Low function (6.69) : it is associated with a polarizability and hence cannot be negative in the region where the theory is meaningful. We note that this derivation may be refined. We assumed that $A(r) = A(e^2(r))$. In reality, at distances r the quantity A is determined by the charge $e^2(r_1)$, where $r_1 \sim r$. Assume first of all that $\ln r_1 = \ln r + \nu$, where ν is a small correction term. Then we have

$$A(e^2(r_1)) = A(e^2(r)) + \frac{dA}{de^2}\frac{de^2}{d\ln r}\nu = \Phi(e^2(r)).$$

By iterating this operation we conclude that indeed $A = \Phi(e^2(r))$.

The Gell-Mann-Low function is a most important feature of field theory; however, up to now no way to calculate it other than perturbation theory has been found. Perturbation theory will give us the first few coefficients in the expansion of $A(e^2)$, but does not allow us to draw any conclusions about the properties of

$A(e^2)$ in the region of real interest, $e^2 \gtrsim 1$.

In principle we cannot exclude the possibility that the function $A(e^2)$ should tend to zero at some value e_*^2 of e^2. Since the polarizability cannot become negative, the function $A(e^2)$ must in this case be tangent to the e^2 axis at $e^2 = e_*^2$. The question of the occurrence of a zero of the Gell-Mann-Low function is of some theoretical interest, since in this case it is possible to construct a strictly local theory with $r_0 = 0$. To see this, note that for $r_0 \to 0$ the integral in (6.71) must diverge. This can happen if either the upper limit e_0^2 or the lower limit e^2 of the integration should coincide with the zero (e_*^2) of the function $\beta(x)$. The observed charge $e^2 = 1/137$ is sufficiently small that we can trust perturbation theory for the function $\beta(e^2)$; this certainly does not have a zero at $e^2 = 1/137$. If e_0^2 is equal to e_*^2, then we can put $r_0 = 0$, i.e. the theory will be strictly local.

If the Gell-Mann-Low function has no zeros at all (as seems most probable) then for $r_0 \to 0$ the bare charge e_0^2 must be put equal to ∞ for the integral to diverge. Thus in this case the concept of a bare interaction would lose all meaning.

The screening of the interaction (that is, the positiveness of the coefficient A in (6.64)) is a general feature of all renormalizable theories investigated until recently: electrodynamics, the scalar-field (with interaction $\sim \lambda \varphi^4$), Yukawa theory (with interaction $\sim \bar{\Psi} \Psi \varphi$) etc. For twenty years the screening of a bare interaction seemed a property inseparable from renormalizable theories. For any finite interaction at large distances the interaction at short distances would be large or even tend to infinity, which makes the very concept of a bare interaction sterile.

The screening of the interaction has often been put forward as an argument against local field theory. Other theoretical approaches have begun to be intensively developed, in which this difficulty can be avoided; an attempt has been made to formulate the theory in terms of relations which arise only from its general properties, such as unitarity and causality (the so-called S-matrix approach). However, as we have already mentioned, it is not possible to formulate this scheme in the form of a self-contained theory.

Recently a new possibility has been discovered. It was observed that the gauge-invariant theory put forward to Yang and Mills as early as 1954 is renormalizable. Theories of this type are a generalization of electrodynamics ; fermion fields interact with with several types of vector fields, the so-called gluons. In contrast to photons, gluons are charged and hence interact among themselves. The structure of the Lagrangian is uniquely determined by the requirements of renormalizability and gauge invariance; the theory contains only a single dimensionless constant, which determines both the interaction of the fermions with the gluons and the interaction of the gluons among themselves. The bare masses of the gluons are zero, but the fermion masses may be nonzero.

The conserved "charge" in this case is, unlike the electric charge, a vector operator whose components commute among themselves like the components of angular momentum. Thus the law of combination of the polarization "charges" is more complicated than in electrodynamics. Calculation shows that depending on the relation between the numbers of fermion fields and of gluon fields, a "charge" placed in the vacuum may be either screened or "anti-screened". For a not too large number of fermion fields anti-screening occurs.

In such a theory the property of renormalizability also leads to the logarithmic law (6.64) for the effective coupling constant, but with a negative coefficient A. This means that the effective interaction increases as the distance increases, while in the limit of ultra-short distances the interaction vanishes (the so-called "asymptotic freedom" phenomenon). This phenomenon is sometimes thought to be relevant to the solution of a paradox of the physics of strong interactions: it is well known that the mass spectrum of the hadrons, as well as their electromagnetic structure as observed in deep inelastic electron-hadron reactions, is well described by the model of noninteracting quarks*. However, no free quarks have been observed, although they have been sought for many years. This paradox might be resolved by the assumption that the interaction of quarks is described by a Yang-Mills theory, that is, that it vanishes at small distances and allows neither quarks nor gluons to escape to large distances.

Thus, the purely theoretical questions of renormalizability and interactions at ultra-short distances may turn out to be crucial for our understanding of the nature of the strong interactions.

---

* R.P. Feynman, Photon-Hadron Interactions (W.A.Benjamin Inc., Reading, Mass. 1972).

INDEX

Printed in the United States
by Baker & Taylor Publisher Services